Drug Stereochemistry

DRUGS AND THE PHARMACEUTICAL SCIENCE SERIES

Series Executive Editor
James Swarbrick
PharmaceuTech Inc.
Pinehurst, North Carolina, USA

Advisory Board

Larry L. Augsburger
University of Maryland
Baltimore, Maryland, USA

Anthony J. Hickey
University of North Carolina,
School of Pharmacy, Chapel
Hill, North Carolina, USA

Yuichi Sugiyama
University of Tokyo, Tokyo,
Japan

Harry G. Brittain
Center for Pharmaceutical
Physics, Milford,
New Jersey, USA

Jeffrey A. Hughes
University of Florida,
College of Pharmacy,
Gainesville, Florida. USA

Elizabeth M. Topp
Purdue University,
West Lafayette, Indiana, USA

Jennifer B. Dressman
University of Frankfurt,
Institute of Pharmaceutical
Technology, Frankfurt, Germany

Joseph W. Polli
GlaxoSmithKline,
Research Triangle Park,
North Carolina, USA

Geoffrey T. Tucker
University of Sheffield,
Royal Hallamshire Hospital,
Sheffield, UK

Robert Gurny
University of Geneva, Geneva,
Switzerland

Kinam Park
Purdue University,
West Lafayette, Indiana, USA

Peter York
University of Bradford,
School of Pharmacy,
Bradford, UK

Recent Titles in Series
For information on other volumes in the *Drugs and Pharmaceutical Science* Series, please visit www.informahealthcare.com

211. *Drug Stereochemistry: Analytical Methods and Pharmacology, Third Edition*; Krzysztof Jóźwiak, W. John Lough, Irving W. Wainer, ISBN 978-1-4200-9238-7, 2012
210. *Pharmaceutical Stress Testing: Predicting Drug Degradation, Second Edition*; Steven W. Baertschi, Karen M. Alsante, and Robert A. Reed, ISBN 978-1-4398-0179-6, 2011
209. *Pharmaceutical Process Scale-Up, Second Edition*; Michael Levin, ISBN 978-1-61631-001-1, 2011
208. *Sterile Drug Products: Formulations, Packaging, Manufacturing, and Quality*; Michael K. Akers, ISBN 978-0-8493-3993-6, 2010
207. *Advanced Aseptic Processing Technology*; James Agalloco, James Akers, ISBN 978-1-4398-2543-3, 2010
206. *Freeze-Drying/Lyophilization of Pharmaceutical and Biological Products, Third Edition*; Louis Rey, Joan May, ISBN 978-1-4398-2575-4, 2010
205. *Active Pharmaceutical Ingredients; Development, Manufacturing, and Regulation, Second Edition*; Stanley Nusim, ISBN 978-1-4398-0336-3, 2009
204. *Generic Drug Product Development: Specialty Dosage Forms*; Leon Shargel, Isadore Kanfer, ISBN 978-08493-7786-0, 2010

Drug Stereochemistry

Analytical Methods and Pharmacology

Third edition

Krzysztof Jóźwiak
W. John Lough
Irving W. Wainer

New York London

This edition published in 2012 by Informa Healthcare, 119 Farringdon Road, London EC1R 3DA, U.K.
Simultaneously published in the USA by Informa Healthcare, 52 Vanderbilt Avenue, 7th Floor, New York, NY 10017, USA.

First published in 1993 by Marcel Dekker, Inc., New York, New York.

Informa Healthcare is a trading division of Informa UK Ltd. Registered Office: 37–41 Mortimer Street, London W1T 3JH, U.K. Registered in England and Wales number 1072954.

© 2012 Informa Healthcare, except as otherwise indicated.

No claim to original U.S. Government works.

Reprinted material is quoted with permission. Although every effort has been made to ensure that all owners of copyright material have been acknowledged in this publication, we would be glad to acknowledge in subsequent reprints or editions any omissions brought to our attention.

All rights reserved. No part of this publication may be reproduced, stored in a retrieval system, or transmitted, in any form or by any means, electronic, mechanical, photocopying, recording, or otherwise, unless with the prior written permission of the publisher or in accordance with the provisions of the Copyright, Designs and Patents Act 1988 or under the terms of any licence permitting limited copying issued by the Copyright Licensing Agency Saffron House, 6-10 Kirby Street, London EC1N 8TS UK, or the Copyright Clearance Center, Inc., 222 Rosewood Drive, Danvers, MA 01923, USA (http://www.copyright.com/ or telephone 978-750-8400).

Product or corporate names may be trademarks or registered trademarks, and are used only for identification and explanation without intent to infringe.

This book contains information from reputable sources and although reasonable efforts have been made to publish accurate information, the publisher makes no warranties (either express or implied) as to the accuracy or fitness for a particular purpose of the information or advice contained herein. The publisher wishes to make it clear that any views or opinions expressed in this book by individual authors or contributors are their personal views and opinions and do not necessarily reflect the views/opinions of the publisher. Any information or guidance contained in this book is intended for use solely by medical professionals strictly as a supplement to the medical professional's own judgement, knowledge of the patient's medical history, relevant manufacturer's instructions and the appropriate best practice guidelines. Because of the rapid advances in medical science, any information or advice on dosages, procedures, or diagnoses should be independently verified. This book does not indicate whether a particular treatment is appropriate or suitable for a particular individual. Ultimately it is the sole responsibility of the medical professional to make his or her own professional judgements, so as appropriately to advise and treat patients. Save for death or personal injury caused by the publisher's negligence and to the fullest extent otherwise permitted by law, neither the publisher nor any person engaged or employed by the publisher shall be responsible or liable for any loss, injury or damage caused to any person or property arising in any way from the use of this book.

A CIP record for this book is available from the British Library.

Library of Congress Cataloging-in-Publication Data

Drug stereochemistry : analytical methods and pharmacology / edited by Krzysztof Jóźwiak, W. John Lough, Irving W. Wainer. -- 3rd ed.
  p. ; cm. -- (Drugs and the pharmaceutical science series ; 211)
Includes bibliographical references and index.
Summary: "Updated to reflect modern advances in the techniques and methodology of drug stereochemistry, the Third Edition comprehensively presents all aspects of chiral drugs from scientific, academic, governmental, industrial, and clinical points of view. This stand-alone text covers the lifespan of stereochemistry, from its early history, including an overview of terms and concepts, to the current drug development process, legal and regulatory issues, and the new stereoisomeric drugs."--Provided by publisher.
  ISBN 978-1-4200-9238-7 (hardback : alk. paper)
  I. Jóźwiak, Krzysztof, 1971- II. Lough, W. J. (W. John) III. Wainer, Irving W. IV. Series: Drugs and the pharmaceutical sciences ; v. 211.
  [DNLM: 1. Chemistry, Pharmaceutical--methods. 2. Molecular Conformation. 3. Stereoisomerism. QV 744]
  615'.19--dc23
                                                                                      2011044110

UK/USA ISBN: 978-1-4200-9238-7; eISBN:978-1-4200-9239-4

India Edition ISBN: 978-1-84214-553-1

Orders may be sent to: Informa Healthcare, Sheepen Place, Colchester, Essex CO3 3LP, UK
Telephone: +44 (0)20 7017 6682
Email: Books@Informa.com
Website: http://informahealthcarebooks.com

For corporate sales please contact: CorporateBooksIHC@informa.com
For foreign rights please contact: RightsIHC@informa.com
For reprint permissions please contact: PermissionsIHC@informa.com

Typeset by MPS Limited, India. Printed and bound in India by Replika Press Pvt. Ltd.

About the Editors

Professor Krzysztof Jóźwiak is Head of Laboratory of Medicinal Chemistry and Neuroengineering of Medical University of Lublin, Lublin, Poland. Following graduation in 2000 he was a postdoctoral fellow in the Gerontology Research Center, National Institute on Aging/National Institutes of Health in Baltimore, Maryland, under the supervision of Irving W. Wainer; and in 2004 assumed Associate Professor position at the Medical University of Lublin. Prof. Jóźwiak's main research interests focus on elucidation of molecular mechanisms of interactions between medicinal molecules and their protein targets, development of new methods for both experimental and theoretical characterization of drug-receptor interactions and their applications in medicinal chemistry projects. Topics of particular interest are molecular modeling of chiral substances and mechanisms of chiral recognition of molecules on protein selectors.

Dr W. John Lough is Reader in Pharmaceutical Analysis in the Department of Pharmacy, Health and Well-Being at the University of Sunderland, U.K. From an ICI-sponsored PhD, over seven years spent with Beecham Pharmaceuticals, to pharmaceutical collaborations during his time in academia, Dr Lough's research interests have always been orientated toward industrial applications. In the general area of pharmaceutical and biomedical analysis, this has included a varied range of funded studies including the exploitation of achiral derivatization in chiral separations, studies in low dispersion chromatography, use of on-column sample focusing in drug bioanalysis, chiral drug bioanalysis, biomedical applications of capillary electrophoresis, pharmaceutical applications of capillary electrochromatography, and the evaluation and exploitation of orthogonal stationary phase selectivity in liquid chromatography. His experience of chiral separations, much of which was gained in the U.K. pharmaceutical industry, dates to the late 1970s. His early research in this field involved chiral ion-pair chromatography and the development of an immobilized chiral metal-diketonate catalyst and a hexahelicene chiral stationary phase for liquid chromatography (LC). His work as a separation sciences leader and chiral separations specialist with Beecham Pharmaceuticals in the United Kingdom in the 1980s came at a time when breakthroughs were being made in LC chiral stationary phases that had a major impact on how chiral drugs were developed. His more recent interests are in chiral drug bioanalysis, screening strategies for chiral method development, and chiral capillary electrophoresis (CE).

Dr Lough has published extensively, including editing *Chiral Liquid Chromatography*, and coediting three other texts. He has been a member of the Executive Committee of The Chromatographic Society for the past 20 years (serving as President from 2007 to 2009), and of the British Pharmacopoeia, Group of Experts A (Medicinal Chemicals) for over 10 years. He chaired the International Symposium on Chiral Discrimination in Edinburgh in 1996 and since then has served on the committees of several international symposia, currently as the Secretary of the Permanent Scientific Committee of the International Symposium on Chromatography. Dr Lough was involved in founding the journal *Chromatography Today*, for which he is currently a contributing editor.

ABOUT THE EDITORS

Irving W. Wainer, PhD, is Senior Investigator in the Bioanalytical Chemistry and Drug Discovery Section, Laboratory for Clinical Investigation, National Institute of Aging/National Institutes of Health. Dr Wainer received his BS in chemistry from Wayne State University, and then received his PhD in chemistry from Cornell University. After conducting postdoctoral doctoral studies in molecular biology at the University of Oregon and clinical pharmacology at Thomas Jefferson Medical School, he worked for the Food and Drug Administration (FDA) as a research chemist. Subsequent posts were Director of Analytical Chemistry, Clinical Pharmacokinetics Lab, and Associate Member, Pharmaceutical Division, St. Jude Children's Research Hospital in Memphis; Professor and Head of the Pharmacokinetics Laboratory, Department of Oncology, McGill University—and remains an Adjunct Professor at McGill; Professor of Pharmacology, Georgetown University, Washington, D.C.

Dr Wainer has published over 350 scientific papers and eight books. He was founding editor of the journal *Chirality* and senior editor of the *Journal of Chromatography B: Biomedical Sciences and Applications* for 11 years. His awards include the Harry Gold Award (corecipient with Dr John E. Stambaugh) from the American College of Clinical Pharmacologists; Sigma Xi Science Award, FDA Sigma Xi Club; and A. J. P. Martin Medal presented by the Chromatographic Society for contributions to the development of chromatographic science. Dr Wainer is an Elected Fellow of the American Academy of Pharmaceutical Sciences and Elected Member United States Pharmacopeial Convention Committee of Revision for 1995–2000. In June 2006, he was awarded an honorary doctorate in medicine from the Medical University of Gdańsk, Poland. His research interests include clinical pharmacology, bioanalytical chemistry, the development of online high-throughput screens, and drug discovery in the areas of oncology, neuropharmacology, and cardiovascular disease.

Contents

About the Editors v
Contributors ix

PART I: INTRODUCTION

1. The early history of stereochemistry: From the discovery of molecular asymmetry and the first resolution of a racemate by Pasteur to the asymmetrical chiral carbon of van't Hoff and Le Bel 1
   Dennis E. Drayer

2. Stereochemistry—basic terms and concepts 17
   Krzysztof Jóźwiak

3. Molecular basis of chiral recognition 30
   Krzysztof Jóźwiak

PART II: THE SEPARATION, PREPARATION, AND IDENTIFICATION OF STEREOCHEMICALLY PURE DRUGS

4. Separation and resolution of enantiomers and their dissociable diastereomers through direct crystallization 48
   Harry G. Brittain

5. Indirect methods for the chromatographic resolution of drug enantiomers 69
   Władysław Gołkiewicz and Beata Polak

6. HPLC chiral stationary phases for the stereochemical resolution of enantiomeric compounds: The current state of the art 95
   W. John Lough

7. Preparative and production scale chromatography in enantiomer separations 113
   Geoffrey B. Cox

8. Enantioselective separations by electromigration techniques 147
   Michał J. Markuszewski

9. Alternative analytical techniques for determination or isolation of drug enantiomers 167
   W. John Lough

PART III: PHARMACOKINETIC AND PHARMACODYNAMIC DIFFERENCES BETWEEN DRUG STEREOISOMERS

10. Stereoselective transport of drugs 171
    Prateek Bhatia and Ruin Moaddel

11. Enantioselective binding of drugs to plasma proteins 182
    Thomas H. Kim

12. Clinical pharmacokinetics and pharmacodynamics of stereoisomeric drugs 206
    Scott A. Van Wart and Donald E. Mager

PART IV: PERSPECTIVES ON THE DEVELOPMENT AND USE OF SINGLE ISOMER DRUGS

13. Regulatory perspective on the development of new stereoisomeric drugs 240
    Sarah K. Branch and Andrew J. Hutt

14. Molecular analysis of agonist stereoisomers at β_2-adrenoceptors 274
    Roland Seifert and Stefan Dove

15. Development of chiral drugs from a U.S. legal patentability perspective: Enantiomers and racemates 294
    Svetlana M. Ivanova

16. The importance of chiral separations in single enantiomer patent cases 304
    Charlotte Weekes

Index 313

Contributors

Prateek Bhatia National Institute on Aging/National Institutes of Health, Baltimore, Maryland, USA

Sarah K. Branch Medicines and Healthcare products Regulatory Agency, London, UK

Harry G. Brittain Center for Pharmaceutical Physics, Milford, New Jersey, USA

Geoffrey B. Cox PIC Solution Inc., West Chester, Pennsylvania, USA

Stefan Dove Department of Pharmaceutical and Medicinal Chemistry II, University of Regensburg, Germany (S. D.), Hannover, Germany

Dennis E. Drayer Retired from Department of Pharmacology, Cornell University Medical College, New York, USA

Władysław Gołkiewicz Retired from Department of Physical Chemistry, Medical University of Lublin, Lublin, Poland

Andrew J. Hutt Division of Pharmaceutical Chemistry, School of Pharmacy, University of Hertfordshire, Hatfield, Hertfordshire, UK

Svetlana M. Ivanova United States Patent and Trademark Office, Alexandria, VA, USA

Krzysztof Jóźwiak Laboratory of Medicinal Chemistry and Neuroengineering, Medical University of Lublin, Lublin, Poland

Thomas H. Kim Department of Anesthesiology, Division of Clinical and Translational Research, Washington University School of Medicine, Washington, USA

W. John Lough Department of Pharmacy, Health and Well-Being, University of Sunderland, Sunderland, UK

Donald E. Mager Department of Pharmaceutical Sciences, University at Buffalo, SUNY, Buffalo, New York, USA

Michał J. Markuszewski Department of Biopharmaceutics and Pharmacodynamics, Medical University of Gdańsk, Gdańsk, Poland; Department of Toxicology, Ludwik Rydygier Collegium Medicum, Nicolaus Copernicus University, Bydgoszcz, Poland

Ruin Moaddel National Institute on Aging/National Institutes of Health, Baltimore, Maryland, USA

Beata Polak Department of Physical Chemistry, Medical University of Lublin, Lublin, Poland

Roland Seifert Department of Pharmacology, Medical School of Hannover, Germany (R. S.), Hannover, Germany

Scott A. Van Wart Department of Pharmaceutical Sciences, University at Buffalo, SUNY, Buffalo, New York, USA

Charlotte Weekes Pinsent Masons LLP, London, UK

1 The early history of stereochemistry

From the discovery of molecular asymmetry and the first resolution of a racemate by Pasteur to the asymmetrical chiral carbon of van't Hoff and Le Bel

Dennis E. Drayer

The first half of the nineteenth century was the great age of geometrical optics. Several French scientists studied diffraction, interference, and polarization of light. In particular, linear polarization of light and rotation of the plane of polarization very quickly attracted attention because of the possible relationship between these phenomena and the structure of matter. Optical activity, the ability of a substance to rotate the plane of polarization of light, was discovered in 1815 at the College de France by the physicist Jean-Baptiste Biot. In 1848 at the Ecole Normale in Paris, Louis Pasteur made a set of observations that led him a few years later to make this proposal, which is the foundation of stereochemistry: Optical activity of organic solutions is determined by molecular asymmetry, which produces nonsuperimposable mirror-image structures. A logical extension of this idea occurred in 1874 when a theory of organic structure in three dimensions was advanced independently and almost simultaneously by Jacobus Henricus van't Hoff in Holland and Joseph Achille Le Bel in France. By this time it was known from the work of Kekule in 1858 that carbon is tetravalent (links up with four other groups or atoms). van't Hoff and Le Bel proposed that the four valances of the carbon atom were not planar, but directed into three-dimensional space. van't Hoff specifically proposed that the spatial arrangement was tetrahedral. A compound containing a carbon substituted with four different groups, which van't Hoff defined as an asymmetric carbon (*asymmetrisch koolstof-atoom*), would therefore be capable of existing in two distinctly different nonsuperimposable forms. The asymmetric carbon atom, they proposed, was the cause of molecular asymmetry and therefore optical activity.

The purpose of this chapter is to describe the observations and reasoning that led Pasteur, van't Hoff, and Le Bel to make these epochal discoveries. In several instances the protagonists will speak for themselves. More detailed accounts of their work are presented in Weyer (1), Partington (2), and Riddell and Robinson (3). Also, the three methods discovered by Pasteur to resolve for the first time an optically inactive racemate into its optically active components (enantiomers) will be discussed. To truly appreciate the contributions of these three chemists, one should remember that during their time even the existence of atoms and molecules was questioned openly by many scientists, and to ascribe shape to what seemed like metaphysical concepts was too much for many of their contemporaries to accept.

Ordinary tartaric acid has been known since the eighteenth century and is a by-product of alcoholic fermentation obtained in great quantities from the tartar deposited in the barrels. This acid has been especially important in medicine and dyeing. Paratartaric acid (also called racemic acid), discovered

2 DRUG STEREOCHEMISTRY: ANALYTICAL METHODS AND PHARMACOLOGY

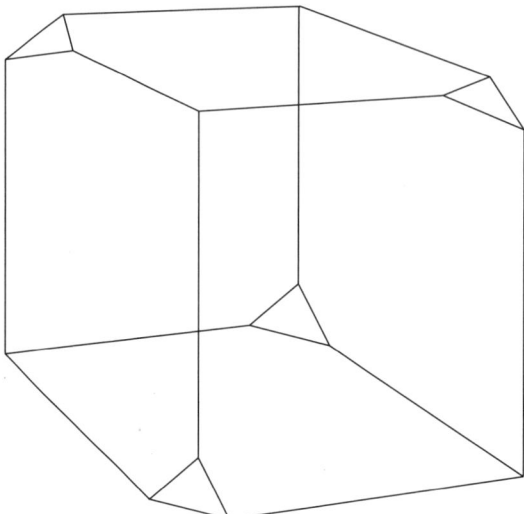

FIGURE 1.1 Hemihedral cube.

in certain industrial processes in the Alsace region of France, came to the attention of chemists only in the 1820s, when Gay-Lussac established that it possessed the same chemical composition as ordinary tartaric acid. Because of their importance for the emerging concept of isomerism, the two acids thereafter attracted considerable notice. On January 20 and February 3, 1860, Pasteur gave lectures before the Council of the Société Chimique of Paris describing the principal results of his research (done from 1848 to 1850) on tartaric acid and paratartaric acid, from which evolved his proposals on the molecular asymmetry of organic products. The excerpts below are taken, with permission, from an English translation made by the Alembic Club (5). An English translation is also found in Pasteur (6). Additional insight is found in Mauskopf (7). The headings and interspersed comments below are mine. To better understand what follows, ordinary tartaric acid is now called dextro-tartaric acid and paratartaric acid is the racemate, (d,l)-tartaric acid.

HEMIHEDRAL CRYSTAL STRUCTURE
Pasteur begins his first lecture by discussing the precedents that led up to his research and then defines hemihedral crystals. These are cubical crystals with four little facets inclined at the same angle to the adjacent surfaces and arranged alternately so the same edge of the cube does not contain two facets (Fig. 1.1). Under these conditions, no point or plane of symmetry exists in the cube.

MOLECULAR ASYMMETRY AND OPTICAL ACTIVITY
Pasteur now describes the research that led to his conclusion about the causal relationship between molecular asymmetry and optical activity.

> When I began to devote myself to special work, I sought to strengthen myself in the knowledge of crystals, foreseeing the help that I should draw from this

in my chemical researches. It seemed to me to be the simplest course, to take, as a guide, some rather extensive work on the crystalline forms; to repeat all the measurements, and to compare my determinations with those of the author. In 1841, M. de la Provostaye, whose accuracy is well known, had published a beautiful piece of work on the crystalline forms of tartaric and paratartaric acids and their salts. I made a study of this memoir. I crystallized tartaric acid and its salts, and investigated the forms of the crystals. But, as the work proceeded, I noticed that a very interesting fact had escaped the learned physicist. All the tartrates which I examined gave undoubted evidence of hemihedral faces.

This peculiarity in the forms of the tartrates was not very obvious. This will be readily conceived, seeing that it had not been observed before. But when, in a species, its presence was doubtful, I always succeeded in making it manifest by repeating the crystallization and slightly modifying the conditions.

The German chemist Eilhard Mitscherlich published a note in 1844 in the *Reports of the Academy of Science* on the subject of the tartrate and paratartrate of sodium and ammonia. The importance of this note is now acknowledged by Pasteur.

I must first place before you a very remarkable note by Mitscherlich which was communicated to the *Academie des Sciences* by Biot. It was as follows:

"The double paratartrate and the double tartrate of soda and ammonia have the same chemical composition, the same crystalline form with the same angles, the same specific weight, the same double refraction, and consequently the same inclination in their optical axes. When dissolved in water their refraction is the same. But the dissolved tartrate deviates the plane of polarisation, while the paratartrate is indifferent, as has been found by M. Biot for the whole series of those two kinds of salts. Yet" adds Mitscherlich, "here the nature and the number of the atoms, their arrangement and distances, are the same in the two substances compared."

This note of Mitscherlich's attracted my attention forcibly at the time of publication. I was then a pupil in the Ecole Normale, reflecting in my leisure moments on these elegant investigations of the molecular constitution of substances, and having reached, as I thought at least, a thorough comprehension of the principles generally accepted by physicists and chemists. The above note disturbed all my ideas. What precision in every detail! Did two substances exist which had been more fully studied and more carefully compared as regards their properties? But how, in the existing condition of the science, could one conceive of two substances so closely alike without being identical? Mitscherlich himself tells us what was, to his mind, the consequence of this similarity:

The nature, the number, the arrangement, and the distance of the atoms are the same. If this is the case what becomes of the definition of chemical species, so rigorous, so remarkable for the time at which it appeared, given by Chevreul in 1823? In compound bodies a species is a collection of individuals identical in the nature, the proportion, and the arrangement of their elements.

In short, Mitscherlich's note remained in my mind as a difficulty of the first order in our mode of regarding material substances.

You will now understand why, being preoccupied, for the reasons already given, with a possible relation between the hemihedry of the tartrates and their rotative property, Mitscherlich's note of 1844 should recur to my memory. I thought at once that Mitscherlich was mistaken on one point. He had not observed that his double tartrate was hemihedral while his

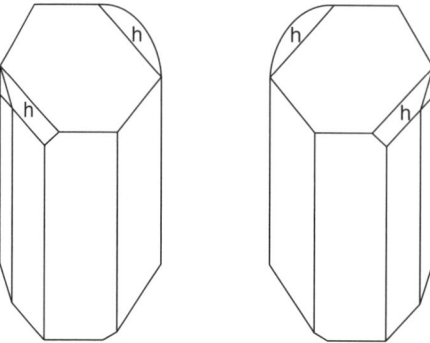

FIGURE 1.2 Paratartrate of soda and ammonia formed by an equal mixture of hemihedral crystals of levo-tartrate (*on left*) and dextro-tartrate (*on right*). The anterior hemihedral facet "h" is on the left side of the observer in the levo-tartrate and on his or her right in the dextro-tartrate. *Source*: From Ref. 4.

paratartrate was not. If this is so, the results in his note are no longer extraordinary; and further, I should have, in this, the best test of my preconceived idea as to the inter-relation of hemihedry and the rotatory phenomenon.

I hastened therefore to re-investigate the crystalline form of Mitscherlich's two salts. I found, as a matter of fact, that the tartrate was hemihedral, like all the other tartrates which I had previously studied, but, strange to say, the paratartrate was hemihedral also. Only, the hemihedral faces which in the tartrate were all turned the same way were in the paratartrate inclined sometimes to the right and sometimes to the left. In spite of the unexpected character of this result, I continued to follow up my idea. I carefully separated the crystals which were hemihedral to the right from those hemihedral to the left, and examined their solutions separately in the polarising apparatus. I then saw with no less surprise than pleasure that the crystals hemihedral to the right deviated the plane of polarisation to the right, and that those hemihedral to the left deviated it to the left (Fig. 1.2); and when I took an equal weight of each of the two kinds of crystals, the mixed solution was indifferent towards the light in consequence of the neutralisation of the two equal and opposite individual deviations.

Thus, I start with paratartaric acid; I obtain in the usual way the double paratartrate of soda and ammonia; and the solution of this deposit, after some days, crystals all possessing exactly the same angles and the same aspect. To such a degree in this case that Mitscherlich, the celebrated crystallographer, in spite of the most minute and severe study possible, was not able to recognise the smallest difference. And yet the molecular arrangement in one set is entirely different from that in the other. The rotatory power proves this, as does also the mode of asymmetry of the crystals. The two kinds of crystals are isomorphous, and isomorphous with the corresponding tartrate. But the isomorphism presents itself with a hitherto unobserved peculiarity; it is the isomorphism of an asymmetric crystal with its mirror image. This comparison expresses the fact very exactly. Indeed, if, in a crystal of each kind, imagine the hemihedral facets produced till they meet, I obtain two symmetrical tetrahedra, inverse, and which cannot be superposed, in spite of the perfect identity of all their respective parts. From this I was justified in concluding that, by crystallisation of the double paratartrate of soda and ammonia, I had separated two symmetrically isomorphous atomic groups, which are intimately united in paratartaric acid. Nothing is easier to show than that these two species of crystals represent two distinct salts from which two different acids can be extracted.

The announcement of the above facts naturally placed me in communication with Biot, who was not without doubts regarding their accuracy. Being charged with giving an account of them to the Academy, he made me come to him and repeat before his eyes the decisive experiment. He handed over to me some paratartaric acid which he had himself previously studied with particular care, and which he had found to be perfectly indifferent to polarised light. I prepared the double salt in his presence, with soda and ammonia which he had likewise desired to provide. The liquid was set aside for slow evaporation in one of his rooms. When it had furnished about 30 to 40 grams of crystals, he asked me to call at the College de France in order to collect them and isolate them before him, by recognition of their crystallographic character, the right and the left crystals, requesting me to state once more whether I really affirmed that the crystals which I should place at his right would deviate to the right, and the others to the left. This done, he told me that he would undertake the rest. He prepared the solutions with carefully measured quantities, and when ready to examine them in the polarising apparatus, he once more invited me to come into his room. He first placed in the apparatus the more interesting solution, that which ought to deviate to the left. Without even making a measurement, he saw by the appearance of the tints of the two images, ordinary and extraordinary, in the analyser, that there was a strong deviation to the left. Then, very visibly affected, the illustrious old man took me by the arm and said:

"My dear child, I have loved science so much throughout my life that this makes my heart throb."

Indeed there is more here than personal reminiscences. In Biot's case the emotion of the scientific man was mingled with the personal pleasure of seeing his conjectures realized. For more than thirty years Biot had striven in vain to induce chemists to share his conviction that the study of rotatory polarisation offered one of the surest means of gaining a knowledge of the molecular constitution of substances.

Let us return to the two acids furnished by the two sorts of crystals deposited in so unexpected a manner in the crystallisation of the double paratartrate of soda and ammonia. I have already remarked that nothing could be more interesting than the investigation of these acids.

One of them, that which comes from crystals of the double salt hemihedral to the right, deviates to the right, and is identical with ordinary tartaric acid. The other deviates to the left, like the salt which furnishes it. The deviation of the plane of polarisation produced by these two acids is rigorously the same in absolute value. The right acid follows special laws in its deviation, which no other active substance had exhibited. The left acid exhibits them, in the opposite sense, in the most faithful manner, leaving no suspicion of the slightest difference.

The paratartaric acid is really the combination, equivalent for equivalent, of these two acids, is proved by the fact that, if somewhat concentrated solutions of equal weights of each of them are mixed, as I shall do before you, their combination takes place with disengagement of heat, and the liquid solidifies immediately on account of the abundant crystallisation of paratartaric acid, identical with the natural product. (This beautiful experiment called forth applause from the audience.)

Pasteur ends the first lecture with the following summary:

1. When the elementary atoms of organic products are grouped asymmetrically, the crystalline form of the substance manifests this molecular asymmetry in nonsuperposable hemihedry.
   The cause of this hemihedry is thus recognised.
2. The existence of this same molecular asymmetry betrays itself, in addition, by the optical rotative property.
   The cause of rotatory polarisation is likewise determined.
3. When the nonsuperposable molecular asymmetry is realised in opposite senses, as happens in the right and left tartaric acids and all their derivatives, the chemical properties of these identical and inverse substances are rigorously the same.

In the second lecture, Pasteur gives a further discussion of his fundamental idea that optical activity of organic solutions is related to molecular geometry. This insight was far ahead of the organic structural theory of the time.

> We saw in the last lecture that quartz possesses the two characteristics of asymmetry—hemihedry in form, observed by Hauy, and the rotative phenomenon discovered by Arago! Nevertheless, molecular asymmetry is entirely absent in quartz. To understand this, let us take a further step in the knowledge of the phenomena with which we are dealing. We shall find in it, besides, the explanation of the analogies and differences already pointed out between quartz and natural organic products.
>
> Permit me to illustrate roughly, although with essential accuracy, the structure of quartz and of the natural organic products. Imagine a spiral stair whose steps are cubes, or any other objects with superposable images. Destroy the stair and the asymmetry will have vanished. The asymmetry of the stair was simply the result of the mode of arrangement of the component steps. Such is quartz. The crystal of quartz is the stair complete. It is hemihedral. It acts on polarized light in virtue of this. But let the crystal be dissolved, fused, or have its physical structure destroyed in any way whatever; its asymmetry is suppressed and with it all action on polarized light, as it would be, for example, with a solution of alum, a liquid formed of molecules of cubic structure distributed without order.
>
> Imagine, on the other hand, the same spiral stair to be constructed with irregular tetrahedra for steps. Destroy the stair and the asymmetry will still exist, since it is a question of a collection of tetrahedra. They may occupy any positions whatsoever, yet each of them will nonetheless have an asymmetry of its own. Such are the organic substances in which all the molecules have an asymmetry of their own, betraying itself in the form of the crystal. When the crystal is destroyed by solution, there results a liquid active towards polarised light, because it is formed of molecules, without arrangement, it is true, but each having an asymmetry in the same sense, if not of the same intensity in all directions.

RESOLUTION OF RACEMATES

Pasteur devised three methods to resolve paratartaric acid: the first was manual, the second was chemical, and the third could be considered biological or physiological. Because paratartaric acid (also called racemic acid) was the first inactive compound to be resolved into optical isomers (enantiomers), an equimolar mixture of two enantiomers is now called a racemate.

THE EARLY HISTORY OF STEREOCHEMISTRY 7

Manual Separation

As indicated in the first lecture, Pasteur, using a hand lens and pair of tweezers, laboriously separated a quantity of the sodium ammonium salt of paratartaric acid into two piles, one of left-handed crystals and the other of right-handed crystals, and in this way accomplished the first resolution of a racemate. After purifying the free tartaric acids from the separate salt solutions, he found one acid to be identical to the previously characterized ordinary tartaric acid (which was dextrorotatory) and the other acid to be the previously unknown levorotatory isomer. Pasteur was extremely fortunate in this area of his research. The tartrate used by him is one of the very few substances that undergo a spontaneous separation into enantiomeric (hemihedral) crystals, thereby allowing resolution by hand. That is, most enantiomers do not form enantiomeric crystals. Moreover, this separation takes place only below 27°C (8). If Pasteur had been working in southern France during a torrid Mediterranean summer, rather than in Paris, we may have praised another chemist as being the first to resolve a racemate.

Chemical Formation of Diastereomers

The physical properties of enantiomers are identical in an achiral environment. However, chemical reactions that add another asymmetric center create a diastereomeric pair, each of which has physical properties that are not completely the same. Therefore, although an enantiomeric pair cannot be separated by ordinary chromatographic means or fractional recrystallization, the diastereomeric pair can often be separated easily by these means, as is indicated later in this book (see chap. 5). After separation, the pure enantiomers can then be regenerated by chemical means. Even today this is a common way of resolving a racemate.

Pasteur, in his second lecture, gives the following account, in which the optically active basic alkaloids quinicine or cinchonicine were used to convert the two enantiomeric tartaric acids into diastereomers:

> We have seen that all artificial or natural chemical compounds, whether mineral or organic, must be divided into two great classes: non-asymmetric compounds with superposable image and asymmetric compounds with non-superposable image.
>
> Taking this into account, the identity of properties above described in the case of the two tartaric acids and their similar derivatives, exists constantly, with the unchangeable characters which I have referred to, whenever these substances are placed in contact with any compound of the class with superposable image, such as potash, soda, ammonia, lime, baryta, aniline, alcohol, ethers—in a word, with any compounds whatever which are non-asymmetric, non-hemihedral in form, and without action on polarised light.
>
> If, on the contrary, they are submitted to the action of products of the second class with non-superposable image—asparagine, quinine, strychnine, brucine, albumen, sugar, etc., bodies asymmetric like themselves—all is changed in an instant. The solubility is no longer the same. If combination takes place, the crystalline form, the specific weight, the quantity of water of crystallisation, the more or less easy destruction by heating, all differ as much as in the case of the most distantly related isomers.

Here, then, the molecular asymmetry of a substance obtrudes itself on chemistry as a powerful modifier of chemical affinities. Towards the two tartaric acids, quinine does not behave like potash, simply because it is asymmetric and potash is not. Molecular asymmetry exhibits itself henceforth as a property capable by itself, in virtue of its being asymmetry, of modifying chemical affinities. I do not believe that any discovery has yet made so great a step in the mechanical part of the problem of combination....

Here is a very interesting application of the facts which have just been explained.

Seeing that the right and left tartaric acids formed such dissimilar compounds simply on account of the rotative power of the base, there was ground for hoping that, from this very dissimilarity, chemical forces might result, capable of balancing the mutual affinity of the two acids, and thereby supply a chemical means of separating the two constituents of paratartaric acid. I sought long in vain, but finally succeeded by the aid of two new bases, quinicine and cinchonicine, isomers of quinine and cinchonine, which I obtained very easily from the latter without the least loss.

I prepare the paratartrate of cinchonicine by neutralising the base and/then adding as much of the acid as necessary for the neutralisation, I allow the whole to crystallise, and the first crystallisations consist of perfectly pure left tartrate of cinchonicine. All the right tartrate remains in the mother liquor because it is more soluble. Finally this itself crystallises with an entirely different aspect, since it does not possess the same crystalline form as the left salt. We might also believe that we were dealing with the crystallisation of two distinct salts of unequal solubility.

Use of Living Organisms

Pasteur also discovered a method for resolving paratartaric acid while he was deeply involved in the study of fermentation. In essence, it depends on the capacity of certain microorganisms to discriminate between enantiomers and selectively to metabolize one instead of the other. This method is obviously less desirable than the chemical method since, at best, only one pure enantiomer can be obtained. The particular example described below by Pasteur in his second lecture grew out of his study of the fermentation of ammonium paratartrate.

Knowing this, I set the ordinary right tartrate of ammonia to ferment in the following manner. I took the very pure crystallised salt, dissolved it, adding to the liquor a clear solution of albumenoid matter. One gram of albumenoid matter was sufficient for one hundred grams of tartrate. Very often it happens that the liquid ferments spontaneously when placed in an oven. I say very often; but it may be added that this will always take place if we take care to mix with the liquid a very small quantity of one of those liquids with which we have succeeded in obtaining spontaneous fermentation.

So far there is nothing peculiar; it is a tartrate fermenting. The fact is well known.

But let us apply this method of fermentation to paratartrate of ammonia, and under the above conditions it ferments. The same yeast is deposited. Everything shows that things are proceeding absolutely as in the case of the right tartrate. Yet if we follow the course of the operation with the help of the polarising apparatus, we soon discover profound differences between the two operations. The originally inactive liquid possesses a sensible rotative power to the left, which increases little by little and reaches a maximum. At this point the fermentation is suspended. There is no longer a trace of the right acid in the liquid. When it is evaporated and mixed with an equal volume of alcohol it gives immediately a beautiful crystallisation of left tartrate of ammonia.

Let us note, in the first place, two distinct things in this phenomenon. As in all fermentation properly so called, there is a substance which is changed chemically, and correlatively there is a development of a body possessing the aspect of a mycodermic growth. On the other hand, and it is this which it is important to note, the yeast which causes the right salt to ferment leaves the left salt untouched, in spite of the absolute identity in physical and chemical properties of the right and left tartrates of ammonia as long as they are not subjected to asymmetric action.

Here, then, the molecular asymmetry proper to organic substances intervenes in a phenomenon of a physiological kind, and it intervenes in the role of a modifier of chemical affinity. It is not at all doubtful that it is the kind of asymmetry proper to the molecular arrangement of left tartaric acid which is the sole and exclusive cause of the difference from the right acid, which it presents in relation to fermentation.

Thus we find introduced into physiological principles and investigations the idea of the influence of the molecular asymmetry of natural organic products, of this great character which establishes perhaps the only well marked line of demarcation that can at present be drawn between the chemistry of dead matter and the chemistry of living matter.

Later qualified, modified, and generalized by others, Pasteur's new method became applicable to the separation of a number of other racemates (9). Pasteur then ends his second lecture with the following:

Such, gentlemen, are in co-ordinated form the investigations which I have been asked to present to you.

You have understood, as we proceeded, why I entitled my exposition, "On the Molecular Asymmetry of Natural Organic Products." It is, in fact, the theory of molecular asymmetry that we have just established, one of the most exalted chapters of the science. It was completely unforeseen, and opens to physiology new horizons, distant, but sure.

I hold this opinion of the results of my own work without allowing any of the vanity of the discoverer to mingle in the expression of my thought. May it please God that personal matters may never be possible at this desk. These are like pages in the history of chemistry which we write successively with that feeling of dignity which the true love of science always inspires.

Although popularly known chiefly for his great work in bacteriology and medicine, Pasteur was by training a chemist, and this work in chemistry alone would have earned him a position as an outstanding scientist.

The development of stereochemical ideas entered a new stage in 1858 when August Kekule introduced the idea of the valence bond and the pictorial representation of molecules as atoms connected by valence bonds. His main thesis was that the carbon atom is tetravalent, and that a carbon atom can form valence bonds with other carbon atoms to form open chains and that sometimes the carbon chains can be closed to form rings (10). This led directly to his proposal for the structure of benzene. On the occasion of celebrations held in his honor, Kekule in 1890 delivered a speech before the German Chemical Society describing the origin of his idea of the linking of atoms (10).

> During my stay in London I resided for a considerable time in Clapham Road in the neighborhood of the Common. I frequently, however, spent my evenings with my friend Hugo Muller at Islington, at the opposite end of the giant town.... One fine summer evening I was returning by the last omnibus, "outside," as usual, through the deserted streets of the metropolis, which are at other times so fully of life. I fell into a reverie and lo, the atoms were gambolling before my eyes! Whenever, hitherto, these diminutive beings had appeared to me, they had always been in motion; but up to that time I had never been able to discern the nature of their motion. Now, however, I saw how, frequently, two smaller atoms united to form a pair; how a larger one embraced two smaller ones; how still larger ones kept hold of three or even four of the smaller; whilst the whole kept whirling in a giddy dance. I saw how the larger ones formed a chain, dragging the smaller ones after them, but only at the ends of the chain.... The cry of the conductor: "Clapham Road," awakened me from my dreaming; but I spent a part of the night in putting on paper at least sketches of these dream forms. This was the origin of the *Structurtheorie*.

Then he related a similar experience of how the idea for the structure of benzene occurred to him.

> I was sitting writing at my textbook, but the work did not progress; my thoughts were elsewhere. I turned my chair to the fire and dozed. Again the atoms were gambolling before my eyes. This time the smaller groups kept modestly in the background. My mental eye, rendered more acute by repeated visions of this kind, could now distinguish larger structures of manifold conformations; long rows, sometimes more closely fitted together; all twisting and turning in snake-like motion. But look! What was that? One of the snakes had seized hold of its own tail, and the form whirled mockingly before my eyes. As if by a flash of lightning I awoke; and this time also I spent the rest of the night working out the consequences of the hypothesis. Let us learn to dream, gentlemen, and then perhaps we shall find the truth ... but let us beware of publishing our dreams before they have been put to the proof by the waking understanding.

In speculating on the kind of atomic arrangements that could produce molecular asymmetry, Pasteur, as already indicated, suggested tentatively in 1860 that the atoms of a right-handed compound, for example, might be "arranged in the form of a right-handed spiral, or situated at the corners of an irregular tetrahedron." But he never developed these suggestions. The solution to this problem of what is the cause of molecular asymmetry was presented in the publications of van't Hoff and Le Bel. On September 5,1874, van't Hoff, while he was still a student at the University of Utrecht and only 22 years of age, published a pamphlet entitled "Proposal for the extension of the structural formulae now in use in chemistry into space, together with a related note on the relation between the optical active power and the chemical constitution of organic compounds" (11). An English translation is presented in van't Hoff (12). Starting with the ideas of August Kekule on the tetravalency of carbon, van't Hoff states, at the beginning of his pamphlet: "It appears more and more that the present constitutional formulas are incapable of explaining certain cases of isomerism; the reason for this is perhaps the fact that we need a more definite statement about the actual positions of the atoms." He then proposed that the four valences of a carbon atom are directed toward the corners of a tetrahedron with the carbon atom at

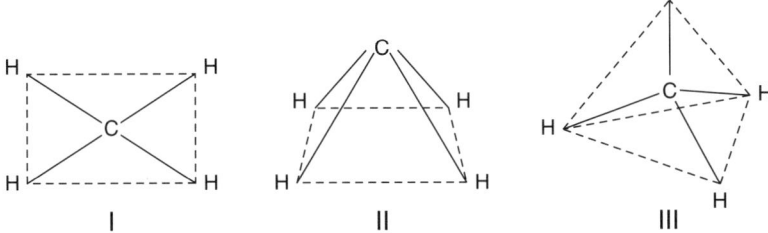

FIGURE 1.3 Spatial models for methane where the four hydrogen atoms are equivalent. I, planar; II, pyramidal; III, tetrahedral.

the center, based on the concept of the isomer number, which is illustrated below.

For any atom Y, only one substance of formula CH_3Y has ever been found. For example, chlorination of methane yields only one compound of formula CH_3Cl. Indeed, the same holds true if Y represents, not just an atom, but a group of atoms (unless the group is so complicated that in itself it brings about isomerism); there is only one CH_3OH, and only one CH_3CO_2H. This suggests that every hydrogen atom in methane is equivalent to every other hydrogen atom, so that replacement of any one of them gives rise to the same product. If the hydrogen atoms of methane were not equivalent, then replacement of one would yield a different compound than replacement of another, and isomeric substitution products would be obtained. In what ways can the atoms of methane be arranged so that the four hydrogen atoms are equivalent? There are three such arrangements (Fig. 1.3): a planar arrangement (I) in which carbon is at the center of a rectangle (or square) and a hydrogen atom is at each corner; a pyramidal arrangement (II) in which carbon is at the apex of a pyramid and a hydrogen atom is at each corner of a square base; a tetrahedral arrangement (III) in which carbon is at the center of a tetrahedron and a hydrogen atom is at each corner. By then comparing the number of isomers that have been prepared for di-, tri- and tetrasubstituted methanes with the number predicted by the above three spatial arrangements, it is possible to decide which one is correct.

For example, with a disubstituted compound CH_2R_2 (Fig. 1.4); (*i*) if the molecule is planar, then two isomers are possible. This planar configuration can be either square or rectangular; in each case, there are two isomers only. (*ii*) If the molecule is pyramidal, then two isomers are also possible. There are only two isomers, whether the base is square or rectangular. (*iii*) If the molecule is tetrahedral, then only one form is possible. The carbon atom is at the center of the tetrahedron. In actuality, only one disubstituted isomer is known. Therefore, only the tetrahedral model for a disubstituted methane agrees with the evidence of the isomer number.

For tetrasubstituted compounds of the type $CR_1R_2R_3R_4$ (Fig. 1.5); (*i*) if the molecule is planar, then three isomers are possible. (*ii*) If the molecule is pyramidal, then six isomers are possible. Each of the forms in Figure 1.5, top, drawn as a pyramid, is not superimposable on its mirror image. Thus, three pairs of enantiomers are possible (one of which is shown in Fig. 1.5, middle).

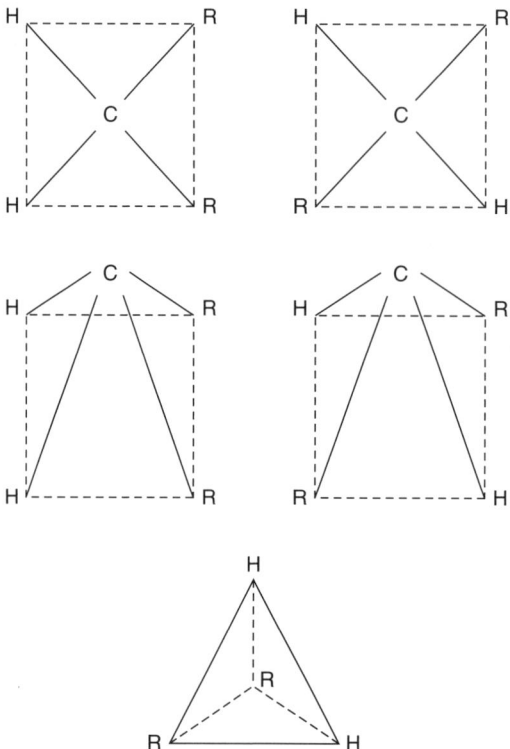

FIGURE 1.4 Spatial models for a disubstituted methane. Top, planar; middle, pyramidal; bottom, tetrahedral.

(*iii*) If the molecule is tetrahedral, two isomers are possible, related to one another as object to mirror image. In actuality, only two tetrasubstituted isomers of methane are known (pair of enantiomers). This is strong evidence for the tetrahedral model for the carbon atom. Similar reasoning leads to the same conclusion for trisubstituted methanes.

The tetrahedral model for the carbon atom has withstood the test of time very well. Hundreds of thousands of organic compounds have been synthesized since it was first proposed. The number of isomers obtained has always been consistent with the concept of the tetrahedral carbon atom.

van't Hoff then introduced the concept of the asymmetric carbon atom as follows: "When the four affinities of the carbon atom are satisfied by four univalent groups differing among themselves, two and not more than two different tetrahedrons are obtained, one of which is the reflected image of the other, they cannot be superposed; that is, we have to deal with two structural formulas isomeric in space." van't Hoff proposed that all carbon compounds that in solution rotate the plane of polarization possess an asymmetric carbon atom. He illustrated this for a great number of compounds: ethylidene lactic acid (now called α-hydroxypropionic acid), aspartic acid, asparagine, malic acid, glutaric acid, tartaric acid, sugars and glucosides, camphor, borneol, and camphoric acid.

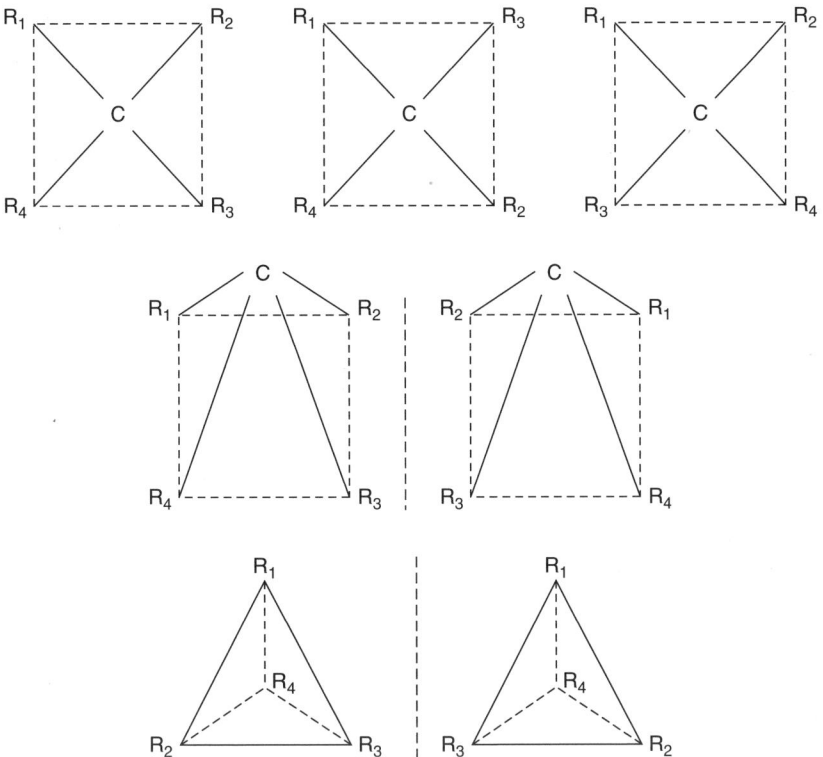

FIGURE 1.5 Spatial models for a tetrasubstituted methane. Top, planar; middle, pyramidal; bottom, tetrahedral.

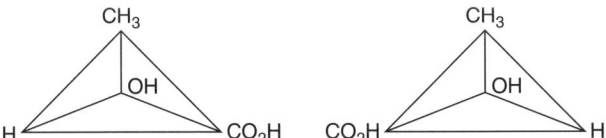

FIGURE 1.6 Tetrahedral model for lactic acid enantiomers (carbon atom is at the center of the tetrahedron) as envisioned by van't Hoff.

Two compounds from this list are worthy of note: lactic acid (Fig. 1.6) and tartaric acid (Fig. 1.7). Wislicenus extensively investigated the isomers of lactic acid between 1863 and 1873, and was convinced that the number of isomers exceeded that allowed by the existing structural theory (13). However, due to experimental difficulties in obtaining pure samples of the isomers, in addition to the limits of the structural theory then known to him, he ended up going around in circles, van't Hoff studied the publications of Wislicenus on lactic acids and they led him to his own stereochemical ideas. In fact, lactic acid was the first concrete example of an optically active compound that van't Hoff discussed after his theoretical introduction. He pointed out that ethylidene lactic acid

14 DRUG STEREOCHEMISTRY: ANALYTICAL METHODS AND PHARMACOLOGY

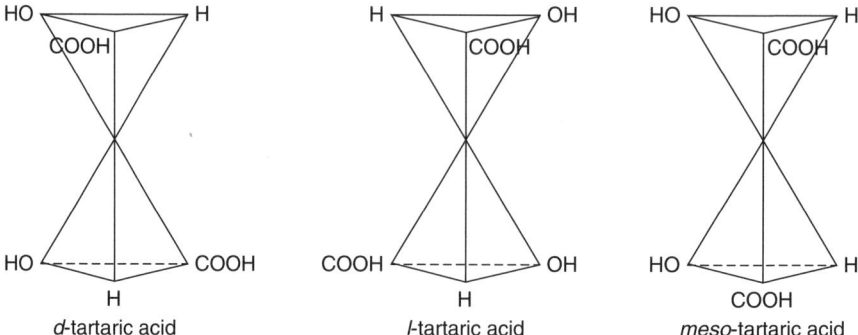

FIGURE 1.7 Structures for three tartaric acid isomers that are representative of the tetrahedral models used by van't Hoff.

contains an asymmetric carbon. Therefore, it can exist as two pure enantiomers or a racemic mixture, which nicely cleared up the confusion surrounding the lactic acid isomers. In a lecture, held much later in Utrecht on May 16, 1904, van't Hoff said the following:

> Students, let me give you a recipe for making discoveries. In connexion with what has just been said about libraries, I might remark that they have always had a mind-deadening effect on me. When Wislicenus' paper on lactic acid appeared and I was studying it in the Utrecht library, I therefore broke off my study half-way through, to go for a walk; and it was during this walk, under the influence of the fresh air, that the idea of asymmetric carbon first struck me.

These proposals of van't Hoff's came as a breath of fresh air to Wislicenus. No wonder that he was the first to welcome it enthusiastically, or that he sponsored the German translation that made it widely known, or that he was the first to make significant further use of the hypothesis, in his work on geometrical isomers of unsaturated compounds (14).

The other example of note is the optically active tartaric acids (Fig. 1.7). Tartaric acid contains two asymmetric carbon atoms. The dextro- and levo-tartaric acids are enantiomers. However, a third isomer is possible in which the two rotations due to the two asymmetric carbon atoms compensate and the molecule is optically inactive as a whole. That is, the molecule contains a plane of symmetry. This form, meso-tartaric acid, was also discovered by Pasteur, differs from the two optically active tartaric acids in being internally compensated, and is not resolvable. Thus, the tetrahedral model for carbon and the asymmetric carbon atom proposed by van't Hoff were able to completely explain the observations of Pasteur relating to the three isomers of tartaric acid.

Le Bel published his stereochemical ideas two months later, in November 1874, under the title, "The relations that exist between the atomic formulas of organic compounds and the rotatory power of their solutions" (15). An English translation is presented in Le Bel (16). Le Bel approached the problem from a different direction from van't Hoff. His hypothesis was based on neither the tetrahedral model of the carbon atom nor the concept of fixed valences between the atoms. He proceeded purely from symmetry arguments; he spoke of the asymmetry, not of individual atoms, but of the entire molecule, so that his views would nowadays be classed under the heading of molecular asymmetry. Only once does he mention the tetrahedral carbon atom, which he regarded as not a general principle but a special case. Today, substituted allenes, spiranes, and biphenyls are but a few examples of asymmetric molecules that do not contain any asymmetric carbons, thus confirming Le Bel's views on molecular asymmetry. The reason for the different approaches by van't Hoff and le Bel is easy to understand. van't Hoff came from the camp of structural chemists and he wished his hypothesis to be understood as an extension of the structural theory to spatial relationships. The tetravalent atomic models used by Kekule in his lectures presumably also prompted his pupil van't Hoff, possibly unconsciously, in the conception of the asymmetric carbon atom. Le Bel, on the other hand, was trained in the tradition of Pasteur (whose investigations he also mentioned expressly in his article), that is, he started out from Pasteur's considerations of the connections between optical rotation and molecular structure.

In 1877 Hermann Kolbe, one of the most distinguished of the older German chemists, published a diatribe in the *Journal für Praktische Chemie* after reading the work of van't Hoff (which had been translated into German by Felix Herrmann at the suggestion of Wislicenus). An English translation of this abusive attack is presented completely in Riddell and Robinson (3). Those individuals interested in seeing an example of the great personal attacks by editors that appeared in journals of the nineteenth century should read this translation. Although defamatory, this criticism served a useful purpose, since it made a decisive contribution to the dissemination of these ideas of van't Hoff. This was fortunate, since van't Hoff soon turned his genius away from stereochemistry to physical chemistry, for which he received the Nobel Prize.

We can now end this historical journey. We have walked through the early days of stereochemistry in the company of giants. In 1949, almost exactly 100 years after the first resolution of (d,l)-tartaric acid by Pasteur, the Dutchman Bijvoet (17), using X-ray diffraction, determined the actual arrangement in space of the atoms of the sodium rubidium salt of (+)-tartaric acid, and thus made the first determination of the absolute configuration about an asymmetric carbon. To further complete the link with the past, Bijvoet did this work while the Director of the van't Hoff Laboratory at the University of Utrecht.

In the intervening years since the first resolution of a racemate by Pasteur, many chromatographic and non-chromatographic methods have been developed for the resolution of racemic compounds. These methods are the subject of many of the other chapters in this book.

REFERENCES

1. Weyer J. A hundred years of stereochemistry—the principal development phases in retrospect. Angew Chemie Internat Ed1974; 23:591–598.
2. Partington JR. A History of Chemistry. Vol. 4. London: Macmillan and Co., Ltd., 1964:749–764.
3. Riddell EG, Robinson MJT. J. H. van't Hoff and J. A. Le Bel—their historical context. Tetrahedron 1974; 30:2001–2007.
4. Vallery-Radot R. The Life of Pasteur (Devonshire RL, transl.). New York: Garden City Publishing Co., Inc., 1926.
5. Pasteur L. Researches on the molecular asymmetry of natural organic products. Alembic Club Reprints, No. 14, reissue edition. Edinburgh: F. and S. Livingstone, Ltd., 1948.

6. Pasteur L. On the asymmetry of naturally occurring organic compounds. In: Richardson GM, ed. The Foundations of Stereo Chemistry: Memoirs by Pasteur, Van't Hoff, Le Bel, and Wislicenus. New York: American Book Co., 1901:1–33.
7. Mauskopf SH. Crystals and Compounds: Molecular Structure and Composition in Nineteenth-Century French Science. Philadelphia: American Philosophical Society, 1976:68–80.
8. van't Hoff JH. The Arrangement of Atoms in Space (Eiloart A, transl. ed.). New York: Longmans, Green, and Co., 1898:34–40.
9. van't Hoff JH, The Arrangement of Atoms in Space (Eiloart A, transl. ed.). New York: Longmans, Green, and Co., 1898:30–33.
10. Japp FP. Kekule memorial lecture. Chem Soc 1989; 73:97–138.
11. van't Hoff JH. Voorstel tot uitbreiding der tegenwoordig in de scherkundegebruiktestructuur-formules in de ruimte. Greven: Utrecht, 1874.
12. van't Hoff JH. A suggestion looking to the extension into space of the structural formulas at present used in chemistry and a note upon the relation between the optical activity and the chemical constitution of organic compounds. In: Richardson GM, ed. The Foundations of Stereo Chemistry: Memoirs by Pasteur, Van't Hoff, Le Bel, and Wislicenus. New York: American Book Co., 1901:37–46.
13. Fisher NW. Wislicenus and lactic acid: the chemical background to van't Hoff's hypothesis. In: Bertrand OB, ed. van't Hoff-Le Bel Centennial. ACS Symp Ser 1975; 12:33–54.
14. Wislicenus J. The space arrangement of the atoms in organic molecules and the resulting geometric isomerism in unsaturated compounds. In: Richardson GM, ed. The Foundations of Stereo Chemistry: Memoirs by Pasteur, Van't Hoff, Le Bel, and Wislicenus. New York: American Book Co., 1901:61–132.
15. Le Bel JA. Sur les relations qui existent entre les formulesatomiques des corps organiques, et le pouvoirrotatoire de leur dissolutions. Bull Soc Chim Paris 1874; 22:337.
16. Le Bel JA. On the relations which exist between the atomic formulas of organic compounds and the rotatory power of their solutions. In: Richardson GM, ed. The Foundations of Stereo Chemistry: Memoirs by Pasteur, Van't Hoff, Le Bel, and Wislicenus. New York: American Book Co., 1901:49–59.
17. Bijvoet JM, Peerdeman AF, van Bommei AJ. Determination of the absolute configuration of optically active compounds by means of X-rays. Nature (Lond.) 1951; 268:271–272.

2 Stereochemistry—basic terms and concepts

Krzysztof Jóźwiak

INTRODUCTION

According to the International Union of Pure and Applied Chemistry (IUPAC) definition (1), stereoisomerism is a type of isomerism that arises from the differences in the spatial arrangement of atoms without any differences in connectivity or bond multiplicity between the isomers. One of the most important branches of stereochemistry is related to molecular dissymmetry and to the study of chiral molecules. The terminology used to describe stereochemical relationships is often a maze of interchangeable terms (capital D's and L's, lowercase *d*'s and *l*'s, mixed in with *R*'s, *S*'s, (+)'s, and (−)'s, to name a few). It is therefore appropriate to address basic stereochemical terms and concepts to lay a foundation for the more technical discussions that follow. This is not meant to be an in-depth treatment of this topic; there are many fine texts on the subject (2,3), which may be consulted if more detailed understanding is required. The chapter is the update to work previously published by I.W. Wainer and A.A. Marcotte in the 2nd edition of this book.

SYMMETRY AND DISSYMMETRY

Symmetry or the lack of it is one of the interesting features of geometric figures with two or more dimensions. Dissymmetry is very common in real life: it may often be confronted without it being immediately apparent. On the other hand, dissymmetry may become painfully apparent when, for example, someone has to switch from driving on one side of the road to the other while passing the English Channel from France to England or vice versa. The Latin alphabet is very good example containing both symmetrical and asymmetrical two-dimensional letters, some of which have different appearances when they are reflected in a mirror. Six letters and their mirror images are presented in Figure 2.1. There is no difference between each of the symmetrical letters A, H, or Y and their corresponding mirror images; however, the mirror images of the asymmetrical letters G, R, or F appear reversed. It is important to note that the first three letters have an internal plane of symmetry (a vertical line bisecting the letter into two parts), while no such plane of symmetry can be found for the letters G, R, and F.

Actually, two-dimensional objects like letters always have a plane of symmetry, which is the plane of the paper. The mirror image of the letter R can be lifted out of the plane of the paper, turned over, and placed exactly on top of the original figure. The real dissymmetry leading to the presence of non-identical mirror image exists for objects with three dimensions. The human body is an example of an asymmetrical three-dimensional figure. The mirror images of the right hand, foot, or ear appear to be the left hand, foot, or ear. These mirror images cannot be made identical by simple spatial manipulations. A mirror image of a left hand cannot be placed exactly on top of a right hand (with both palms up or both palms down). Thus, three-dimensional figures may exist as nonsuperimposable mirror images.

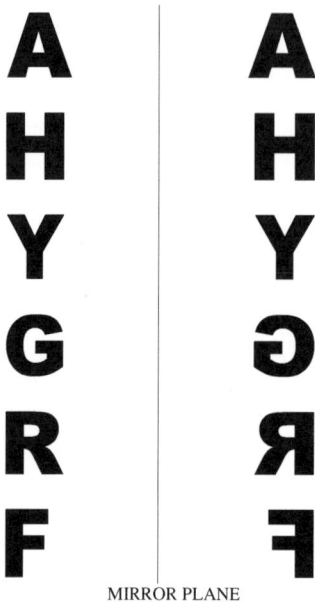

FIGURE 2.1 Examples of symmetrical and dissymmetrical letters in the Latin alphabet and their mirror images.

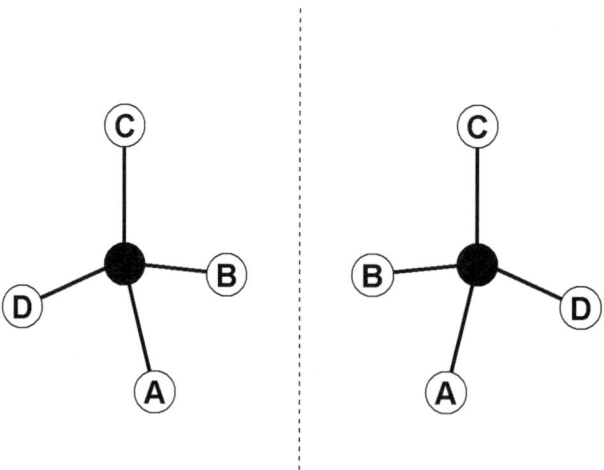

FIGURE 2.2 An asymmetric tetrahedral carbon atom and its mirror image.

The spatial relationships that exist for the human body also exist for molecules as three-dimensional symmetrical and dissymmetrical objects. One of the simplest dissymmetric molecules is a tetrahedral carbon atom with four different groups attached to it. The spatial arrangement of the atoms in this molecule is shown in Figure 2.2. The carbon atom from Figure 2.2 is the dissymmetric center of the molecule and the molecule is a chiral stereoisomer.

STEREOCHEMISTRY—BASIC TERMS AND CONCEPTS

Rigorous symmetry group theory classifies a chiral object or a molecule as the one having no symmetry elements of the second kind (a mirror plane, a center of inversion, and a rotation-reflection axis) (1) described in detail in chapter 4. If the molecule and its mirror image are nonsuperimposable, the relationship between the two molecules is enantiomeric, and the two stereoisomers are enantiomers. Carbon is not the only atom that can act as an asymmetric center. Phosphorus, sulfur, and nitrogen are among some of the other atoms that can form chiral molecules; some examples are presented in section "Types of Stereoisomers" of this chapter.

OPTICAL ROTATION

Because each member of a pair of enantiomeric molecules differs from the other only in the spatial arrangement of the moieties attached to the chiral center, most of their physical properties like melting and boiling points, density, refractive index, etc., are identical. The major difference between the isomers of an enantiomeric pair was first observed by Biot in 1815 when he noted that one form of tartaric acid rotated plane-polarized light, whereas another form did not (see chap. 1).

Light is a form of electromagnetic radiation and is composed of electric and magnetic fields that oscillate in all directions perpendicular to each other and to the direction from which the beam is propagated. In plane-polarized light, the component electric and magnetic fields oscillate as in ordinary light, except that they are contained within two perpendicular planes. When the electric component of light interacts with an asymmetric molecule, the direction of the field is altered or rotated because of the dissymmetry of the molecule. The substance that rotates the plane-polarized light is said to be optically active. Because enantiomers exist as mirror images, they interact with electric component of light to an equal but opposite extent. This situation is depicted in Figure 2.3; in viewing this figure, the observer is looking directly at the beam of plane-polarized light, which is initially at position 0. One of the isomers rotates this beam in a counterclockwise direction. This isomer is defined as the levorotatory or

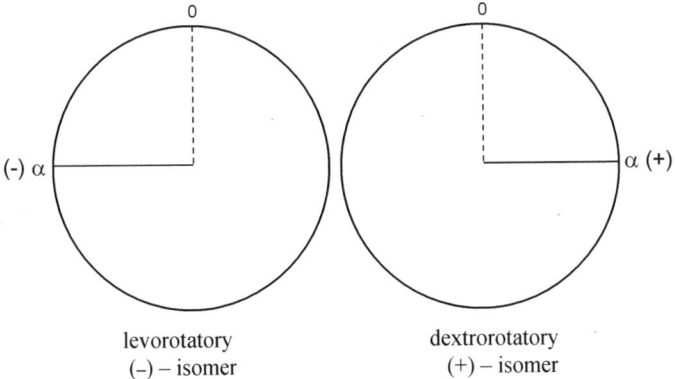

FIGURE 2.3 The rotation of plane-polarized light by the (−) and (+) enantiomers of an optically active substance. In the illustration, the polarized light is being propagated in a plane perpendicular to the plane of the page.

l-enantiomer, and the angle of the rotation α is defined as a negative (–) rotation. The other isomer rotates the beam of plane-polarized light in a clockwise direction and is defined as dextrorotatory or *d*-enantiomer, and α is defined as a positive (+) rotation (4). *d*- and *l*- notations for enantiomers used to be popular at the beginning of stereochemistry. However, it must be underlined that due to confusion with the Fischer naming convention (see section "Naming of Stereoisomers"), the use of *d*- and *l*- notations is strongly discouraged by IUPAC. Now *d*- and *l*- nomenclature is entirely replaced by (+)– and (–)- in describing plane-polarized light rotation direction. The number and type of molecules through which the beam of light passes influence the magnitude of α. Macroscopically, observed rotation is dependent on the concentration of the chiral substance in the solution, the distance through which the light travels, the temperature at which the measurements are made, and the wavelength of light used. There are some known cases where ionization state of chiral substance influences the magnitude and the sign of rotation of plane-polarized light. For example, an aqueous solution of naproxen sodium salt shows negative (–) rotation, which turns into positive (+) rotation in acidic media, where naproxen occurs as a neutral molecule (5).

The optical activity in respect to plane-polarized light was the central feature of interest of chiral substances in the first century of studies in stereochemistry. Nowadays, the key focus of research involving stereochemistry has moved into other directions and the main hallmarks are unequal interactions of chiral isomers with the dissymmetric environment, mainly of biological origin in biological systems. Dissymmetric enzymes or receptors specifically recognize dissymmetric substrate or drug molecule. In this way, a mixture of enantiomers can be resolved using an asymmetric selector in a separation technique, etc. Such modern aspects of stereochemistry are the focal point of this volume.

TYPES OF STEREOISOMERS

When molecules composed of the same constituents have the same structural formulae but differ only with respect to the spatial arrangement of certain atoms or groups of atoms, they are defined as stereoisomers. Chiral stereoisomers are those that are optically active. A group of chiral stereoisomers constitutes at least one pair of enantiomers: nonsuperimposable molecules, which relates to each other as their mirror images. Stereoisomers that are not related to each other as enantiomers are diastereomers or diastereoisomers. In some cases, a diastereoisomer may acquire an element of symmetry forbidden for a chiral molecule and, thus, lose its optical activity.

Enantiomers

The largest class of chiral molecules is compounds in which the asymmetric center (called a center of chirality) is a tetravalent carbon atom, as in Figure 2.2. The tetrahedral orientation of the bonds to a tetravalent carbon is such that when four nonidentical ligands are present, the mirror image of the molecule is nonsuperimposable, the molecule is enantiomeric and chiral. When two of the moieties are identical, the mirror image is superimposable, and the molecule is achiral. Several examples of chiral drug molecules are presented in Figure 2.4.

STEREOCHEMISTRY—BASIC TERMS AND CONCEPTS

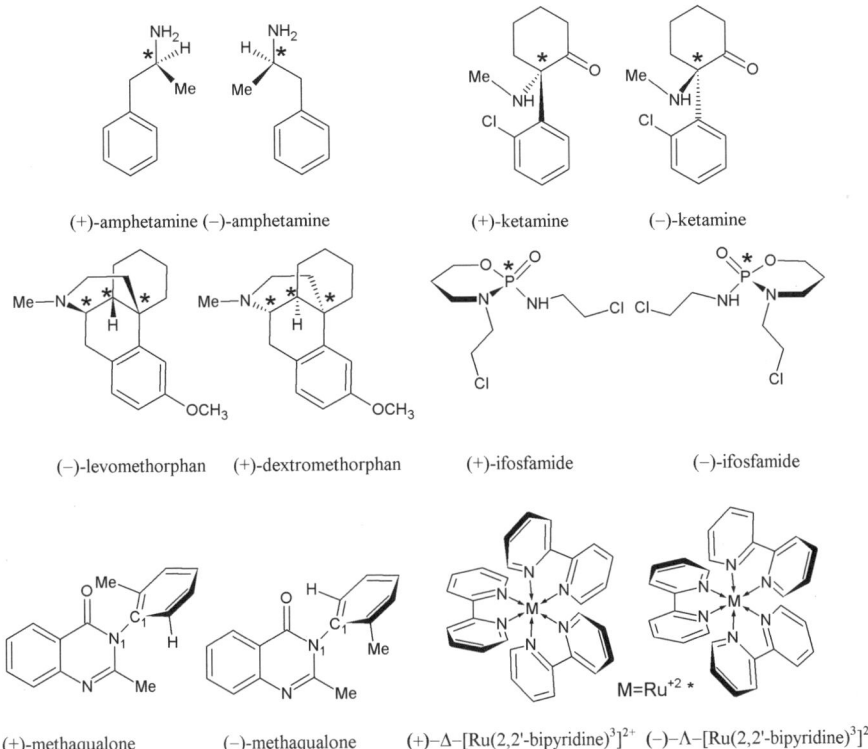

FIGURE 2.4 Examples of enantiomers. Atoms constituting centers of chirality are labeled with asterisks.

Chirality is not necessarily associated with tetrahedral hybridization of valencies. Interesting example is the metal-complex compound tris(2,2'-bipyridine)ruthenium(II) (Fig. 2.4). In this case the octahedral Ru^{2+} cation is a center of chirality. It coordinates three planar and achiral bipyridine molecules and the spatial arrangement of these planes predicates dissymmetry of the complex. The structure resembles a three-winged propeller with two possible rotations: clockwise (the Δ enantiomer) or counterclockwise (the Λ enantiomer).

Molecules that do not possess an asymmetric center may still have nonsuperimposable mirror images and exist as enantiomers. These molecules contain a chiral axis or chiral plane and are dissymmetric with respect to either that plane or axis. The structures of the enantiomers of the sedative-hypnotic methaqualone are presented in Figure 2.4. In this molecule there is a chiral axis between the nitrogen atom N-1 and carbon atom at the phenyl ring C-1. The dissymmetry of the two forms of the molecule is a result of steric hindrance, which prevents rotation around this axis (4). Some other examples of axially dissymmetric molecules include allenes, biaryls, or alkylidenecyclohexanes. Higher level of axial dissymmetry is represented by molecules with helical twist in a structure (e.g., hexahelicene). A planar dissymmetry occurs in many cyclic molecules (e.g., *trans*-1,2-dichlorocyclopropane or *trans*-cyclooctene).

Diastereomers

A compound with one stereogenic center may exist as two stereoisomers, which are mirror image enantiomers. A molecule with n asymmetric centers can exist in maximally 2^n stereoisomeric forms. Enantiomeric pairs are these among them, which have all centers of chirality in reversed configurations and consequently are related to each other as mirror images. Isomers with other combinations of chiral configurations are not related to each other as an object and its mirror image and are called diastereoisomers or diastereomers. Unlike enantiomers, the physical and chemical properties of diastereomers can differ and it is not unusual for them to have different melting and boiling points, refractive indices, solubilities, chemical reactivities, etc. Their optical rotations can differ in both sign and magnitude.

Figure 2.5A shows four stereoisomers of sympathomimetic amines originated from *Ephedra* plants ephedrine and pseudoephedrine (with two asymmetric carbon atoms). By convention, two molecules in which hydroxyl- and methyl- groups attached to two chiral centers point in the same direction are enantiomers of ephedrine. Consequently, molecules that point these two moieties into the opposite directions are enantiomers of pseudoephedrine. In these molecules, different relation of configurations of two chiral centers result in a non-mirror-image relationship between ephedrine and pseudoephedrine. Another example of two diastereomers of pharmaceutical interest are natural alkaloids, quinine and quinidine, molecules that contain in total five stereogenic centers (including one asymmetric nitrogen atom) (Fig. 2.5B). These two diastereomers differ by configuration of only one asymmetric carbon atom (denoted as C_1).

Sometimes a combination of configurations of stereogenic centers in a molecule ends up in a diastereoisomer, which acquires a plane of internal symmetry. Such a symmetric system is called a *meso*-diastereomer. Figure 2.5C illustrates three stereoisomers of tartaric acid. Two of them are enantiomers and the configuration of the third one is associated with the internal plane of symmetry, which implies that the mirror image of the molecule will be the same as the original. This is *meso*-tartaric acid.

Cis-trans Isomers

Molecules that contain a double bond, for example, C=C or C=N, can exist as diastereoisomers. Such a double bond is planar as associated with sp^2 hybridization of atoms forming it; thus, a system contains a plane of symmetry and such stereoisomers are usually not chiral. Stereoisomers differ by the spatial configuration of moieties attached to the double bond. A set of such isomers used to be classified as geometrical isomers, but this nomenclature is no longer recommended (1). Currently, these types of compounds are referred to as "*cis-trans* isomers." In these molecules the source of stereoisomerism is the location of moieties attached to the double bond in respect to the axis of this bond. In *cis*-isomer two moieties are placed on the same side of the axis of the double bond, while in *trans*- isomer these moieties are on the opposite sides. When a molecule contains four different substituents attached to the double bond a *cis-trans* naming may appear unambiguous, in this case the atomic mass priority rule (see section "The Cahn-Ingold-Prelog Convention") is applied to assign substituent order and a molecule containing two heavier moieties on one side (i.e., a

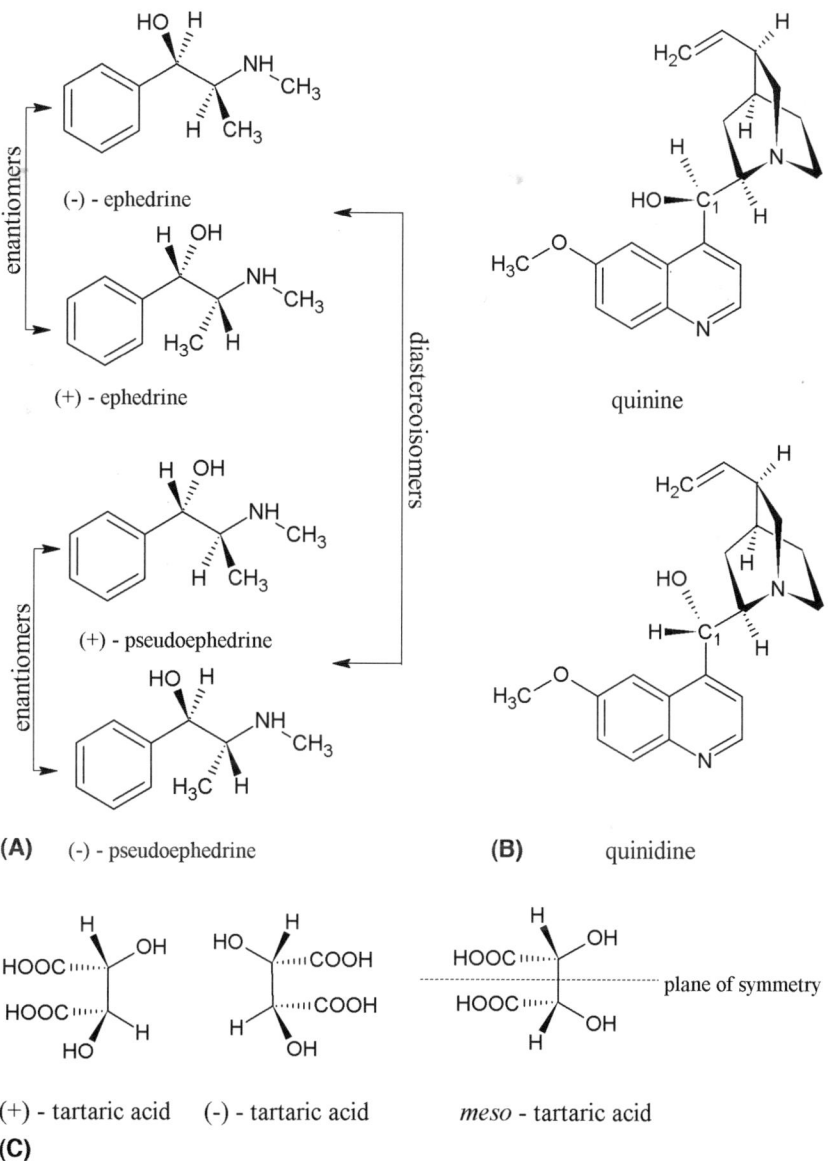

FIGURE 2.5 Examples of diastereoisomers. **(A)** 2-(methylamino)-1-phenylpropan-1-ol may exist as four stereoisomers: two ephedrine enantiomers are diastereoisomers for two pseudoephedrine enantiomers, **(B)** quinine and quinidine are diastereoisomers and differs by the stereo configuration of C_1 carbon atom, and **(C)** three stereoisomers of tartaric acid, the achiral *meso*-form contain internal plane of symmetry (denoted as dashed line).

cis- variant) is defined as Z stereoisomer while *trans*- variant is denoted as E isomer (from German words *Zusammen* and *Entgegen*, meaning together and opposite, respectively). The *cis-*, *trans-*, and Z, E relation is illustrated by two stereoisomers of flupentixol (Fig. 2.6A).

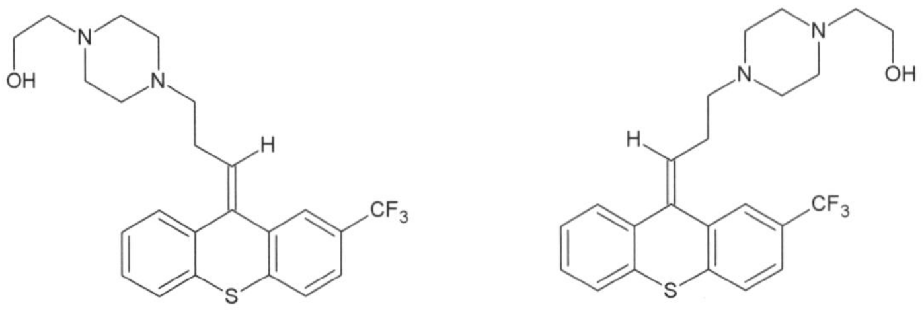

trans-(*E*)-flupentixol

cis-(*Z*)-flupentixol

(A)

trans-1,4-dichlorocyclohexane

cis-1,4-dichlorocyclohexane

(B)

FIGURE 2.6 (**A**) *Cis-* and *trans-* stereoisomers of flupentixol. In this molecule, a relative orientation of CF_3 group in respect to the moiety containing piperazine determines the configuration. The *cis-*(*Z*)-isomer is marketed as an antipsychotic drug. (**B**) *cis-* and *trans-* isomers of 1,4-dichlorocyclohexane.

However, *cis-trans* isomerism is not necessarily associated with a double bond. Similar relationship can be found in many molecules with substituted cycloaliphatic systems. One of the simplest examples is 1,4-dichlorocyclohexane, which exists as two (achiral) stereoisomers depending on the mutual orientation of chlorine atoms in the molecule (Fig. 2.6B).

NAMING OF STEREOISOMERS

The earliest method of naming enantiomers was based on their optical activity, that is, the (+)- and the (–)- forms. This, however, does not describe the actual spatial arrangement of ligands about the chiral center, that is, the absolute configuration. Two conventions are currently in use to assign the absolute configuration of a molecule with tetrahedral chiral center(s): the Fischer convention assuming chemical transformation of the chiral molecule to an arbitrarily chosen standard, (+)-glyceraldehyde, and the Cahn-Ingold-Prelog (C.I.P.) convention employing rules of priority of substituents about the chiral center. In stereochemistry of drugs, the latter method is primarily used.

The Fischer Convention

The Fischer convention was elaborated by Emil Fischer in 1919 based on his method of two-dimensional projection of three-dimensional molecules (4,6). The configuration at the asymmetric center of the analyzed molecule is related to (+)-glyceraldehyde, which was arbitrarily assigned the D- configuration (Fig. 2.7). Interestingly, while defined by Fischer, the actual configuration of (+)-glyceraldehyde was unknown. Several decades later it became possible to

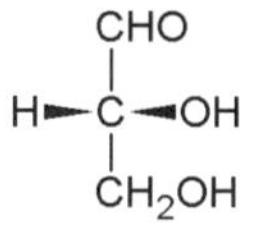

STEREOCHEMISTRY—BASIC TERMS AND CONCEPTS

```
        CHO                      CHO
         |                        |
    H──C──OH                 HO──C──H
         |                        |
       CH₂OH                    CH₂OH

  D-(+)-glyceraldehyde    L-(-)-glyceraldehyde


       C¹HO                     C¹HO
         |                        |
    H──C²──OH                HO──C²──H
         |                        |
    H──C³──OH                 H──C³──OH
         |                        |
       C⁴H₂OH                   C⁴H₂OH

      D-erythrose              D-threose
```

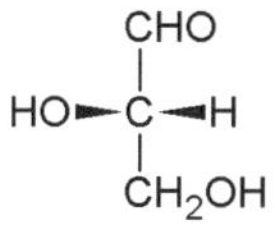

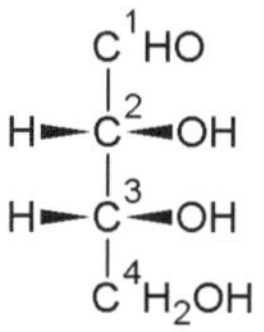

FIGURE 2.7 The configurations of L-(+)- glyceraldehyde, L-(–)-glyceraldehyde, D-erythrose, and D-threose, according to the Fischer convention.

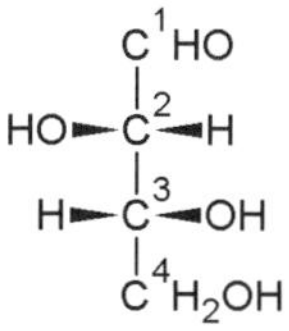

establish the absolute configuration, and X-ray crystallography studies determined that the assigned configuration for D-glyceraldehyde was, in fact, correct (6). To assign a configuration as D- or L-, consideration must be given to how the molecule is chemically related to D- or L-glyceraldehyde or another enantiomer of known configuration. The sign of rotation of polarized light cannot be used a priori to assign a configuration, because they do not always correspond. For example, L-alanine has a (+) sign of rotation, whereas the sign of rotation for L glyceraldehyde is (–) (4).

The Fischer convention is used in sugar and amino acid chemistry. For hydrocarbons, which contain a number of asymmetric centers, the convention assigns D- or L- according to the absolute configuration at the highest numbered asymmetric center. In Figure 2.7, the D- configuration is assigned to the erythrose and threose series because of the D- configuration at C^3 atom.

The Fischer convention is often inexact and difficult to use, especially when complicated chemical transformations are required to relate the molecule into a reference molecule of known configuration. The assigned configuration, D- or L-, used to be confused with the observed sign of rotation, d (dextrorotatory) or l (levorotatory). Because of these difficulties, the Fischer convention has been almost entirely replaced by the C.I.P. convention(4).

The Cahn-Ingold-Prelog Convention

The C.I.P. convention was designated as the "priority rule" since it assigns the priority of substituents around the asymmetric center (7). In this method, the substituents at the chiral center are first ordered according to their atomic number from the largest to the smallest. This is illustrated in Figure 2.8, where the order of substituent is L (large), M (medium), S (small), and S' (smallest). The molecule is oriented in such a manner that the smallest (S') substituent is directed away from the viewer. The configuration is then determined by whether the sequence L-M-S goes in a clockwise or counterclockwise sense of

26 DRUG STEREOCHEMISTRY: ANALYTICAL METHODS AND PHARMACOLOGY

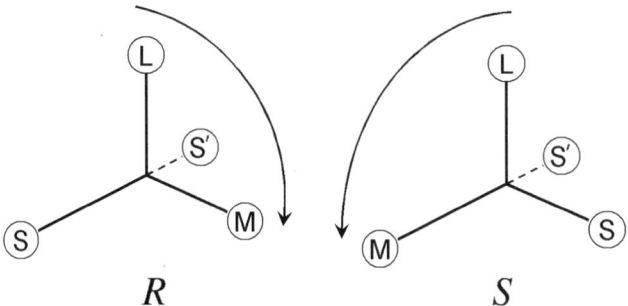

FIGURE 2.8 The Cahn-Ingold-Prelog sequence rule, where L is the largest group attached to the chiral center, followed in size by M, S, and S'.

rotation. A center of chirality with clockwise rotation is specified as *R*, whereas the counterclockwise sense of rotation designates the *S* description of the stereocenter. The names are derived from the Latin terms **Rectus** and **Sinister** for right and left, respectively.

In order to assure ambiguous assignment of priority in C.I.P. rule, two groups are first compared by atomic number of atoms directly attached to center of chirality; the moiety having this primary atom of higher atomic number receives higher priority. If primary atoms are the same, lists of secondary atoms (i.e., bonded to the one directly attached to the stereocenter) arranged in order of decreasing atomic number are compared between two moieties. The lists are analyzed atom by atom and, at the earliest difference, the group containing the atom of higher atomic number receives higher priority. If two lists are still the same, each secondary atom is replaced with a sublist of the tertiary atoms bonded to it and arranged in decreasing order of atomic number; the entire list for one moiety is again compared atom by atom with an analogous list for the other moiety. If necessary, the process is repeated until the earliest difference is found. For usage of priority rule in other cases (like ring systems, unsaturated moieties, etc.), readers are referred to original articles (7,8).

This convention can be used to rapidly and unambiguously specify the configuration of a chiral center. As would be expected, for a chiral molecule the direction of the sequence for one enantiomer is reversed for the other enantiomer. The D-enantiomer of glyceraldehyde has the *R*- configuration; its mirror image (the L-enantiomer) is in the *S*- configuration.

The C.I.P convention is also extremely useful for describing diastereomers. In this case, each chiral center is designated independently and the configuration of the whole molecule can be easily assigned. For example, (+)- and (−)- ephedrine and pseudoephedrine stereoisomers (Fig. 2.5A) have absolute configurations assigned as (*R*,*S*)- and (*S*,*R*)-ephedrine and (*R*,*R*)- and (*S*,*S*)- pseudoephedrine, respectively. Enantiomeric relationships within the ephedrine and pseudoephedrine pairs and diastereomeric relationship between ephedrines and pseudoephedrines can be easily inferred. The absolute C.I.P. configuration for (+)- and (−)- tartaric acid (Fig. 2.5C) is (*R*,*R*)- and (*S*,*S*)-, respectively, while a *meso*-diastereomer has (*R*,*S*)- configuration. A dextromethorphan molecule (Fig. 2.4) has all three centers of chirality in *S*- configuration, while its enantiomer, levomethorphan, has all three centers in *R*- configuration. As seen in examples of diastereomers with multiple chiral centers, two enantiomers are these isomers, which have all chiral centers in reversed configurations (i.e., each R transformed to S and vice versa).

The C.I.P. convention can be easily digitalized (9) and many computer programs for molecular visualization can quickly assign absolute configurations of chiral centers. This is currently recommended standard for naming stereoisomers of drug molecules and their metabolites. The convention, however, has one drawback that in contrary was a strong point of the Fischer system in that it does not necessarily follow the chemical transformation of a molecule. For example, cytochrome P450 transformation of ifosfamide (Fig. 2.4) in vivo produces two dechloroethylated metabolites. Removing this moiety from the ring system (position 3) causes downgrading of the nitrogen atom located there in the size order of C.I.P. convention, and as a result, the (R)-ifosfamide molecule (the (+) isomer in Fig. 2.4) is transformed into the (S)-3-dechloroethylifosfamide metabolite (10).

MIXTURES OF STEREOISOMERS

Racemic mixture or a racemate is an equimolar mixture of a pair of enantiomers. The chemical name of a racemic mixture is distinguished from single enantiomer by a prefix *rac-*, (±)-, (D,L)-, or (R,S)-. In contrast to enantiomers, a racemate does not show optical activity. Interestingly, some other common physicochemical properties of a racemate may significantly differ from the properties of a single enantiomer.

A scalemic mixture is a proposed term for any nonequimolar mixture of two enantiomers (11). Composition of such mixtures can be quantitatively described by the term enantiomeric excess (*ee*), which is defined as an absolute difference between the mole fraction of each enantiomer in a mixture. Conventionally, this figure is expressed in percent, and %*ee* value for racemic mixture is 0% while %*ee* value for a single enantiomer is 100%. Enantiomeric excess was introduced in 1971 as very effective measure to deal with optical purity of scalemic mixtures. Nowadays, with growing numbers of techniques allowing direct determination of each enantiomer in a mixture, it was suggested that the *ee* concept should be replaced with enantiomeric ratio (*er*) or *S/R*, which is a relative ratio of enantiomers' molar fractions (12). Analogous measures can be applied to quantify composition of mixtures of diastereoisomers. Terms diastereomeric excess (*de*) or diastereomeric ratio (*dr*) are used in this case.

In stereochemistry, *ee, er, de,* or *dr* are employed to describe enantioselectivity/stereoselectivity. These terms are defined as a preferential involvement of one stereoisomer over the other(s) in a chemical or enzymatic reaction. In chiral separation techniques, enantioselectivity is used to quantify the degree to which two enantiomers can be resolved from the racemic or scalemic mixtures. A related term is enantiospecificity/stereospecificity: the reaction is termed stereospecific if starting materials differing only in their configuration are converted into stereoisomeric products. According to this definition, a stereospecific process is necessarily stereoselective but not all stereoselective processes are stereospecific. Stereospecificity may be total (100%) or partial. The term is also applied to situations where a reaction can be performed with only one stereoisomer (1,13).

SUMMARY

A summary of the basic nomenclature used in stereochemistry is presented in Table 2.1. For further rules and definitions, a publication by the IUPAC (1) is very useful.

TABLE 2.1 Commonly Used Stereochemical Terms

Stereoisomers	Molecules with the same constitution and atom connectivities, but which differ in respect to the spatial arrangement of certain atoms or groups.
Enantiomeric molecule	A molecule that is not superimposable on its mirror image.
Chiral molecule	A molecule having at least one pair of enantiomers.
Enantiomers	Stereoisomers that are related as nonsuperimposable mirror images.
Diastereomers	Stereoisomers that are not related as an object and its mirror image.
Optical activity	A property of a chiral molecule—the ability to rotate a beam of plane-polarized light.
Optical rotation	The angle that a beam of plane-polarized light is rotated by a chiral molecule.
Dextrorotary or (+) rotation	A clockwise rotation of a beam of plane-polarized light by a stereoisomer, usually used to denote a specific enantiomer of a chiral molecule, that is, the (+)-enantiomer.
Levorotatory or (-) rotation	A counterclockwise rotation of a beam of plane-polarized light by a stereoisomer, usually used to denote a specific enantiomer of a chiral molecule, that is, the (-)-enantiomer.
Configuration	The description of the spatial arrangement about a chiral center.
Fischer convention (D,L)	The assignment of configuration about a chiral atom by comparison to a standard, (D)-(+)-glyceraldehyde, usually by actual chemical transformation of the molecule under investigation into the standard.
Cahn-Ingold-Prelog convention (C.I.P.)(R,S)	The assignment of configuration about a chiral atom by designation of the sequence of substituents from the largest (L), medium (M), small (S) to the smallest (S'); a clockwise direction of the L-M-S sequence rotation is assigned the R- configuration and a counterclockwise direction is assigned the S- configuration.
Racemate, racemic mixture	A mixture of two enantiomers where concentration of both components is equal (50:50). To comprise the fact that both enantiomers exist in a racemate, a mixture is denoted rac-, ($\pm$), (D,L), or (R,S).
Scalemic mixture	A mixture of two enantiomers where concentration of both components is not equal.
Enantiomeric excess	Absolute difference between molar fractions of each enantiomer in a mixture. $$\%ee = \frac{x_R - x_S}{x_R + x_S} \times 100\%$$
Enantiomeric ratio	Relative ratio of molar fractions of each enantiomer in a mixture. $$\%er = \frac{x_R}{x_S} \times 100\%$$
Eutomer	The enantiomer of a chiral compound that is the more potent for a particular action (e.g., biological activity or curing potency).
Distomer	The enantiomer of a chiral compound that is the less potent for a particular action. It does not exclude the possibility of other effects where this enantiomer shows greater potency.

REFERENCES
1. McNaught AD, Wilkinson A, eds. IUPAC. Compendium of Chemical Terminology. 2nd ed. (the "Gold Book"). Oxford: Blackwell Scientific Publications, 1997. Available at: http://goldbook.iupac.org/. Accessed June 2009.
2. Morris DC. Stereochemistry. New York: Wiley-Interscience/Royal Society of Chemistry, 2002.
3. Carrey FA, Sundberg RJ. Advanced Organic Chemistry. Part A: Structure and Mechanism. New York: Springer, 2007.
4. Wainer IW, Marcotte AA. Stereochemical terms and concepts, an overview. In: Wainer IW, ed. Drug Stereochemistry: Analytical Methods and Pharmacology. 2nd ed. New York: Marcel Dekker Inc., 1993.
5. Vargek M, Freedman TB, Lee E, et al. Experimental observation of resonance Raman optical activity. Chem Phys Lett 1998; 287:359–364.
6. Kunz H. Emil Fischer—unequalled classicist, master of organic chemistry research, and inspired trailblazer of biological chemistry. Angew Chem Int Ed Engl 2002; 41:4439–4451.
7. Cahn RS, Ingold CK, Prelog V. Specification of molecular chirality. Angew Chem Int Ed Engl 1966; 5:385–415.
8. Prelog V, Helmchem G. Basic principles of the CIP-system and proposals for a revision. Angew Chem Int Ed Engl 1982; 21:567–583.
9. Cieplak T, Wisniewski JL. A new effective algorithm for the unambiguous identification of the stereochemical characteristics of compounds during their registration in databases. Molecules 2001; 6:915–926.
10. Oliveira RV, Onorato JM, Siluk D, et al. Enantioselective liquid chromatography–mass spectrometry assay for the determination of ifosfamide and identification of the N-dechloroethylated metabolites of ifosfamide in human plasma. J Pharm Biomed Anal 2007; 45:295–303.
11. Mislow K. Molecular chirality. In: Denmark SE, ed. Topics in Stereochemistry. Vol. 22. New York: John Wiley and Sons, 1999.
12. Gawle RE. Do the terms "% ee" and "% de" make sense as expressions of stereoisomer composition or stereoselectivity? J Org Chem 2006; 71:2411–2416.
13. Davankov VA. The nature of chiral recognition: is it a three-point interaction? Chirality 1997; 9:99–102.

Molecular basis of chiral recognition

Krzysztof Jóźwiak

INTRODUCTION

Two enantiomers of a chiral substance have all their main physicochemical properties the same. Three-dimensional (3D) structures of enantiomers are nearly identical; the only difference lies in the fact that two molecules are related to each other as mirror images. Molecules are dissymmetric and any differences in the properties of enantiomers may be expressed in unequal interactions with a dissymmetric environment. Such interactions are particularly an issue in biological systems where biomacromolecules (polypeptides, polynucleotides, sugars, etc.) are asymmetric and can disproportionately recognize enantiomers of a chiral substance. Molecular mechanism of such chiral recognition processes is a matter that has drawn researchers' attention right from the earliest studies of stereochemistry. Recently, with the emergence of many new structural and computational techniques that may be employed in these studies, much more is known about the molecular basis of interactions between chiral molecules and (selector) macromolecules that lead to resolution. The purpose of this chapter is to present the introduction to various models describing molecular chiral recognition mechanisms and to overview several techniques that assist in deciphering these processes.

MODELS OF CHIRAL RECOGNITION

Considering the molecular level, when molecules representing two enantiomers, (+) and (–), interact with a dedicated binding site on a selector asymmetric macromolecule, two types of molecular complexes are formed. These complexes, (+)-selector and (–)-selector, are actually a pair of diastereoisomers and may express different properties to each other. Under certain conditions, this difference can lead to enantioselectivity or enantiospecificity of a reaction in which complexes are formed. A selector, while forming a complex with ligand molecule tends to recognize the latter in all three dimensions by a network of mutual interactions. Covalent bonding rarely occurs in such processes and a variety of attractive and repulsive non-covalent intermolecular interactions are predominant here (1). Interactions that are commonly found in ligand-selector binding are listed and characterized in Table 3.1. From the point of view of stereochemistry, it is evident that unequal interactions of enantiomers with asymmetric binding site(s) will arise when there is unequal distribution of the ligands' moieties along the three axes of the space (2). It produces a difference of interaction patterns between enantiomers that is required to observe enantioselective or enantiospecific conditions of a reaction. Starting from the early thirties of the twentieth century, a number of different models were proposed to explain molecular mechanisms of these conditions and these are presented later in this chapter.

TABLE 3.1 Molecular Interactions: Strength, Working Distance, and Vector of Direction

Type of interaction	Strength	Distance	(+), Attraction (−), Repulsion
Coulombic	Very strong	Long range	(+) or (−)
Hydrogen bond	Very strong	Long range	(+)
Steric hindrance	Very strong	Very short range	(−)
π-π interaction	Strong	Medium range	(+) or (−)
Ion-dipole	Strong	Short range	(+)
Dipole-dipole	Intermediate (weak)	Short range	(+)
Dipole–induced dipole	Weak (very)	Very short range	(+)
Van der Waals	Very weak	Very short range	(+)

Three-Point Models

As illustrated by the Cahn-Ingold-Prelog naming convention (described in detail in chapter 2), it is apparently sufficient to localize three reference points (valencies) on the asymmetric tetrahedral system in order to distinguish the object from its mirror image (2,3). Such a simple assumption was the basis of the first structural explanation of enantioselective interactions proposed in as early as 1933 by Easson and Steadman (4). In this model, a common asymmetric site on the enzyme or a receptor needed three nonequivalent points linked to a chiral carbon atom of a ligand molecule in order to distinguish differential binding of two enantiomers. This model is schematically depicted in Figure 3.1A where points A, B, and C located on the 3D surface of the macromolecule interact with a, b, and c moieties attached to a chiral tetrahedral. In this classical explanation, only one enantiomer [in Fig. 3.1A indicated as (+)] is able to saturate complementary interactions in all three points. The other isomer [indicated as (−)] while bound to the site cannot assume the orientation to meet all three of these interactions (Fig. 3.1A). The (−) enantiomer, due to a spatial restriction, forms only a limited number of interactions. If the interactions are positive, then the overall interaction will be weaker and as a consequence its binding is not favored by the site to the extent observed for the (+).

In 1948, Ogston (unaware of the previous Easson and Steadman deliberations) developed an analogous explanation to deal with enantiospecificity of enzymatic reactions in the citric acid cycle (5). In this version of a "three-point" model (Fig. 3.1B), a prochiral molecule containing two chemically equivalent moieties (a' and a'' in the original report -CH_2-COOH groups) can be stereospecifically transformed into a chiral product. The reason for that is the fact that two other moieties of the substrate, b and c, interacting with complementary points located on the binding site (B and C, respectively) sterically determine which a group is processed by the A component of the catalytic site (Fig. 3.1B).

Such three-point models as originally defined are conceptually valid. However, their use in pharmacology is limited to situations where drug-receptor relationships are relatively simple (6). The illustrative examples, which is considered as a classical three-point interaction model, are complexes between (R)-propranolol or (S)-propranolol and β-cyclodextrin developed by Armstrong et al. in 1986 (7). Figure 3.2 shows complex configurations for both enantiomers overlaid exactly to the point of the chiral atom and allows direct analysis of the network of propranolol-cyclodextrin interactions.

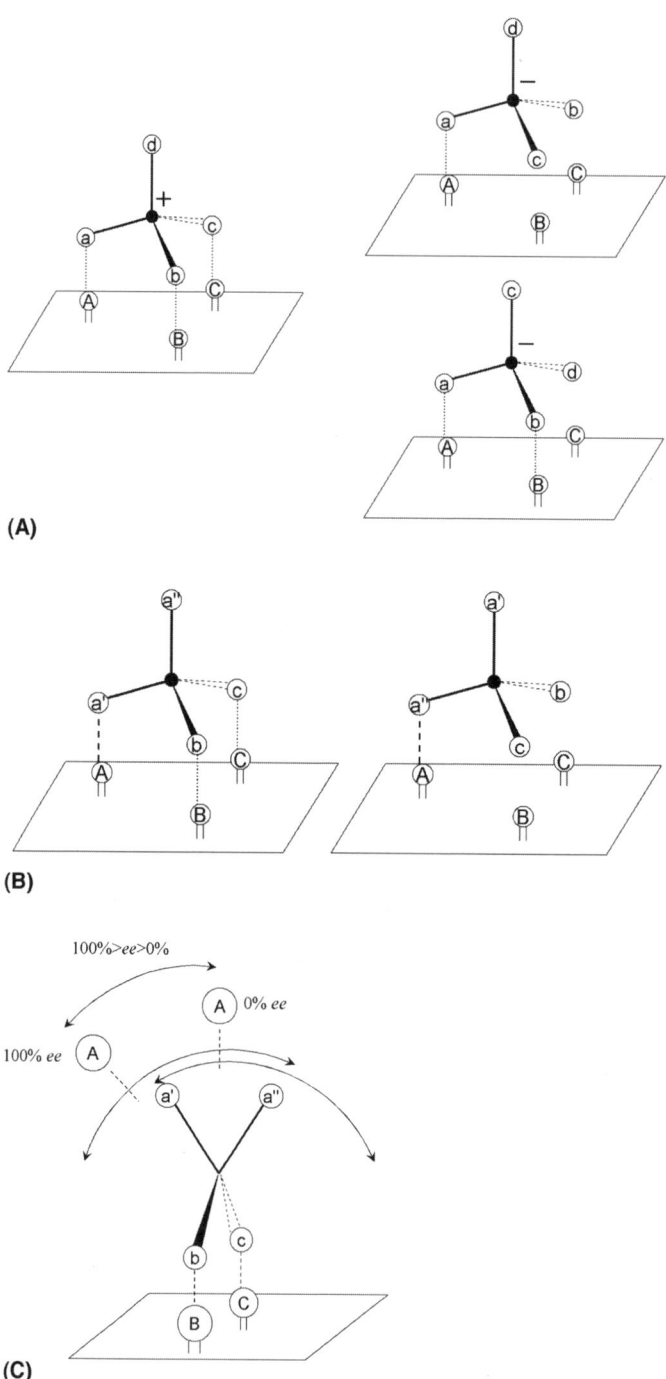

FIGURE 3.1 Three-point models. (**A**) Easson and Steadman model; (**B**) Ogston model; (**C**) Sokolov and Zefirov conformationally driven model.

MOLECULAR BASIS OF CHIRAL RECOGNITION

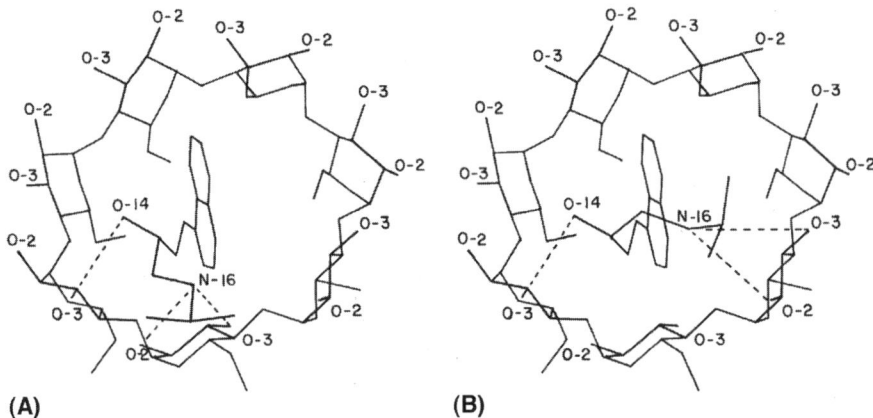

FIGURE 3.2 Computer projections of inclusion complexes between (S)-propranolol (**A**) or (R)-propranolol (**B**) and β-cyclodextrin. The dotted lines represent potential hydrogen bonds. *Source*: From Ref. 7.

Analysis indicates that locations of naphthyl moieties and β-hydroxyl moieties are equivalent for both enantiomers: the lipophilic naphthyl moiety is tightly interacting to the hydrophobic interior of the cyclodextrin molecule (interaction A-a), while the hydroxyl group (denoted in the figure as O-14) interacts with 3-OH moiety of the selector located close to the mouth of the molecule (interaction B-b). Such distribution of these two fragments determines the nonequivalent location of the third interacting fragment for two enantiomers, the aminoalkyl chain (the amino group is denoted in the figure as N-16). In the (S)-propranolol complex, the N-16 atom is optimally placed for hydrogen bonding with both O-2 and O-3 atoms of the selector with respective bond distances 3.3 Å and 2.8 Å (interaction C-c). The amine in the (R)-propranolol complex is positioned less favorably for hydrogen bonding; the distances to the closest 2- and 3-hydroxyl groups are 3.8 Å and 4.5 Å, respectively (7). Thus, the difference in C-c interaction determines that the (S)-enantiomer can form a stronger network of attracting interactions with the β-cyclodextrin in the orientation, which cannot be attained by the (R)-enantiomer within the interior of the host molecule.

The above example of propranolol interacting with a cyclodextrin molecule demonstrates that in reality the selector's contact points are not located on the planar surface as was idealized on the illustrations of the Easson and Steadman model or the Ogston model. Another important aspect is that, originally, the contacts in these models were defined as point attractions [i.e., linkages A-a, B-b, and C-c consisted of strictly defined intermolecular attractive forces (2)]. With time, the concept has been redefined as a three-interactions rule where the type of interaction is immaterial; it could be either attractive or repulsive. Steric hindrance by a bulky group or any loose type of interaction involving a contour of a larger part of the molecule may play an important role in defining the orientation of an enantiomer within the binding site (6). As illustrated by the Ogston model, in the case of stereospecific/stereoselective enzymes the residues performing a catalytic reaction incorporate point interaction(s) to the system (e.g., the interaction A-a' in Fig. 3.1B).

Four-Point Models

Three-point models are very simplistic and assume several conditions, which in real life are not frequently fulfilled. Recalling once again the analogy to the Cahn-Ingold-Prelog convention, determining the rotation order of three largest valencies around the tetrahedral center of chirality is indeed sufficient to establish the configuration. However, a central requirement needs to be additionally assumed: all three valencies are directed into the viewer's eye and the smallest fourth valency must always point away from the viewer. Similarly, the central (and implicit) requirement for three-point models is the fact that a chiral molecule can approach the binding surface of the selector only from one side. The opposite enantiomer molecule is not permitted to flip into the interior of the macromolecule of the selector in order to make the network of interactions with A, B, and C points equivalent to the network of interactions made by the original enantiomer approaching the surface from the "regular" side. At first glance, this seems to be a very reasonable assumption; however, many situations may be hypothesized where such conditions could actually occur. If it is assumed, for example, that binding to the internal surface of the channel where A, B, and C points are distributed in a planar fashion in three angles of a triangle perpendicular to the channel axis. If the distance is appropriate, each a, b, and c valency of the tetrahedral object can form optimized interactions regardless of chiral configuration. The difference will be manifested by the fourth nonbonded valency pointing in opposite directions in both enantiomeric tetrahedrals (a scheme is depicted in Fig. 3.3A). In such a situation, the network made by A-a, B-b, and C-c contacts will be the same for both enantiomers and additional contact point(s) will be required in order to observe any enantioselectivity.

This topological condition to the three-point models was first conceptually formulated in 1950 by Wilcox et al. (8), but the idea tended to be overlooked for a long time (6). Yet in 1989, Topiol and Sabio (9) in a theoretical paper addressing simulation of energetics of short intermolecular interactions between two chiral tetrahedrons defined that interactions between all eight valencies play a role in the mutual interactions of two tetrahedrons. In other words, four contact points are necessary for chiral recognition. They concluded that for the object to be chiral, 3D space is required: thus "chirality discriminative forces" cannot be represented by the interaction of objects reduced to two dimensions (as is presented by a 2D triangle in the classical illustration of three-point model, Fig. 3.1). This concept was elegantly illustrated by Bentley (6), who showed that if a three-point model is drawn in 3D space (instead of 2D), both enantiomeric molecules can be positioned in a manner allowing equal A-a, B-b, and C-c interactions for both molecules (the scaffold of such an orientation is presented in Fig. 3.3B). In this case the three-point system is not enantioselective and an additional point of interaction is required to observe chiral discrimination.

Such theoretical elucidations found a vivid experimental verification in 2000 in an X-ray crystallographic study of isocitrate dehydrogenase by Mesecar and Koshland (10,11). The enzyme is stereoselective and preferentially processes *threo*-D_s-isocitrate [(1-R,2-S)-(+)-1-hydroxypropane-1,2,3-tricarboxylate] into α-ketoglutarate; a magnesium ion is essential component of the active site to perform the reaction. The authors cocrystallized the enzyme with a racemate of *threo*-isocitrate with and without the presence of Mg^{2+}. X-ray crystallography revealed that the magnesium-free form of the enzyme binds exclusively the inactive L-enantiomer, whereas when crystals are produced in the presence of

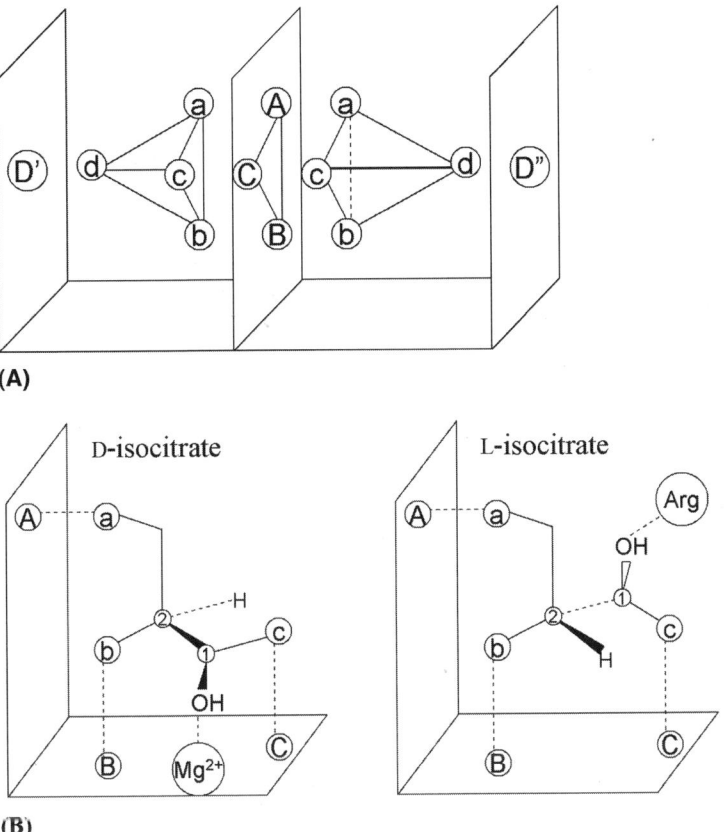

FIGURE 3.3 Four-point models of chiral recognition. **(A)** The schematic model of enantiomer discrimination proposed initially by Mesecar and Koshland for their investigations of the isocitrate dehydrogenase system (10) is very simplistic as it assumes coplanar orientation of A, B, and C locations. The d points (for isocitrate the −OH groups) directing the opposite locations within the binding site is a source of enantioselectivity. **(B)** A more precise scheme of interactions between threo-D_s-isocitrate (the (1-R,2-S)-enantiomer) (*left*) or its inactive enantiomer, (1S,2R)-isocitrate (*right*), and isocitrate dehydrogenase enzyme based on the noncoplanar representation (6). a, b, and c are carboxylate moieties, which interact with their respective A, B, and C contacts distributed in 3D space. This orientation results in the opposite positioning of the −OH groups and their interaction with Mg^{2+} ion (for the D-enantiomer) or Arg[119] (for the L-enantiomer).

magnesium ion, only the active substrate, D-enantiomer is seen in the electron density maps. Superposition of the two molecular complexes indicates that three of the four groups attached to the asymmetric C2 atom of both enantiomers bind to the same three locations within the enzyme active site. The difference is a hydroxyl group, the fourth group attached to this center of chirality. The −OH group of the inactive (1-S,2-R) form interacts with the Arg[119] residue in the Mg^{2+}-free enzyme, in contrast to the −OH group of the active substrate, which binds to the catalytic magnesium ion associated with two aspartate residues and water molecules (11). This differential in the positions of −OH moiety is essential for isocitrate dehydrogenase to distinguish between the two enantiomers. The schematic modes of chiral recognition proposed for this system are depicted in Figure 3.3A and B. *a*, *b*, and *c* fragments occupy the same protein locations, the consequence of which is that the *d* fragment (a hydroxyl group) points into the opposite directions and interacts with different protein locations (Mg^{2+} or Arg^{119} residue for D- and L-enantiomer, respectively). The authors concluded that a similar "four-location model" is capable of explaining other stereoselective enzyme systems including mandelate racemase (10).

The four-point model was found important in further studies of stereoselective binding. For example, Kato et al. (12) used this explanation to elucidate the mechanism of selective enclathration of (2-*R*,3-*S*)-3-methyl-2-pentanol by 3-epideoxycholic acid. During crystallization, the acid forms microscopic channels that can preferentially accommodate alcohol molecules; the crystalline inclusion method is used to resolve the (2-*R*,3-*S*)-3-methyl-2-pentanol from a racemic mixture. The authors established the crystal structure of the enclathrate and concluded that the alcohol molecule interacts with the channel-like cavity by the mechanism described earlier: three asymmetric valencies of the selectand bind to the selector's moieties at the 3D surface of the channel and the fourth contact is needed for the host to distinguish between enantiomers. The authors identified the fourth contact as a C*–H···O hydrogen bond formed between the hydrogen atom attached to the asymmetric carbon atom of the alcohol molecule and the oxygen atom from the hydroxyl group of the host molecule. The mechanism of the chiral recognition in this system can be schematically illustrated by the model proposed in Figure 3.3A where the *d*-D contact for (2-*R*,3-*S*)-3-methyl-2-pentanol is performed by C*H···O interaction.

Multiple-Point Model

Further considerations of point models and their generalization for molecules containing multiple centers of chirality led Sundaresan and Abrol (13) to elaborate the *Stereocenter Recognition* model based on the analysis of the topology of a ligand's stereocenters. The model rigorously defines a minimum number of locations distributed on the ligand surface that need to enter into interaction (either attractive or repulsive) with the receptor sites in order to observe stereoselectivity. It solely addresses interactions between the receptor and distinct stereoisomers of a ligand, but not enantioselective enzyme action at the substrate prochiral atom. According to the model, stereoselectivity toward an acyclic ligand with n stereocenters distributed along the single chain requires interaction involving a minimum of $n + 2$ ligand locations distributed over all stereocenters in the ligand molecule. So for a long multistereocenter chiral substance far more than three points or four points (as discussed above) may be required. Interestingly, the model considers that more points of interactions are required for the receptor to distinguish between different diastereoisomers than to distinguish between two enantiomers of acyclic multistereocenter molecule (13).

Conformationally Driven Models

Mechanisms presented so far still hold several conditions and assumptions that are not always realistic in the molecular world. First of all, the notion of point models implies that molecules of enantiomers are rigid objects and interact with point locations fixed in 3D space. On the other hand, both a macromolecule and ligand molecule(s) undergo conformational changes upon binding and the dynamics of this process plays a role in enantiorecognition. Such aspects were elegantly illustrated in 1991 by Sokolov and Zefirov in the so-called rocking tetrahedron model (14). The model proposed a dynamic explanation to the enzymatic reaction, which, in contrast to the Ogston reaction model, may be not only fully but also partially stereospecific. As seen in Figure 3.1C, the model assumes that two moieties (b and c) of the tetrahedral prochiral molecule are attached to the enzyme-binding site that determines the fixed location of the two other equivalent substituents undergoing further catalytic reaction (a' and a''). Both substituents are flexible and possess significant conformational freedom. As a result a' and a'' arms sweep certain defined angles around the prochiral center and each moiety sweeps different spaces that may partially overlap. The location of the enzymatic reaction center (A) determines the extent of enantiospecificity of the reaction performed by this system. If A point is located in the area that is swept equally by both a' and a'' moieties (top position, Fig. 3.1C), both enantiomers have equal likeliness of being produced and the reaction is not enantiospecific (enantiomeric excess $ee = 0\%$). The more left the A point is located in respect to "equality" point, the more enantiospecific is the reaction, and ee increases. The extreme situation can be expected when the A point is in the location that is reachable by only the a' moiety and $ee = 100\%$ should be expected in such systems. An analogous situation occurs when the A point is located to the right of the equality point; obviously this time the enantiospecificity is reversed and the a'' moiety is preferably processed by the enzyme.

The described model is an example of a conformationally driven chiral recognition process. Booth and Wainer (15,16) in further considering this topic suggested that flexible molecules under certain conditions may adopt the conformation that allows saturating interactions with all three points required by classical models on the selector surface regardless of the configuration of the stereocenter. As an example, the interaction between benoxaprofen and the amylose tris(3,5-dimethylphenylcarbamate) chiral stationary phase was presented. The authors concluded that the only difference leading to the observed enantioselectivity in this case was the entropic difference in internal energies resulting from different conformational restraints for the two enantiomers adopting the desired orientations within the binding site (16).

Other Diastereoisomeric Models

It is not only the selectand molecule that can undergo conformational changes. Specific stereoconfiguration of the ligand may also induce a unique conformational shift of the recognizing macromolecule (a selector). Binding of the substrate can induce small local or even significantly large global changes in a receptor's 3D structure (17,18). A very good illustration of this effect is recent findings in the stereospecific interaction between β2-adrenergic receptor and its agonists (see chap. 14). It was found, for example, that some agonists depending on their stereoconfiguration are able to preferentially activate different classes of G protein as a downstream effect of binding to the β2-adrenergic receptor (19). Further investigations revealed that two isomers can induce different protein conformations of the β2 receptor within the agonist binding site. As a result, each isomer stabilizes distinct and unique active conformations of the receptor that is further manifested as a significant difference in functional assays (reviewed in Ref. 20). The details of the mechanism of chiral recognition of agonists in β2-adrenergic receptor system are presented in chapter 14 of this book.

In the light of these considerations, any model of individual points of interactions common to both enantiomers will be too simplistic to be of much practical value in the context of ligands binding to macromolecules. After all, it must be borne in mind that a central requirement (but not always sufficient condition) to differentiate between enantiomers is to form a diastereoisomeric pair of complexes, (+)-selector and (−)-selector. This is achieved by different binding modes of each enantiomer to the selector; in the most extreme case, two enantiomers of the chiral substance do not have to share the same binding site at all. In this situation, three-point or four-point attachment is not a unique condition to observe enantioselectivity. As pointed by Bentley, "dissymetric treatment of a substrate by an enzyme can occur whenever the enzyme imposes, whether actively by binding or passively by obstruction, a particular orientation at the site of reaction. According to this general hypothesis the attachment of the substrate to the enzyme by one group alone would suffice for a completely stereospecific reaction" (6).

A good illustration of this hypothesis may be seen in X-ray crystallographic studies of chiral recognition in alcohol dehydrogenase from different species (21). The enzymes can be characterized by distinct enantiospecificity depending on from which particular species they originated (i.e., depending on the protein sequence.). Their active site cavities are lined with different hydrophobic residues, which vary between species. In this case, it is hard to indicate individual contact points that are responsible for the observed stereospecificity. It appears that the overall size and the shape of the cavity, hydrophobicity distribution, and steric hindrance within the cavity are responsible for controlling the stereospecificity of the reaction rather than specific contact points defined within the binding site (6).

THERMODYNAMIC CONSIDERATIONS

In terms of thermodynamics, a complex between a chiral isomer and a selector macromolecule can be described by standard Gibbs energy differences (ΔG°) using the equation:

$$\Delta G^\circ = -RT \ln K \qquad (1)$$

where R is the gas constant (8.3145 J mol^{-1}K^{-1}), T is temperature of the experiment in degrees Kelvin, and K is the equilibrium constant of complex association. When two enantiomers of a chiral substance, the (+)-isomer and the (−)-isomer, form diastereoisomeric pairs of complexes, with a chiral selector, the energy difference between the two complexes is described by the $\Delta(\Delta G^\circ)$ function:

$$\Delta(\Delta G^\circ) = \Delta G^\circ_{(+)} - \Delta G^\circ_{(-)} = -RT \ln K_{(+)} + RT \ln K_{(-)} = -RT \ln \frac{K_{(+)}}{K_{(-)}} = -RT \ln \alpha \qquad (2)$$

Where $K_{(+)}$ and $K_{(-)}$ are association equilibrium constants for the (+)-isomer and (−)-isomer complexes, respectively, and α is the system stereoselectivity (the ratio of equilibrium constants).

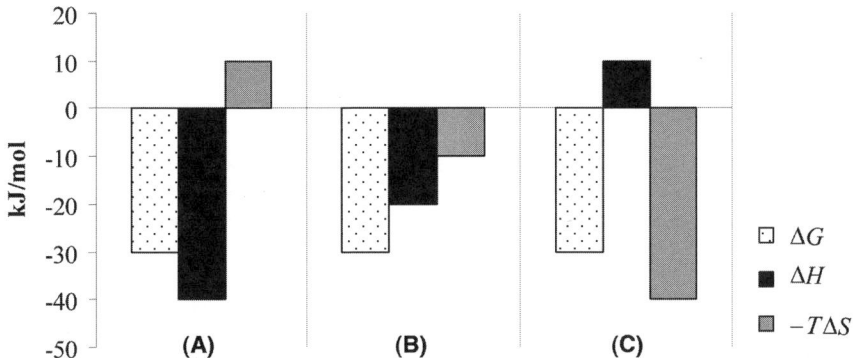

FIGURE 3.4 Thermodynamic parameters for three different patterns of ligand binding: (**A**) an enthalpy-driven process—good polar interactions and a conformational change; (**B**) a process driven by both enthalpy and entropy—favorable polar interactions and hydrophobic interactions; and (**C**) an entropy-driven process—bonding dominated by hydrophobic interaction. *Source*: Adapted from Ref. 22.

Standard free energy of binding can be in turn split into two components, an enthalpic factor ($\Delta H°$) and an entropic factor ($T\Delta S°$), according to the equation:

$$\Delta G° = \Delta H° - T\Delta S \quad (3)$$

These two thermodynamic functions can be used to provide the insight into the nature of interactions between a chiral molecule and a selector. Also, differences between these values characterizing two diastereoisomeric complexes [i.e., $\Delta(\Delta H°)$ and $\Delta(\Delta S°)$] give a description of the type of interactions and processes involved in chiral recognition of the studied system. How thermodynamic factors can be interpreted in terms of molecular effects of governing the ligand-binding process is illustrated in Figure 3.4.

Several direct techniques can be used to determine thermodynamic characteristics of the ligand binding (23) but still classical van't Hoff analysis is widely employed in these studies. In this approach, combining of Eq. 1 and 3 allows assessment of enthalpy and entropy by measurement of the equilibrium constant as a function of the reciprocal of:

$$\ln K = \frac{\Delta S°}{R} - \frac{\Delta H°}{R} \times \frac{1}{T} \quad (4)$$

Although several exceptions were reported (15), the relationship (4) is usually linear and when carried out separately for two enantiomers, it allows determining enthalpic, $\Delta(\Delta H°)$, and entropic, $\Delta(-T\Delta S°)$, contributions of the $\Delta(\Delta G°)$ factor, which can further be used to characterize the nature of molecular interactions governing a chiral recognition process. More negative enthalpic contribution reflects the preferential strength of a stereoisomer-selector interaction, primarily due to attractive polar and nonpolar interactions. Entropic contribution corresponds to the relative strength of hydrophobic interactions (associated greatly with reorganization of water molecules within the binding site) as well as to a change in conformational freedom of the binding partners (24).

In one of the examples of the thermodynamic analysis, the enantiomeric recognition of dextromethorphan and levomethorphan by the nicotinic acetylcholine receptor system was studied (25). In affinity chromatography experiments, using the column with immobilized α3β4 subtype of nicotinic receptor, dextromethorphan shows significantly stronger retention than its enantiomer, levomethorphan. Van't Hoff analysis was used to establish thermodynamic characteristics of binding in this system. Linear regression of relation (4) for both enantiomers allowed to determine $\Delta H°$ and $\Delta S°$ values for dextromethorphan–nicotinic receptor complex as -6.92 kcal·mol^{-1} and -15.7 cal·mol^{-1}·K^{-1}, respectively, and for levomethorphan–nicotinic receptor complex as -6.59 kcal·mol^{-1} and -15.2 cal·mol^{-1}·K^{-1}, respectively. The data demonstrate that observed enantioselectivity is due to a significant decrease of enthalpy change for the dextromethorphan complex in respect to the levomethorphan complex, $\Delta(\Delta H°) = -0.33$ kcal·mol^{-1}, while there is no significant difference in entropy between the two complexes. This suggests that dextromethorphan forms stronger attractive polar interaction with the α3β4 subtype of nicotinic acetylcholine receptor than levomethorphan, which was further confirmed in molecular modeling studies (26).

Another example of this analysis with very interesting findings was published by Karlsson et al. (27), who studied temperature dependence of chromatographic retention of enantiomers of 5-HT$_4$ selective agonist, mosapride, on a Chiral-AGP column. The authors observed that van't Hoff plots for enantiomers were highly dependent on pH of the mobile phase used in the experiment. A peculiar behavior was observed at pH = 6.0 where two regression lines for two enantiomers crossed at the point corresponding to 303 K; at this temperature no enantioselectivity was found. As a result, the reversal of enantiomers' elution order was shown for this pH: in temperature lower than 30°C (R)-mosapride was eluted before the (S)-enantiomer, while in higher temperature the (S)-mosapride was eluted first.

Very interesting thermodynamic behavior was recently observed for stereoisomers of fenoterol, a β2-adrenergic receptor agonist. The compound 2-(3,5-dihydroxyphenyl)-2-hydroxy-2'-(4-hydroxyphenyl)-1'-methyldiethylamine possesses two stereocenters at positions 2 and 1' and exists as four stereoisomers. The clinically used drug is a racemic mixture of (R,R)- and (S,S)- forms but the (R,R)-enantiomer is demonstrated to be the significantly more active form in either affinity binding and functional studies (28,29). The van't Hoff relations were determined for all four stereoisomers measuring temperature dependence of binding affinity of a drug in a radioligand displacement assay (30). The analysis showed that K_d values determined for (S,S)- and (S,R)-isomers increase with the increase of temperature while the temperature has the opposite effect on (R,R)- and (R,S)-isomers: the K_d values decrease with the increase of temperature of experiment. These distinct trends are a consequence in completely opposite thermodynamic characteristics. While the (S,S)- and (S,R)-stereoisomers have $\Delta H°$ values highly negative and $-T\Delta S°$ values close to zero, the (R,R)- and (R,S)-isomers have $\Delta H°$ values slightly positive and $-T\Delta S°$ highly negative. This thermodynamic data suggest that fenoterol exhibits two distinct binding mechanisms to the β2-adrenergic receptor and the configuration of the first stereogenic center (position 2) of molecule is a critical factor determining these mechanisms. The binding affinity of the fenoterol molecule with the (2-S)-configuration is predominantly enthalpy-driven, indicating favorable polar interactions between the ligand and the receptor, while binding of the molecule with (2-R)- configuration is purely entropy-driven, which suggests that hydrophobic interactions as well as possibly conformational change of the complex are the crucial role ligand-receptor interactions. Interestingly, the configuration of the second chiral center (position 1′) does not influence significantly the mechanism of binding. Thus, both pairs of enantiomers showed a distinct mechanism of interaction in this study; the binding mechanism of (2-R, 1′-R)-fenoterol and (2-R, 1′-S)-fenoterol is different from the mechanism observed for (2-S, 1′-S)-fenoterol and (2-S, 1′-R)-fenoterol, respectively (20).

Using a different approach Kafri and Lancet (31) studied the rules that govern enantioselective separations from a statistical viewpoint. In this work, a comprehensive data of over 72,000 enantioseparation measurements stored in the CHIRBASE database (32,33) was studied. The separation factors were computed using the formula:

$$\alpha = \frac{k'_{(+)}}{k'_{(-)}} \quad (5)$$

where k' are the capacity factors. The free-energy differences $\Delta(\Delta G°)$ were computed from the separation factors using Eq. 2. It was found that the distribution of separation factors (α) followed a power law ($P(\alpha) = \lambda e^{-\lambda}$), which corresponded to an exponential decay for the chiral free-energy differences. A string model for enantiorecognition (SMED) was proposed in this paper to explain this observation on the basis of an extended Ogston three-point interaction model. Partially overlapping molecular interaction domains were analyzed in terms of a string-complementarity model for ligand-receptor complementarity. The results suggested that chiral selection statistics may be interpreted in terms of more general concepts related to biomolecular recognition.

In 1992, Berthod et al. proposed a purely empirical procedure to predict chromatographic enantioselectivity using a representation of the molecular structure of chiral analytes (34). The approach was based on the assumption that the difference in molecular free energy of the chiral interaction between two enantiomers ($\Delta\Delta G_c$) can be broken down into four terms, each term being related to one of the four different substituents attached to the asymmetric carbon of the analyte:

$$\Delta\Delta G_c = (\Delta G_{c11} - \Delta G_{c12}) + (\Delta G_{c21} - \Delta G_{c22}) + (\Delta G_{c31} - \Delta G_{c32}) + (\Delta G_{c41} - \Delta G_{c42}) \quad (6)$$

Each substituent has its own contribution on the $\Delta\Delta G_c$ value, and in order to assess these impacts a cohort of 126 chiral compounds resolved on two chiral stationary phases (based on modified cyclodextrins) were analyzed. A cohort introduced totally 81 different substituents and the chiral free energy contribution for each of them was estimated using the optimization procedure in a calculation spreadsheet. The optimization involved adjusting the contribution of each substituent in order to minimize the sum of differences between experimental and computed enantioselectivity factor ($\Sigma|\alpha_{calc} - \alpha_{exp}|$) and assumed a reference chiral free energy contribution of a hydrogen atom as equal to 0 cal/mol. The authors concluded that the approach could be useful for analysts to predict the enantioseparation of other analytes using studied chiral stationary phases (34). The approach, however, by definition estimates only the absolute value of α; it cannot predict the relative order of eluting enantiomers.

QSAR Studies

The prediction of chemical or biological properties of chemical substances based on their molecular structure is a very feasible task. One of the techniques that has found use in theoretical modeling of enantioselectivity is quantitative structure-activity relationships (QSAR). In its classical form, the approach is based on the linear free energy relationships postulate (LFER) (35,36) and essentially describe the trend of changes of certain property for the group of congeneric compounds as a function of changes of structural descriptors for these substances. Descriptors are supposed to be easily accessible for measurement either by experiment or by computational methods for all tested substances. In the biological world, an interaction of ligand with macromolecule is a very specific and frequently multistep process. Accordingly, several such descriptors taking into account different aspects of modeled biological property are necessary to build realistic QSAR model by multiple linear regression technique.

Originally developed for predicting biological activity, the approach can be adopted to predict the enantioselectivities observed for a group of compounds in a given chiral recognition system. Enantioselectivity can be described for a group of compounds in terms of $\Delta\Delta G$ values (Eqs. 2 and 3), thus the linear free energy relationships can be assumed and regression models explaining trends in enantioselectivity ratios (α) can be developed. This modified analysis is sometimes termed QSER (the acronym of quantitative structure-enantioselectivity relationships) (37).

An example of QSER analysis was presented by Beck et al. (38). The authors focused on characterizing the chiral selectivity properties of lambda–carrageenan, naturally occurring linear polysaccharide—employed as a chiral selector in capillary electrophoresis. A series of 13 structurally related racemic compounds were analyzed by capilary electrophoresis (CE) and observed enantioselectivities were rationalized by QSER technique. It was found that the number of aromatic bonds and the partial charge of the hydrogen atom associated with the amine group of the analyte were the factors contributing the most to the enantioselectivity observed in this system.

Wolbach et al. (37) developed several quantitative models of enantioselectivity observed in capillary electrophoresis using β-cyclodextrin, hydroxypropyl-β-cyclodextrin, and tri-O-methyl-β-cyclodextrin as chiral selectors. In this work, racemic mixtures of 22 compounds were resolved using these three chiral selector systems and such molecular descriptors as log P (lipophilicity), TASA (total apolar surface area), HOMO (highest occupied molecular orbital), and LUMO (lowest unoccupied molecular orbital) were found to be important in constructing the QSER models.

In the work by Kaliszan et al. (39), chiral benzodiazepines were chromatographically separated using immobilized human serum albumin (HSA), and QSER models of these data were constructed. Two types of binding sites were postulated. For benzodiazepines in the P-conformation, binding to HSA involved a hydrophobic region with steric restrictions. For drugs in the M-conformation, a hydrophobic region was also involved, as well as a cationic region that interacted electrostatically with carbon C(3) of the diazepine system and substituents at that atom. These differences were found important to explain different binding patterns for enantiomers and provide a rationalization for the diversified behavior of benzodiazepine analogues.

In the work by Booth and Wainer (15) on the chromatographic enantioseparations of certain α-alkyl arylcarboxylic acids on an amylase tris(3,5-dimethylphenylcarbamate) chiral stationary phase, the QSER models identified that retention of enantiomers depends on three descriptors: the number of hydrogen bond donors (1) and acceptors (2) in the molecule and their degree of aromaticity. The authors postulated a conformationally driven chiral recognition mechanism for this system.

3D QSAR

Chirality is the property that occurs in three dimensions and regular QSAR descriptors usually cannot handle specific (i.e., asymmetric) 3D properties of the molecule. Although several attempts have been made [including definition of the so-called chirality descriptors (40)], more effective in modeling chiral recognition are techniques of 3D QSAR. In these approaches, molecular models of studied ligands are aligned and mapped in 3D space with a specific property probe(s). Obtained grid maps are capable of taking into account any stereochemical features of the molecules including their dissymmetry. These spatial representations of properties of molecules are processed statistically using multiple regression techniques (usually partial least square procedure) and as a result a set specific property fields are generated (41). Fields represent statistically significant areas where a property of the probe is correlated with modeled property/activity for the whole set of molecules. By definition the fields are distributed in all three dimensions of space around the modeled molecules, and thus, they are able to account for enantio/stereoselectivity of a modeled property.

A number of 3D QSAR models of molecular recognition for stereoisomeric compounds were reported (42–47). In one such study, Suzuki et al. (48) modeled chromatographic enantioseparation of 42 arylalkylcarbinols on Pirkle-type chiral stationary phases. Enantioseparation factors (in the form of log α) were modeled using one of the most popular 3D QSAR procedures: comparative molecular fields analysis (CoMFA) (41). The analysis identified both steric and electrostatic fields asymmetrically oriented around aligned molecules, which were found important for enantioseparation. Analysis of the steric fields shows that increased bulkiness of derivatives in certain regions almost exclusively produced a decrease of enantioresolution. Similarly, the analysis of electrostatic fields identifies several regions of derivatives where higher positive charge generally led to increased resolution, whereas only a small region was identified, where lower (negative) charge produced increased enantioresolution (48). In the same article, the authors developed a regular QSER model to explain the molecular level of chiral recognition. It was found that two descriptors, lipophilicity and a sum of partial charges on carbon atoms of the phenyl moiety of each tested ligands, were the factors influencing enantioseparation the most. The combination of both parameters can be interpreted as an effect of face-to-face π-π interactions between aromatic moieties of an analyte and the chiral selector. Further analyses by the same group led to development of 3D QSAR models for other chiral stationary phases as well as nonlinear models using artificial neural networks (48,49).

In 2007, a very illustrative example of CoMFA model of stereoselective interactions of fenoterol congeners was developed by Jozwiak et al. (28). As was mentioned earlier, fenoterol, a β2-adrenergic receptor selective agonist, has two centers of chirality, and exists as four stereoisomers. In this study, seven derivatives of fenoterol, modified at the aminoalkyl part of the molecule were synthesized in all possible stereoisomeric variants totaling in 26 closely related ligands. Binding affinity equilibrium constants toward the β2-adrenergic receptor were determined by radioligand displacement assay and used as input data for CoMFA analysis. Since every stereochemical configuration of chiral centers was probed in studied derivatives, the model clearly accounts for the effect of stereochemistry on modeled binding affinity. As shown in Figure 3.5, CoMFA fields are predominantly associated with the aminoalkyl part of the molecules (far right part of the molecule, Fig. 3.5) as it is the region where most structural modifications take place. However, a certain fraction of the fields asymmetrically surrounds both chiral centers. In case of

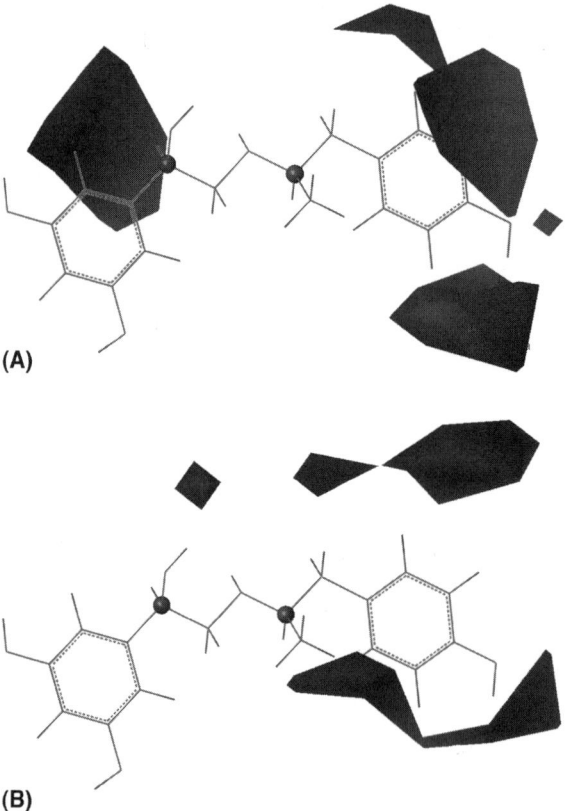

FIGURE 3.5 Comparative molecular field analysis for the series of fenoterol derivatives and analogs (28). Figures depict (**A**) electrostatic fields and (**B**) steric fields overlaid on a structure of R,(R)-fenoterol (centers of chirality are marked as gray balls). The first chiral center (*left*) is accompanied by the asymmetric electropositive field located behind the molecule—the indication of destabilizing hydrogen bond property of the —OH group in the *S*-configuration. The second chiral center (*right*) is accompanied by the asymmetric steric field indicating stabilizing steric properties of the —CH$_3$ group in the *R*-configuration.

first center (associated with β-hydroxyl group), a field of positive electrostatic is located behind the molecule—the indication of hydrogen bond interaction with the oxygen atom stabilizing the binding of molecules but only when the chiral center is in the *R*-configuration. The second chiral center (associated with CH_3 group) is accompanied with the asymmetric steric field located in front of the molecule—the indication of stabilizing steric properties of the methyl group in the *R*-configuration.

CONCLUSION

As can be deduced from above examples, different models of molecular discrimination between stereoisomers find applications for different studies and systems. It is apparent that chiral recognition is a complex phenomenon and a network of various types of static and dynamic interactions may affect binding. Isomers are always dissymmetrical 3D objects, thus the network of these interactions has to be dissymetrically defined in 3D space imposed by the macromolecule on the ligand. It must be borne in mind that the presented explanations are models, which by definition are simplified versions of the actual phenomena that have been modeled. Therefore, some simplified explanations (e.g., three-point model on flat surface, Fig. 3.1A) may appear useful but only under additional precisely defined conditions.

REFERENCES

1. Berthod A. Chiral recognition mechanisms with macrocyclicglycopeptide selectors. Chirality 2009; 21:167–175.
2. Davankov VA. The nature of chiral recognition: is it a three-point interaction? Chirality 1997; 9:99–102.
3. Pirkle WH. On the minimum requirements for chiral recognition. Chirality 1997; 9:103.
4. Easson LH, Stedman E. Studies on the relationship between chemical constitution and physiological action. V. Molecular dissymmetry and physiological activity. Biochem J 1933; 27:1257–1266.
5. Ogston AG. Interpretation of experiments on metabolic processes, using isotopic tracer elements. Nature 1948; 162:963.
6. Bentley R. Diastereoisomerism, contact points, and chiral selectivity: a four-site saga. Arch Biochem Biophys 2003; 414:1–12.
7. Armstrong DW, Ward TJ, Armstrong, RD, et al. Separation of drug stereoisomers by the formation of β-cyclodextrin inclusion complexes. Science 1986; 232:1132–1135.
8. Wilcox PE, Heidelberger C, Van Potter R. Chemical preparation of asymmetrically labelled citric acid. J Am Chem Soc 1950; 72:5019–5024.
9. Topiol S, Sabio M. Interactions between eight centers are required for chiral recognition. J Am Chem Soc 1989; 111:4109–4110.
10. Mesecar AD, Koshland DE. A new model for protein stereospecificity. Nature 2000; 403:614–615.
11. Mesecar AD, Koshland DE. Sites of binding and orientation in a four-location model for protein stereospecificity. IUBMB Life 2000; 49(5):457–466.
12. Kato K, Aburaya K, Miyake Y, et al. Excellent enantio-selective enclathration of (2R,3S)-3-methyl-2-pentanol in channel-like cavity of 3-epideoxycholic acid, interpreted by the four-location model for chiral recognition. Chem Commun (Cambr.) 2003; 23:2872–2873.
13. Sundaresan V, Abrol R. Towards a general model for protein–substrate stereoselectivity. Prot Sci 2002; 11:1330–1339.
14. Sokolov VI, Zefirov NS. Enantioselectivity at a two-point attachment: model of a rocking tetrahedron. Dokl Akad Nauk 1991; 319:1382–1384.

15. Booth TD, Wainer IW. Investigation of the enantioselective separations of a-alkyl arylcarboxylic acids on an amylose tris (3,5-dimethylphenylcarbamate) chiral stationary phase using quantitative structure-enantioselective retention relationships: identification of a conformationally driven chiral recognition mechanism. J Chromatogr A 1996; 737:157–169.
16. Booth TD, Wahnon D, Wainer IW. Is chiral recognition a three point process? Chirality 1997; 9:96–98.
17. Koshland DE. Application of a theory of enzyme specificity to protein synthesis. Proc Natl Acad Sci 1958; 44:98–104.
18. Koshland DE, Nemethy G, Filmer D. Comparison of experimental binding data and theoretical models in proteins containing subunits. Biochemistry 1966; 5:365–385.
19. Woo AY, Wang TB, Zeng X, et al. Stereochemistry of an agonist determines coupling preference of β2-adrenoceptor to different G proteins in cardiomyocytes. Mol Pharmacol 2009; 75:158–165.
20. Jozwiak K, Plazinska A, Toll L, et al. The effect of fenoterol stereochemistry on the β2 adrenergic receptor system–ligand directed chiral recognition. Chirality 2011; 23(1E): E1–E6.
21. Hou CT, Patel R, Barnabe N, et al. Stereospecificity and other properties of a novel secondary-alcohol-specific alcohol dehydrogenase. Eur J Biochem 1981; 119:359–364.
22. Isothermal titration calorimetry and drug design (application note). MicroCal LLC, 2006. Available at: http://www.microcal.com/documents/ITCand-Drug-Design.pdf
23. Holdgate GA, Ward WHJ. Measurements of binding thermodynamics in drug discovery. Drug Discov Today 2005; 10:1543–1550.
24. Perozzo R, Folkers G, Scapozza L. Thermodynamics of protein–ligand interactions: history, presence, and future aspects. J Recept Signal Transduct Res 2004; 24:1–52.
25. Jozwiak K, Hernandez SC, Kellar KJ, et al. Enantioselective interactions of dextromethorphan and levomethorphan with the α3β4-nicotinic acetylcholine receptor: comparison of chromatographic and functional data. J Chromatogr B Analyt Technol Biomed Life Sci 2003; 797:373–379.
26. Jozwiak K, Ravichandran S, Collins JR, et al. Interaction of noncompetitive inhibitors with an immobilized α3β4 nicotinic acetylcholine receptor investigated by affinity chromatography, quantitative-structure activity relationship analysis, and molecular docking. J Med Chem 2004; 47:4008–4021.
27. Karlsson A, Skoog A, Ohlen K. Effect of temperature on the reversal in the retention order of the enantiomers of mosapride on Chiral-AGP. J Biochem Biophys Methods 2002; 31:347–356.
28. Jozwiak K, Khalid C, Tanga MJ, et al. Comparative molecular field analysis of the binding of the stereoisomers of fenoterol and fenoterol derivatives to the β2 adrenergic receptor. J Med Chem 2007; 50:2903–2915.
29. Jozwiak K, Woo A, Tanga MJ, et al. Comparative molecular field analysis of fenoterol derivatives: a platform towards highly selective and effective β2 adrenergic receptor agonists. Bioorg Med Chem 2010; 18:728–736.
30. Jozwiak K, Toll L, Jimenez L, et al. The effect of stereochemistry on the thermodynamic characteristics of the binding of fenoterol stereoisomers to the β2-adrenoceptor. Biochem Pharmacol 2010; 79:1610–1615.
31. Kafri R, Lancet D. Probability rule for chiral recognition. Chirality 2004; 16:369–378.
32. Koppenhoefer B, Nothdurft A, Pierrot-Sanders J, et al. CHIRABSE, a graphical molecular database on the separation of enantiomers by liquid, supercritical fluid, and gas chromatography. Chirality 1995; 5:213–219.
33. Koppenhoefer B, Graf R, Hozschuh H, et al. CHIRBASE, a molecular database for the separation of enantiomers by chromatography. J Chromatogr A 1994; 666:557–563.
34. Berthod A, Chang SC, Armstrong DW. Empirical procedure that uses molecular structure to predict enantioselectivity of chiral stationary phases. Anal Chem 1992; 64:395–404.
35. Kaliszan R. Structure and Retention in Chromatography: a Chemometric Approach. Boca Raton, FL: CRC Press, 1997.

36. Famini GR, Wilson LY. Using theoretical descriptors in quantitative structure activity relationships and linear free energy relationships. Available at: http://www.netsci.org/Science/Compchem/feature08.html.
37. Wolbach JP, Lloyd DK, Wainer IW. Approaches to quantitative structure-enantioselectivity relationship modeling of chiral separations using capillary electrophoresis. J Chromatogr A 2001; 914:299–314.
38. Beck GM, Neau SH, Holder AJ, et al. Evaluation of quantitative structure property relationships necessary for enantioresolution with Lambda- and sulfobutylether Lambda-carrageenan in capillary electrophoresis. Chirality 2000; 12:688–696.
39. Kaliszan R, Doctor TA, Wainer IW. Stereochemical aspects of benzodiazepine binding to human serum albumin. II. Quantitative relationships between structure and enantioselective retention in high performance liquid affinity chromatography. Mol Pharmacol 1992; 42:512–517.
40. Folkers G, Yarim M, Pospisil P. Keywords in chirality modeling molecular modeling of chirality—software and literature research on chirality in modeling, chirality in docking, chiral ligand–receptor interaction and symmetry. In: Francotte E, Lindner W, eds. Chirality in Drug Research. Weinheim: Wiley-VCH, 2006.
41. Kubinyi H. Comparative molecular field analysis (COMFA). In: von Rague Schleyer P, ed. Encyclopedia of Computational Chemistry. New York: John Wiley & Sons, Ltd., 1998. Available at: http://www.wiley.com//legacy/wileychi/ecc/.
42. Park HJ, Choi Y, Lee W, et al. Enantioseparation of aromatic amino acids and amino acid esters by capillary electrophoresis with crown ether and prediction of enantiomer migration orders by a three-dimensional quantitative structure-property relationship/comparative field analysis model. Electrophoresis 2004; 25:2755–2760.
43. Robarge MJ, Agoston GE, Izenwasser S, et al. Highly selective chiral N-substituted 3alpha-[bis(4′-fluorophenyl)methoxy]tropane analogues for the dopamine transporter: synthesis and comparative molecular field analysis. J Med Chem 2000; 43:1085–1093.
44. Haining RL, Jones JP, Henne KR, et al. Enzymatic determinants of the substrate specificity of CYP2C9: role of B′-C loop residues in providing the pi-stacking anchor site for warfarin binding. Biochemistry 1999; 38:3285–3292.
45. Lipkowitz KB, Pradhan M. Computational studies of chiral catalysts: a comparative molecular field analysis of an asymmetric Diels-Alder reaction with catalysts containing bisoxazoline or phosphinooxazoline ligands. J Org Chem 2003; 68:4648–4656.
46. Schefzick S, Lammerhofer M, Lindner W, et al. Comparative molecular field analysis of quinine derivatives used as chiral selectors in liquid chromatography: 3D QSAR for the purposes of molecular design of chiral stationary phases. Chirality 2000; 12:742–750.
47. Altomare C, Cellamare S, Carotti A, et al. Substituent effects on the enantioselective retention of anti-HIV 5-aryl-delta 2-1,2,4-oxadiazolines on R,R-DACH-DNB chiral stationary phase. Chirality 1996; 8:556–566.
48. Suzuki T, Timofei S, Iuoras BE, et al. Quantitative structure-enantioselective retention relationships for chromatographic separation of arylalkylcarbinols on Pirkle type chiral stationary phases. J Chromatogr A 2001; 922:13–23.
49. Fabian WM, Stampfer W, Mazur M, et al. Modeling the chromatographic enantioseparation of aryl- and hetarylcarbinols on ULMO, a brush-type chiral stationary phase, by 3D-QSAR techniques. Chirality 2003; 15:271–275.

4 Separation and resolution of enantiomers and their dissociable diastereomers through direct crystallization

Harry G. Brittain

INTRODUCTION

Molecules whose mirror images cannot be superimposed on each other are identified as being chiral and an entire area of separation science has developed around the resolution of such compounds. Initially, scientists associated the phenomenon of optical rotation with the presence of carbon atoms bound to four different molecular fragments and these asymmetrically substituted carbon atoms became known as "asymmetric carbons." Continuing work showed that compounds incapable of rotating the plane of polarized light, but which were known to contain at least one asymmetric carbon atom, could be separated into chemically identical "optical isomers" that now exhibited the phenomenon of optical rotation. Over time it became clear that optical activity could exist in compounds having no asymmetric atoms and that other compounds existed that contained two or more asymmetric carbons, but still could not be rendered optically active. These findings necessitated a return to the proposal of Pasteur, who held that optical activity is a consequence of molecular dissymmetry. In other words, a molecule superimposable with its mirror image cannot be optically active and any molecule not superimposable with its mirror image will exhibit optical activity.

The fundamental requirement for the existence of molecular dissymmetry is that the molecule cannot possess any improper axes of rotation, the minimal interpretation of which implies additional interaction with light whose electric vectors are circularly polarized. This property manifests itself in an apparent rotation of the plane of linearly polarized light (polarimetry and optical rotatory dispersion) (1–5), or in a preferential absorption of either left- or right-circularly polarized light (circular dichroism) that can be observed in spectroscopy associated with either transitions among electronic (3–7) or vibrational states (6–8). Optical activity has also been studied in the excited state of chiral compounds (9,10). An overview of the instrumentation associated with these various chiroptical techniques is available (11).

When a compound contains a single center of dissymmetry, its mirror images are termed as *enantiomers*. Individual enantiomeric molecules are completely equivalent in their molecular properties, with the exception of their interaction with polarized light. An equimolar mixture of two enantiomers is termed a *racemic mixture*. The generally accepted configurational nomenclature (see chap. 2 for more detailed coverage of nomenclature) for tetrahedral carbon enantiomers was devised by Cahn, Ingold, and Prelog and is based on sequencing rules (12). Enantiomers are identified by their absolute configuration as being either *R* or *S*, depending on the direction (clockwise or counterclockwise) of substituents after they have been arranged according to increasing atomic mass. Compounds containing more than one center of dissymmetry are identified as *diastereomers*, and in compounds containing n dissymmetric centers, the number of diastereomers will equal 2^n. Diastereomers that differ in configuration at only one dissymmetric center only are termed *epimers*.

In order to specify the absolute configuration of an enantiomer containing an asymmetric carbon atom, the four atoms attached to that carbon are first arranged in a sequence of decreasing atomic mass. If two or more of these first atoms have the same atomic mass, one is chosen by comparing the atomic masses of the second group of atoms attached to the first atoms. If ambiguity still persists, the third, fourth, etc., sets (working outward from the asymmetric carbon atom) are compared until a selection can be made. In the second group of atoms, those atoms with the highest, next highest, etc., atomic numbers are always ranked in that order.

By virtue of symmetry constraints, a resolved enantiomer must crystallize in a noncentrosymmetric space group. Racemic mixtures are under no analogous constraint, but over 90% of all racemic mixtures are found to crystallize in a centrosymmetric space group (13,14). This consequence of molecular dissymmetry results in the situation where differing crystal structures can be obtained for the same chemical compound, depending only on the degree of resolution. Over the years, the study of molecular optical activity has been intimately related to corresponding studies of crystal morphology and structure and detailed histories can be found in the literature (15,16).

CRYSTALLOGRAPHY OF CHIRAL COMPOUNDS

A relatively small number of the possible crystallographic lattice symmetries are available for the crystals of separated enantiomers, since these must crystallize in a lattice structure that does not contain inverse elements of symmetry. Out of the 230 possible space groups belonging to the 32 crystal classes, only 66 space groups within 11 crystal classes are noncentrosymmetric and can accommodate homochiral sets of enantiomers (17). Racemates are permitted to crystallize in any of the 230 space groups and are not restricted to crystallizing in a centrosymmetric group. It has been found that most racemates crystallize into a group that possesses some elements of inverse symmetry. As will be discussed later, racemic mixtures occasionally crystallize in an enantiomorphic lattice system and this system results in a spontaneous resolution of the enantiomers upon crystallization.

A symmetry element is defined as an operation that when performed on an object results in a new orientation of that object which is indistinguishable from and superimposable on the original. There are five main classes of symmetry operations: (*i*) the identity operation (an operation that places the object back into its original orientation), (*ii*) proper rotation (rotation of an object about an axis by some angle), (*iii*) reflection plane (reflection of each part of an object through a plane bisecting the object), (*iv*) center of inversion (reflection of every part of an object through a point at the center of the object), and (*v*) improper rotation (a proper rotation combined with either an inversion center or a reflection plane) (18). Every object possesses some element or elements of symmetry, even if this is only the identity operation.

The rigorous group theoretical requirement for the existence of chirality in a crystal or a molecule is that no improper rotation elements be present. This definition is often trivialized to require the absence of either a reflection plane or a center of inversion in an object, but these two operations are actually the two simplest improper rotation symmetry elements. It is important to note that a chiral object need not be totally devoid of symmetry (i.e., be asymmetric), but that it merely be dissymmetric (i.e., containing no improper rotation symmetry elements). The tetrahedral carbon atom bound to four different substituents may be asymmetric, but the reason it represents a site of chirality is by virtue of dissymmetry.

Jacques has evaluated the compilations published by various authors, and has reported that 70% to 90% of homochiral enantiomers crystallize in the $P2_12_12_1$ or $P2_1$ space groups (13). The most frequently encountered chiral space group, $P2_12_12_1$, is orthorhombic, with the unit cell commonly consisting of four homochiral molecules that are related to each other by three binary screw axes. Practically, all of the other homochiral enantiomers crystallize in the monoclinic $P2_1$ space group, which is characterized by a plane of symmetry and a twofold axis. The unit cell generally contains two molecules related by a binary screw axis.

Jacques has also concluded that among the 164 space groups possessing at least one element of inverse symmetry, it is found that 60% to 80% of racemic compounds crystallize in either the $P2_1c$, $C2/c$, or Pi space groups (13). The most common group is monoclinic $P2_1c$ and the unit cell of which contains two each of the opposite enantiomers related to one another by a center of symmetry and a binary screw axis.

A chirality classification of crystal structures, which distinguishes between homochiral (type A), heterochiral (type B), and achiral (type C) lattice types, has been provided by Zorkii et al. (19) and expanded by Mason (20). In the type A structure, the molecules occupy a homochiral system, or a system of equivalent lattice positions. Secondary symmetry elements (e.g., inversion centers, mirror or glide planes, or higher-order inversion axes) are precluded in type A lattices. In the racemic type B lattice, the molecules occupy heterochiral systems of equivalent positions and opposite enantiomers antipodal are related by secondary lattice symmetry operations. In type C structures, the molecules occupy achiral systems of equivalent positions and each molecule is located either on an inversion center, on a mirror plane, or on a special position of a higher-order inversion axis. If there are two or more independent sets of equivalent positions in a crystal lattice, the type D lattice becomes feasible. This structure consists of one set of type B and another of type C, but is not commonly encountered. Of the 5000 crystal structures studied, 28.4% were classified as being type A, 55.6% were of type B, 15.7% were of type C, and only 0.3% were considered as being type D.

A detailed discussion of crystal packing and the resulting space groups has been given by Kitaigorodskii (21). This approach assumes that molecular crystals are assemblages for which compactness tends toward the maximum, which is compatible with the molecular geometry. He defined a packing coefficient as vZ/V, where v is the volume of the molecule, V is the volume of the cell, and Z is the number of molecules in the unit cell. In this way, the space-filling or packing coefficient in crystals always lies between 0.65 and 0.77, which is of the same order as the regular packing of spheres or ellipsoids. Molecular structures of which cause an inability to attain a packing coefficient at least equal to 0.6 are not found to crystallize, and these compounds can only form glasses from the melt. In order to fill space in the most compact manner with objects of indeterminate geometry, the lattice must be populated in a compact fashion.

There are only a limited number of two-dimensional arrays in which an object may reside in contact with six neighbors, which is necessary for the optimal packing of a molecular assembly. It has been concluded that the binary screw axis is highly conducive toward efficient packing. In the final analysis, the limited possible combinations of stacking leads to a limited number of space groups that fulfill the requirements of three-dimensional close-packing.

The close-packing criterion for crystal stability generally yields lower free energies for a structure composed of a racemic assembly over that composed of homochiral molecules. This view has been expressed in the empirical rule of Wallach, which states that the combination of two opposite enantiomers to form a racemate is accompanied by a volume contraction. As might be expected, there are many exceptions to this rule and these have been discussed in a systematic manner (22). The modification to Wallach's rule contributed by Walden is more generally valid, and states that if an enantiomer has a lower melting point than its corresponding racemate, then the crystals of the latter will have the higher density.

As an example of the effect of molecular chirality on the crystal structure of isolated materials, the system studied by Pasteur will be considered (23), who found that the resolved enantiomers of sodium ammonium tartrate could be obtained in a crystalline form that featured nonsuperimposable hemihedral facets (Fig. 4.1). Pasteur was quite surprised to learn that when he conducted the crystallization of racemic sodium ammonium tartrate at temperatures below 28°C, he also obtained crystals that contained nonsuperimposable hemihedral facets. He was able to manually separate the left-handed crystals from the right-handed ones and found that these separated forms were optically active upon dissolution. More surprising was his discovery that when the crystallization was conducted at temperatures exceeding 28°C, he obtained crystals having different morphologies that did not contain the hemihedral crystal facets (also Fig. 4.1).

The explanation to Pasteur's observations is that the sodium ammonium salt prepared from racemic tartaric acid crystallizes in the orthorhombic $P2_12_12_1$

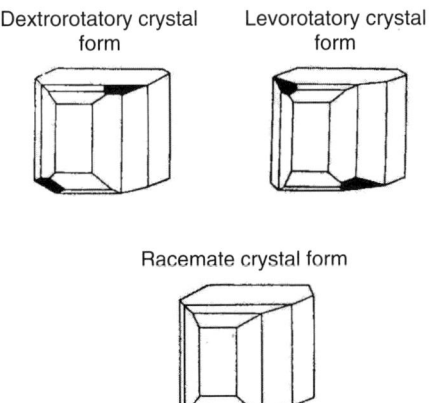

FIGURE 4.1 Crystals of sodium ammonium tartrate, obtained under conditions yielding the hemihedral facets (*darkened crystal faces*) distinctive of the chiral crystalline forms. Also shown is the crystal morphology of racemic sodium ammonium tartrate.

space group, characterized by unit cell parameters of a = 12.173 Å, b = 14.412 Å, and c = 6.235 Å (24). This particular crystal class is noncentrosymmetric and as a result individual crystals are dissymmetric. When formed below a temperature of 28°C, the preferred molecular packing does not permit the intermingling of the enantiomers to yield a true racemic crystal and hence the crystallization results in a spontaneous resolution of the substance into physically separable dissymmetric crystals. On the other hand, when racemic sodium ammonium tartrate is isolated at temperatures higher than 28°C, a different polymorphic form is obtained, that of sodium ammonium tartrate that crystallizes in the monoclinic $P2_1/a$ space group and is characterized by unit cell parameters of a = 15.244 Å, b = 5.066 Å, c = 10.218 Å, and β = 93.60° (24). The racemic crystal form is characterized by a completely different packing pattern and pattern of hydrogen bonding, which allows for the formation of a racemic modification.

PROPERTIES AND RESOLUTION OF RACEMIC MIXTURES OF DISSYMMETRIC SOLIDS

The early work of Pasteur clearly demonstrated that racemic compounds were actually mixtures of the mirror images of the compounds and that these mixtures could be separated into their component enantiomers by various means. He found that while certain compounds could be resolved by mechanical separation of directly crystallized chiral crystals, the majority of organic compounds required some type of reversible chemical reaction in order to effect an enantiomeric separation. Racemic compounds separable by the direct crystallization method were termed *conglomerates*, while racemic compounds that became separable by crystallization subsequent to their derivatization were termed *racemates*.

Many procedures result in the incomplete resolution of a racemic mixture into the component enantiomers, and quantities have been defined that are used to specify the degree of resolution. The *enantiomeric purity* (EP) of a substance will have values between 0 and 1, and for substances where the (*R*)-enantiomer is present in excess, the EP is given by:

$$\text{EP} = 2(X_R) - 1 \qquad (1)$$

where X_R is the mole fractions of the (*R*)-enantiomer. The *enantiomeric excess* (EE) of a substance will have values between 0 and 0.5 and is the quantity of excess enantiomer over that of the racemic mixture. Where the (*R*)-enantiomer is present in excess, the EE is given by

$$\text{EE} = (X_R - 0.5) \qquad (2)$$

For example, in the instance where the mole fraction of excess enantiomer in a substance is 0.65, the EP would equal 0.3 (i.e., 30%) and the EE would equal 0.15 (i.e., 15%).

Historically, optical rotation was used to measure the EP of a substance, giving rise to the term *optical purity* (OP). When measured using optical rotation, the OP of a substance is given by

$$\text{OP} = \frac{\alpha_{\text{OBS}}}{[\alpha]} \qquad (3)$$

Where α_{OBS} is the observed optical rotation of the substance and $[\alpha]$ is the specific rotation of the enantiomerically pure substance. When the enantiomeric composition of a substance is determined by nonoptical means (such as chiral chromatography or nuclear magnetic resonance), the results should not be referred to as optical purities but should be properly identified as enantiomeric purities.

Conglomerate Systems

Conglomerate solids are characterized by the presence of a single enantiomer within the unit cell of the crystal, even when the solid is obtained through crystallization of a racemic or partially resolved mixture. These solids consist of separate crystals, each of which consists entirely of one enantiomer or its mirror image and which may be separated solely on the basis of their physical properties. The small-scale resolution of dissymmetric compounds crystallizing as conglomerates can be straightforward since the resolution step takes place spontaneously with the crystallization step. The key to a successful resolution by direct crystallization lies in the means used to physically separate the crystals containing the opposite enantiomers.

Approximately 10% of chiral compounds inherently form conglomerates by virtue of their crystallization tendencies (13,25), and Jacques and coworkers have provided details of approximately 250 organic compounds forming known conglomerate systems (26,27). The apparent randomness of conglomerate system formation has been explained through considerations of the thermodynamics of these systems. The stability of true racemic solids is defined by the free energy change associated with the process of combining the (R)-enantiomer with the (S)-enantiomer to produce the (R,S)-enantiomeric solid. This process has been calculated to be in the range of 0 to –2 kcal/mol and is roughly proportional to the difference in melting points between the racemate and the resolved enantiomers (28). In most cases, the free energy associated with the formation of racemates is exothermic, owing to the positive nature of the enthalpies and entropies of formation. In those cases where the melting point of the racemic mixture is at least $20°$ lower than the melting points of the separated enantiomers, a conglomerate system will generally be obtained.

The use of crystal structure prediction in the evaluation of systems for their tendency toward spontaneous resolution has been explored (29). Using a gas phase conformational search to locate low-energy conformations, the possible crystal structures of 5-hydroxymethyl-2-oxazolidinone and 4-hydroxymethyl-2-oxazolidinone assembled from these conformations have been predicted. For both compounds, the racemate was predicted to have the lowest energy, with the energy difference between the lowest-energy racemic structure and the lowest-energy separated enantiomer structure being in the range of 0.2 to 0.9 kcal/mole. The authors noted that as the lattice energy calculations become more accurate, it should become possible to predict whether a chiral molecule will crystallize in a space group where its racemate or conglomerate nature would be expressed.

A theoretical explanation has been developed to explain the effect of chiral impurities on the crystallization rates of the enantiomorphic components of a conglomerate system (30). The theory provides the time required to complete the crystallization of the separated enantiomers, suggests that one might be able to obtain an enantiomerically pure product even if the chiral impurity was less than enantiomerically pure and even provides information regarding the particle size distribution. The model was tested on the crystallization of D,L-glutamic acid that was carried out in the presence of a resolved L-lysine impurity.

Inherently Conglomerate Systems

Since a racemic conglomerate system will consist of independently formed, enantiomerically pure crystals, such racemic mixtures are seen to constitute a binary system that consists of physical mixtures of the enantiomer components. Such binary mixtures are easily described by the phase rule and can be characterized by their melting point phase diagrams. Since the components of a conglomerate racemate will melt independently at the same temperature, the material will exhibit the melting phenomena of a pure substance. One would therefore predict the existence of a eutectic point in the melting point phase diagram that would be located at exactly the racemic composition. In principle, the enantiomer present in excess could be separated out in a partially resolved conglomerate system simply by heating the sample to a temperature just above the melting point of the racemate and then collecting the crystalline excess of the residual enantiomer.

An example of the type of melting point phase diagram that is typical for a conglomerate system is shown in Figure 4.2, which illustrates the phase diagram

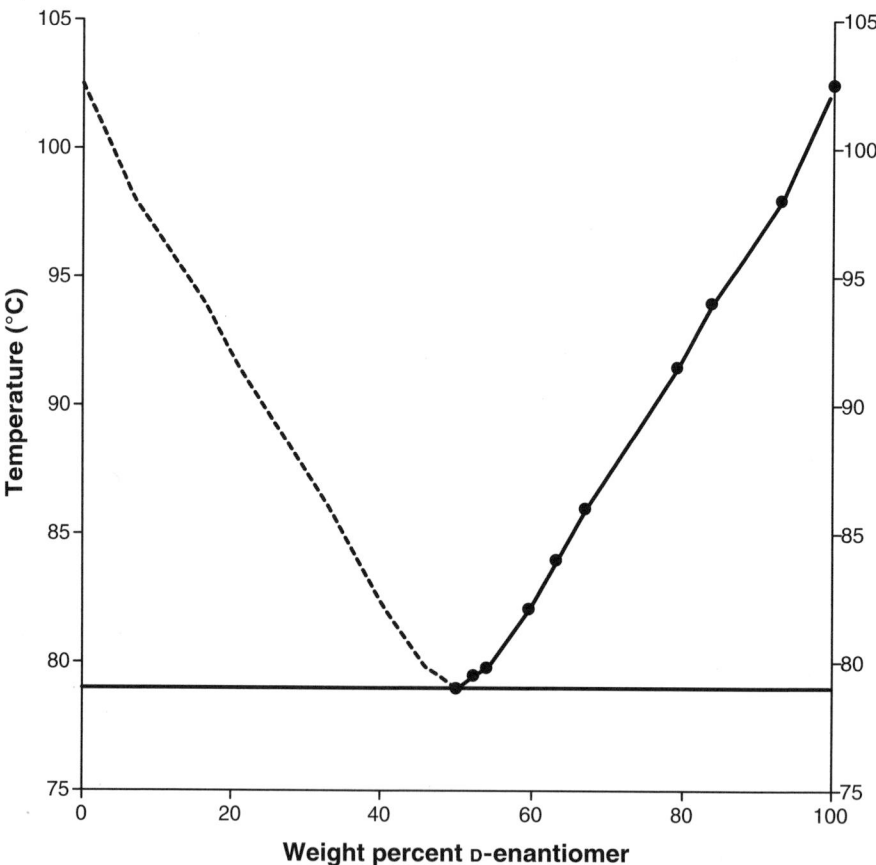

FIGURE 4.2 Melting point phase diagram obtained for methyl diacetyltartrate. The actual reported data points are shown as the filled circles, and the left-hand side of the phase diagram was obtained by taking the mirror image of the right-hand side of the diagram. *Source*: Prepared using data published in Ref. 31.

reported for methyl diacetyltartrate (31). Below the eutectic temperature of 79°C, the system will exist as a mixture of solid D-enantiomer and L-enantiomer. At the exact composition of the racemic mixture ($X = 0.5$), the system will exist entirely in the liquid phase above the eutectic temperature. At mole fractions where the amount of L-enantiomer exceeds that of the D-enantiomer, the system will exist as an equilibrium mixture of racemic liquid and solid L-enantiomer. As required by the phase diagram of a conglomerate, the eutectic temperature is the lowest temperature attainable where any liquid phase can exist in equilibrium with any solid phase.

An alternate approach for the characterization of conglomerate systems is through the use of ternary phase diagrams, where one component is the solvent and the solubility of the substance is used as the observable parameter (10,25). A racemic mixture will be more soluble than are the separated enantiomers and the rule of Meyerhoffer states that a conglomerate will have a solubility that is twice the solubility of the resolved enantiomer (32). This situation arises since ideally the mole fraction of an independent constituent in a liquid depends only on the enthalpy of fusion and melting point of the substance and since the solubility of one enantiomer cannot affect the solubility of the other. Jacques and Gabard have examined the solubilities of a number of conglomerate systems and have largely confirmed the double solubility rule (33). A selection of the results reported for nondissociable substances is presented in Table 4.1.

Jacques and coworkers have provided a comprehensive background regarding methodologies that enable direct crystallization for the resolution of racemic mixtures (13,25), which may be supplemented by other reviews in this area (34,35).

The first method of enantiomeric separation by direct crystallization is the mechanical technique used by Pasteur, where he separated the enantiomorphic crystals that were simultaneously formed while the residual mother liquor remained racemic. Enantiomer separation by this particular method can be extremely time-consuming and not possible to perform unless the crystals form with recognizable chiral features (such as well-defined hemihedral faces). Nevertheless, this procedure can be a useful means to obtain the first seed crystals required for a scale-up of a direct crystallization resolution process. When a particular system has been shown to be a conglomerate and the crystals are not sufficiently distinct so as to be separated, polarimetry or circular

TABLE 4.1 Solubilities of Separated Enantiomers and Racemic Mixtures of Conglomerate Materials

Compound	System	Solubility of separated enantiomer (g/100 mL)	Solubility of racemic mixture (g/100 mL)	Ratio of racemate solubility to enantiomer solubility
Asparagine	Water (25°C)	2.69	5.61	2.16
(α-Naphtoxy)-2-propionamide	Acetone (25°C)	1.38	2.24	2.08
p-Nitrophenylaminopropanediol	Methanol (25°C)	1.63	3.35	2.14
N-acetyl-leucine	Acetone (25°C)	1.86	4.12	2.13
N-acetyl-glutamic acid	Water (25°C)	4.14	8.28	2.12
Diacetyldiamide	Water (35.5°C)	12.4	23.7	2.14
3,5-Dinitrobenzoate-lysine	Water (30°C)	6.96	13.17	1.99

dichroism spectroscopy can often be used to establish the chirality of the enantiomeric solids.

Even a few seed crystals, mechanically separated, can be used to produce larger quantities of resolved enantiomerically pure material. A second method of resolution by direct crystallization involves the localized crystallization of each enantiomer from a racemic, supersaturated solution. With the crystallizing solution within the metastable zone, oppositely handed, enantiomerically pure, seed crystals of the compound are placed in geographically distant locations in the crystallization vessel. These serve as nuclei for the further crystallization of the like enantiomer, and enantiomerically resolved product grows in the seeded locations. This procedure has been used to obtain both enantiomers of methadone, where approximately 50% total yield of enantiomerically pure material can be obtained (36).

Enantiomer separation may be practiced on the large industrial scale using the procedure known as resolution by entrainment (13,25). The method is based on the condition that the solubility of a given enantiomer is less than that of the corresponding racemate. To begin, a solution that contains a slight excess of one enantiomer is prepared. Crystallization is induced (usually with the aid of appropriate seed crystals); whereupon the desired enantiomer is obtained as a solid and the mother liquor is enriched in the other isomer. In a second crystallization step, the other enantiomer is obtained. The method can be applied to any racemic mixture that crystallizes as a conglomerate and the main complication that can arise is when the compound exhibits polymorphism. In that case, the entrainment procedure must be carefully designed so as to generate only the desired crystal form.

Resolution by entrainment is best illustrated through the use of an example, and the laboratory-scale resolution of hydrobenzoin (37) is an appropriate example. Initially, 1100 mg of racemic material was dissolved along with 370 mg of (–)-hydrobenzoin in 85 g of 95% ethanol and then the solution was cooled down to 15°C. Then 10 mg of the (–)-isomer was added in the form of seed crystals and a crop of additional crystalline material was allowed to form. After 20 minutes, 870 mg of (–)-hydrobenzoin was recovered; 870 mg of racemic hydrobenzoin was then dissolved with heating, the resulting solution cooled to 15°C, and seeded with 10 mg of the (+)-isomer. At this time, 900 mg of (+)-hydrobenzoin was recovered. The process was cycled 15 times, and ultimately yielded 6.5 g of (–)-hydrobenzoin and 5.7 g of (–)-hydrobenzoin. Each isomer was obtained as approximately 97% enantiomerically pure.

Racemate Systems Rendered into Conglomerates

Although the number of organic compounds forming conglomerate systems is not large, a significant number of other chiral compounds can be transformed into conglomerate-forming systems through formation of a salt or covalent derivative. For example, substitution of additional functionalities onto the achiral portion of a compound can sometimes also yield a conglomerate system. For example, although mandelic acid and many of its derivatives crystallize as racemic mixtures, *ortho*-chloromandelic acid is obtained as a conglomerate (38). The enantiomers of *sec*-phenethyl alcohol form a conglomerate system upon formation of the 3,5-dinitrobenzoyl ester (39).

While the frequency of conglomerate-forming systems is fairly small for organic compounds, a statistical analysis of more than 500 salts has demonstrated that the probability of observing spontaneous resolution is two or three times higher for salts of the same organic compounds (40). The possibility has been further demonstrated through studies of the crystallization of achiral dicarboxylic acids with α-phenethylamine, where racemates were obtained for the salts with hydrogen malonate and hydrogen phthalate, but a conglomerate was obtained for the hydrogen succinate salt (41). In a subsequent work, it was concluded that conglomerate formation took place when the protonated and deprotonated carboxylic acid groups formed hydrogen-bonded chains rather than forming intermolecular hydrogen-bonded dimers in the crystal (42).

To illustrate how a racemate system can be turned into a conglomerate system, and to demonstrate how one may recognize the existence of a conglomerate system, we will consider the example of ibuprofen, or 2-(4-isobutylphenyl) propanoic acid. The single crystal structures of enantiomerically pure (61) and racemic (62) ibuprofen have been published and the powder diffraction patterns of the enantiomerically pure and racemic forms were calculated from the crystallographic properties of the respective unit cells (Fig. 4.3). The existence

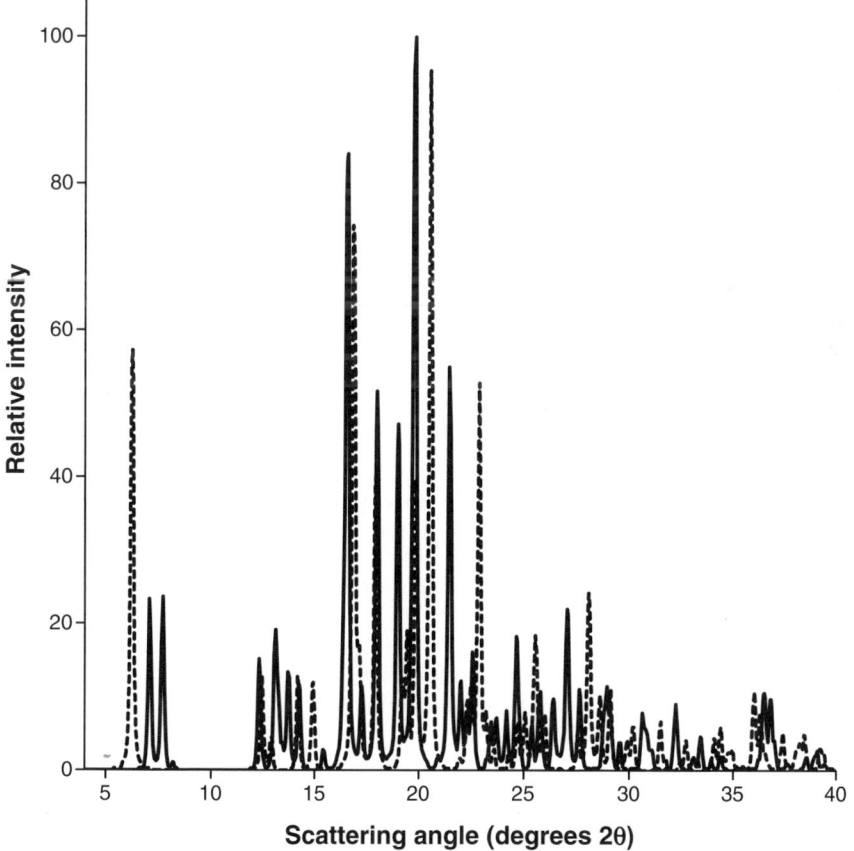

FIGURE 4.3 X-ray powder diffraction patterns calculated from the published crystal structures of (S)-ibuprofen (*solid trace*; derived from Ref. 43) and (RS)-ibuprofen (*dashed trace*; derived from Ref. 44).

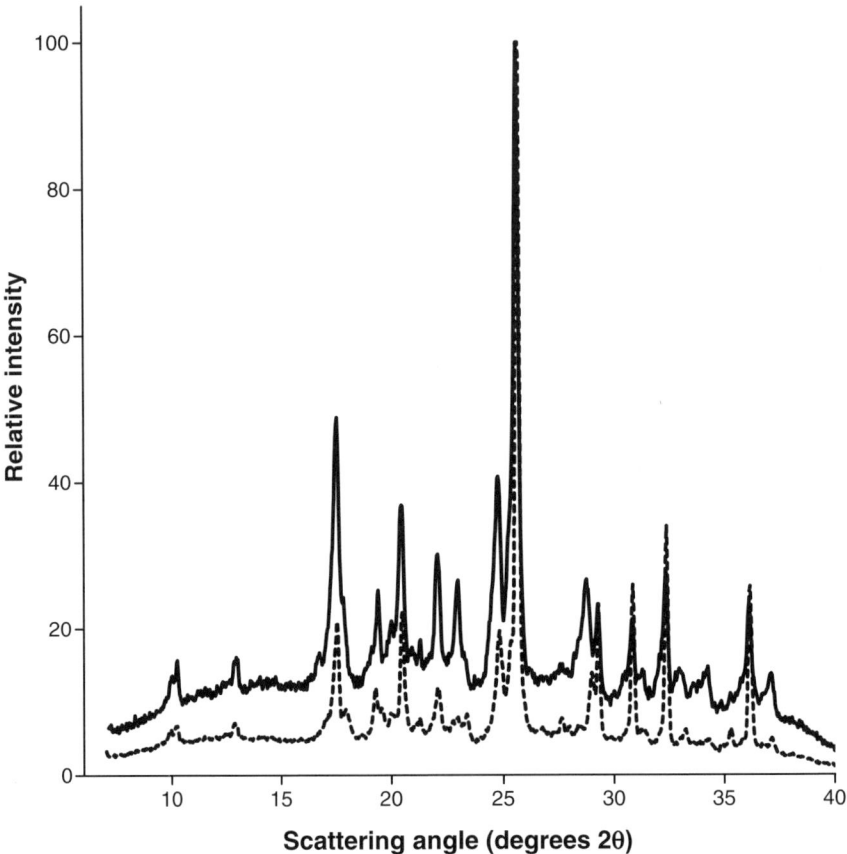

FIGURE 4.4 X-ray powder diffraction patterns obtained for the tromethamine salts with (S)-ibuprofen (*solid trace*) and (RS)-ibuprofen (*dashed trace*).*Source*: Courtesy of H. G Brittain, unpublished results.

of a conglomerate system is indicated if the diffraction patterns of enantiomerically pure and racemic forms of a compound are the same, but this is not the case for ibuprofen. Formation of the tromethamine (2-amino-2-(hydroxymethyl) propane-1,3-diol) salt of ibuprofen was found to result in the generation of a conglomerate system, as demonstrated in Figure 4.4 by the equivalence in X-ray powder diffraction patterns of the (S)-ibuprofen/ tromethamine and (RS)-ibuprofen/tromethamine salts (45). In addition, the differential scanning calorimetry thermograms of the salts are effectively the same, with the (S)-ibuprofen salt exhibiting a melting endotherm at 158.4°C (enthalpy of fusion of 159.1 J/g) and the (RS)-ibuprofen salt having a melting endotherm at 158.9°C (enthalpy of fusion of 160.2 J/g) (45).

The crystal structures of the cinnamic acid salts with (+)-l-phenylethylamine salt and (+)-1-(4-isopropylphenyl)ethylamine were determined using single-crystal X-ray analysis, where it was found that two amine components and two acid components formed a helical column through hydrogen bonding in each crystal (46). On the basis of structural similarity of the two conglomerates, a criterion was proposed for the transformation of a racemic modification of an amine into a conglomerate crystal. For this to occur, the achiral acid should form a helical column by hydrogen bonds with the amine, the acid should be rigid and flat to limit the orientation of the amine, and the sizes of the amine and the acid should be similar to each other. The proposed criteria were supported through crystal structure determinations of the conglomerate systems formed by the p-chlorobenzenesulfonic acid salt of L-alanine, the p-toluenesulfonic acid salt of L-serine, and the benzenesulfonic acid salt of (L)-leucine (47).

Racemate Systems

Racemate solids are characterized by the presence of equimolar amounts of both enantiomers within the unit cell of the crystal, which crystallizes within a centrosymmetric space group. This space group will necessarily be different from the noncentrosymmetric space group characteristic of the separated enantiomers. As a result, the range of physical properties associated with a heterochiral solid will generally be completely different from those of the homochiral solid. For instance, the melting point of the exact racemic mixture may be greater than or less than that of the separated enantiomers and there is no general rule that can be invoked to provide a reliable prediction of melting point behavior.

The melting point phase diagram of a racemate system will contain two eutectic minima, and crystallization at either condition will yield a racemic mixture and not enantiomerically pure product. The phase diagram shown in Figure 4.5 for methyl dipropionyltartrate is typical for a racemate system, where a eutectic temperature of 24°C was detected for a D-enantiomer composition of 75 mole-percent (31). For compounds containing a single center of dissymmetry, construction of the full phase diagram only requires the acquisition of melting point data for one enantiomer in excess to develop one side of the phase diagram, as the other side of the phase diagram must be the mirror image.

Care must be observed in the interpretation of a melting point phase diagram for those instances where the eutectic temperatures are located close to the equimolar composition; as such systems could be erroneously interpreted as indicating the existence of a conglomerate system. An example of this type of behavior is shown in Figure 4.6 for the ethyl diacetyltartrate system, where a eutectic temperature of 42°C was found for a D-enantiomer composition of 55.7 mole-percent (31). On the other hand, when the eutectic temperatures are close to the melting points of the pure enantiomers, the phase diagram could be mistaken for that of a solid solution. An example of this behavior is shown in Figure 4.7 for the methyl dibenzoyltartrate system, where a eutectic temperature of 130.4°C was found for a D-enantiomer composition of 94.5 mole-percent (31). It is clear that the melting points of a sufficient number of compositions must be measured if the proper phase diagram is to be obtained.

In order to separate the enantiomers of a racemate system, the properties of the molecule must be transformed in such a way that the phase diagram containing two eutectic points becomes a phase diagram containing one eutectic

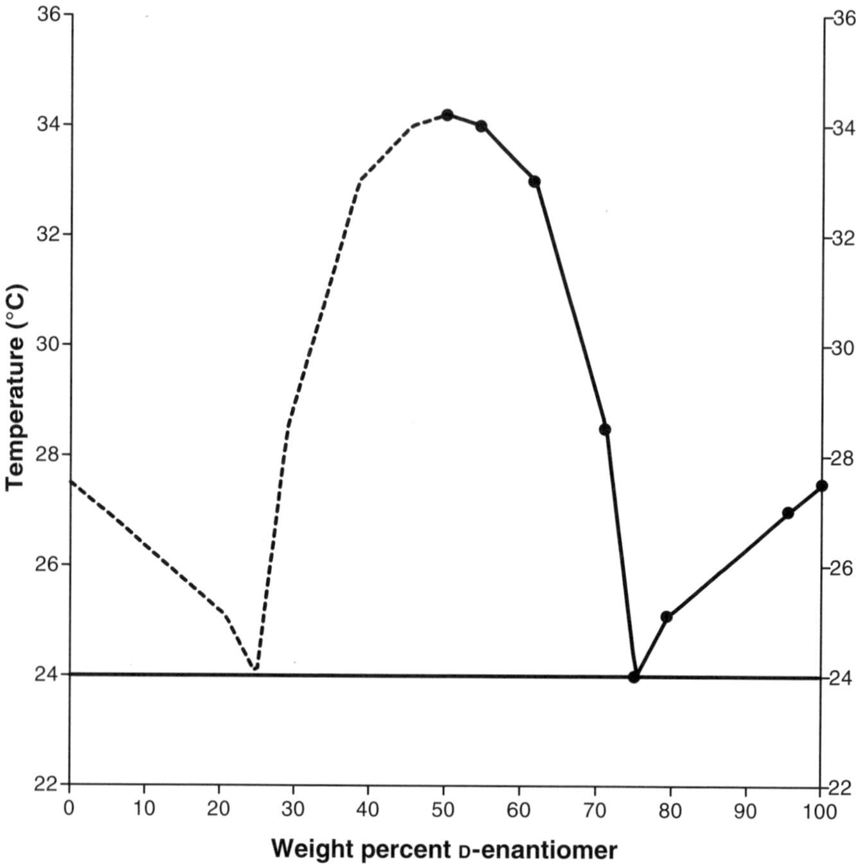

FIGURE 4.5 Melting point phase diagram obtained for methyl dipropionyltartrate. The actual reported data points are shown as the filled circles, and the left-hand side of the phase diagram was obtained by taking the mirror image of the right-hand side of the diagram. *Source*: Prepared using data published in Ref. 31.

point. This can be brought about through performance of a derivatization reaction with a suitable chiral reagent that leads to diastereomers (often identified in the literature as the p-salt and the n-salt) that exhibit a phase diagram containing only a single minimum. In general, this minimum will not be located at the exact racemic composition, but will instead be observed at some other concentration value. In fact, for the p-salt to be separated from the n-salt in a single crystallization step, the position of the eutectic should be substantially removed from the equimolar point.

To illustrate the effect of diastereomer formation on phase diagrams, the example of ibuprofen may again be considered. As evident from the phase diagram shown in Figure 4.8, ibuprofen forms a classic racemate system (45,48), with a eutectic temperature of 47.1°C detected at 81.2 weight-percent (S)-ibuprofen composition. The mirror image eutectic would be predicted to form at the same temperature but at 18.8 weight-percent (S)-ibuprofen

SEPARATION AND RESOLUTION OF ENANTIOMERS

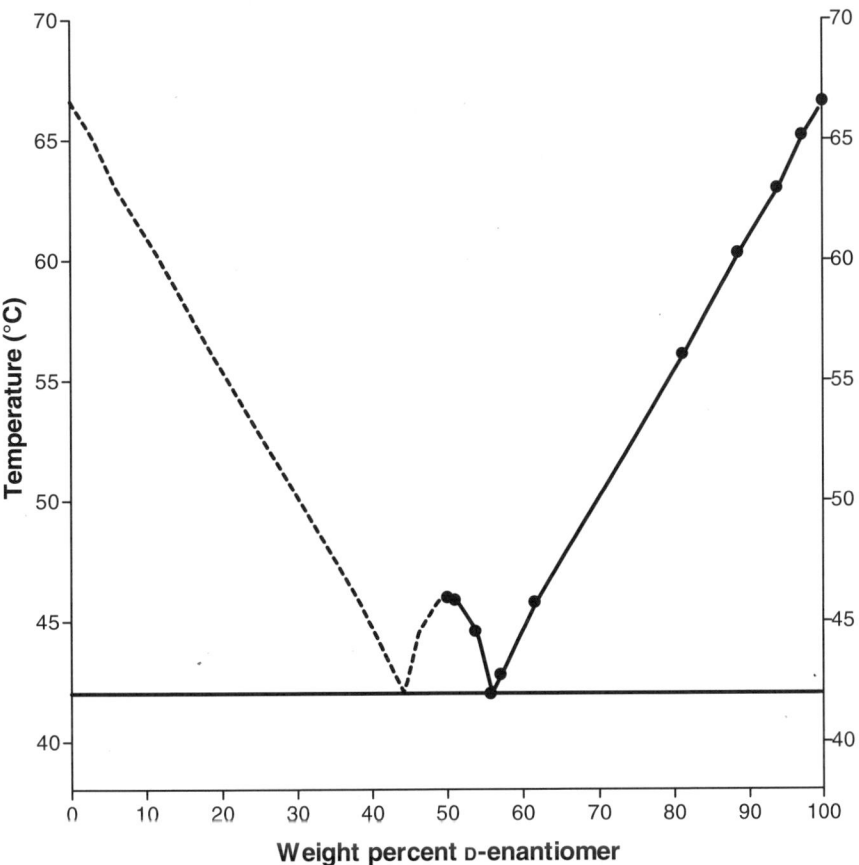

FIGURE 4.6 Melting point phase diagram obtained for ethyl diacetyltartrate. The actual reported data points are shown as the filled circles, and the left-hand side of the phase diagram was obtained by taking the mirror image of the right-hand side of the diagram. *Source*: Prepared using data published in Ref. 31.

composition. However, as shown in Figure 4.9, when the salt of ibuprofen is formed with α-methylbenzylamine (i.e., the resolving agent), the phase diagram is simplified into that of a single eutectic, characterized by a temperature of 147.0°C detected and at 66.5 weight-percent (S)-ibuprofen composition (45,49). A procedure could be designed that would serve to crystallize one diastereomer over the other and then remove the resolving agent to obtain purified enantiomer.

The preceding example demonstrates the general view that the procedure most likely to alter the crystallization thermodynamics of true racemate systems will entail the formation of dissociable diastereomer species (50–54). In most instances, these diastereomers are simple salts formed between proton donors and proton acceptors, or electron-pair donors and electron-pair acceptors. For example, the first resolving agents introduced for acidic enantiomers were alkaloid compounds, and hydroxy acids were used for the resolution of

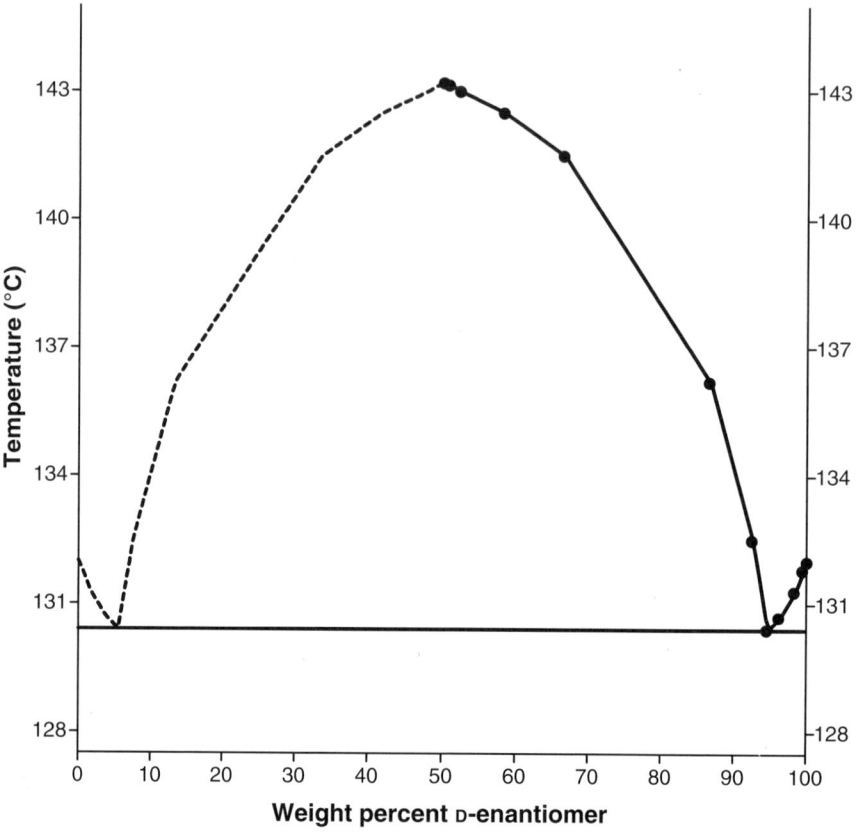

FIGURE 4.7 Melting point phase diagram obtained for methyl dibenzoyltartrate. The actual reported data points are shown as the filled circles, and the left-hand side of the phase diagram was obtained by taking the mirror image of the right-hand side of the diagram. *Source*: Prepared using data published in Ref. 31.

basic enantiomers. This type of resolution procedure has been known since the time of Pasteur and extensive tables of resolving agents and procedures are available (47,55,56).

The general procedure for separation of enantiomers through formation of dissociable diastereomers can be illustrated for the instance where a racemic acid is to be resolved through the use of a basic resolving agent. The first step of the resolution procedure involves formation of the p- and n-diastereomeric salts:

Racemic mixture	+	Resolving agent	→	Diastereomer salts
(+)-R–COOH				(+)-R–COO–NH$_3$–R′–(+)
	+	2 NH$_2$–R′–(+)	→	
(−)-R–COOH				(−)-R–COO–NH$_3$–R′–(+)

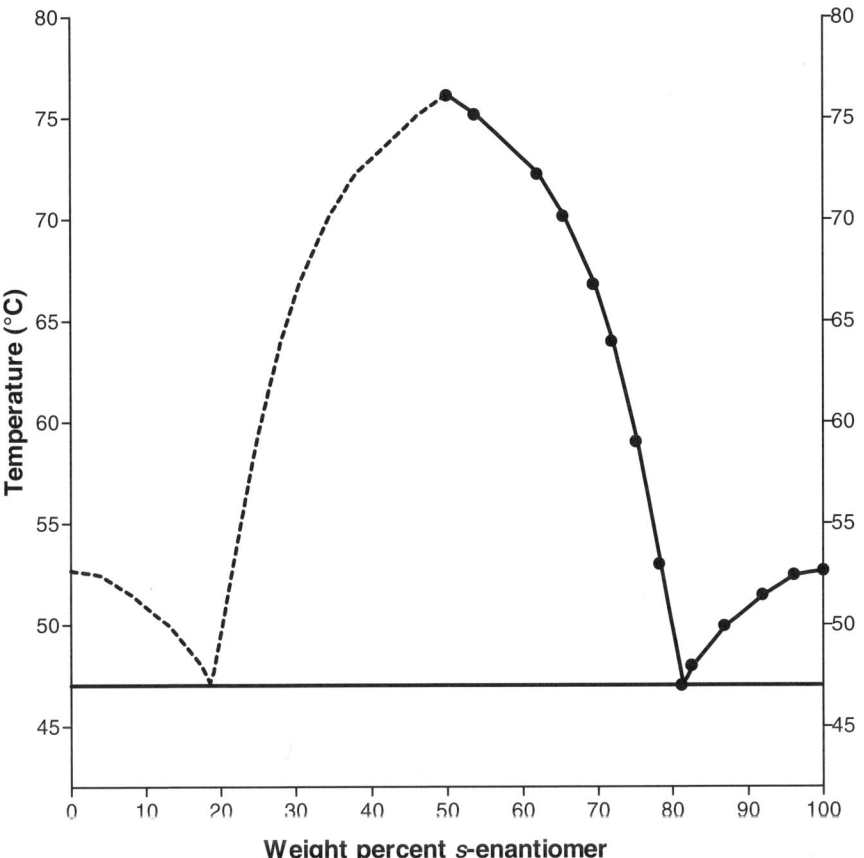

FIGURE 4.8 Melting point phase diagram obtained for racemic ibuprofen free acid. The actual reported data points are shown as the filled circles, and the left-hand side of the phase diagram was obtained by taking the mirror image of the right-hand side of the diagram *Source*: Courtesy of H G Brittain, unpublished results.

According to convention (57), the (R,R) and (S,S) diastereomers are termed the p-salts, and the (R,S) and (S,R) diastereomers are identified as the n-salts. In the example used above, the (+)-R-COO-NH$_3$-R'-(+) diastereomer would correspond to the p-salt, and the (-)-R-COO-NH$_3$-R'-(+) diastereomer would be the n-salt. In the usual practice, the p- and n-salts are separated by fractional crystallization and the success of the resolution process is critically related to this crystallization step. As discussed above, the act of derivatization may be considered as a process that alters the phase diagram from that of a double eutectic into a pseudoconglomerate phase diagram exhibiting only one eutectic.

Since the relative solubilities of the diastereomeric species play such an important role in their separation by crystallization, it follows that ternary phase diagrams incorporating solvent can play a large role in understanding a resolution process (58). A study of the separation of the enantiomers of

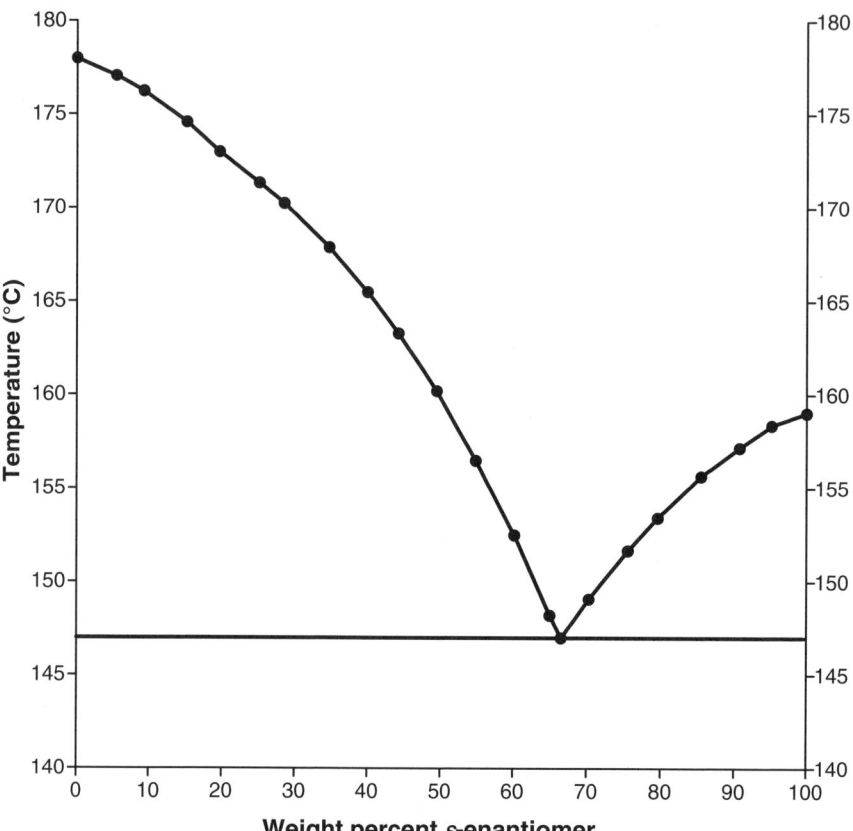

FIGURE 4.9 Melting point phase diagram obtained for the salt formed between racemic ibuprofen and the (S)-enantiomer of α-methylbenzylamine. *Source*: Courtesy of H G Brittain, unpublished results.

phenylsuccinic acid after formation of the phenylsuccinate/proline salt demonstrated the utility of phase diagrams in the design of enantiomeric separations (59). In this work, it was found that the degree of resolution for the salt having the 1:2 stoichiometry was only minimally affected by temperature, while the separation of the 1:1 stoichiometric phenylsuccinate/proline salt was strongly affected by temperature.

The chiral discrimination existing in the diastereomer systems formed by ephedrine and various mandelic acids has been studied in great detail using a variety of spectroscopic and structural tools. Although significant differences in the solubilities of the n- and p-salts obtained after reaction of ephedrine with mandelic acid were known to exist, both salts were found to crystallize in the same monoclinic space group with crystal structures that were isosteric (60). In subsequent work, it was shown that the crystalline diastereomer salts formed by (R)-mandelic acid formed a more compact repeating pattern (with higher melting points and enthalpies of fusion) than did analogous salts formed by (S)-mandelic acid (61). It was deduced that the subtle chiral discriminations led to the existence of different hydrogen-bonding modes, which in turn became manifest in a variety of other physical properties.

There is no doubt that the crystallographic differences between the diastereomers generated during derivatization directly affect the outcome of the resolution process. For example, when D,L-phenylglycine was resolved with (S)-camphor-10-sulfonic acid, the n-salt was isolated in 45.7% yield and had an EP of 98.8% (62). The less soluble n-salt exhibited a higher melting point and enthalpy of fusion than did the p-salt, which was found to be freely soluble. The crystal structure of the n-salt exhibited a dense structure containing alternating phenylglycine cation and camphor sulfonate anion layers, while the structure of the p-salt exhibited a coarser structure with vacancy layers between the anion and cation planes. These differences in crystal structure directly led to the differing physical properties of the diastereomeric salts that caused the enantiomeric resolution to be so successful.

Depending on the details of the lattice dynamics, there is no restriction regarding the ability of a given pair of p- and n-salts to crystallize in the same or in different space groups. For example, in the diastereomeric system formed by α-methylbenzylamine and hydratropic acid, the p-salt was found to crystallize in the $P2_1$ space group, while the n-salt crystallized in the $P2_12_12_1$ space group (63). Although the conformations of the constituent species in the two crystal forms were reported to be similar, the mode of molecular packing that led to the structure of the individual crystals was very different.

This latter behavior differs from that observed in the crystalline diastereomers of mandelic acid with α-phenylethylamine (64). Here the n-salt crystallized in the triclinic P1 space group, while the p-salt crystallized in the monoclinic $P2_1$ space group. The crystallographic structures of the two diastereomers revealed the existence of fairly equivalent hydrogen-bonding patterns, but at the same time substantially different conformations for the two molecular ions making up the donor-acceptor complex were noted. These structural differences became manifest in the relative solubilities of the two diastereomers, where the aqueous solubility of the p-salt was much less than that of the n-salt.

Once either the p-salt or the n-salt (or both) is successfully crystallized, the diastereomer salt is dissociated and the resolving agent separated. In the specific instance of acid resolution, the diastereomeric salt is efficiently cleaved by means of a hydrolysis reaction:

Separated diastereomers + Hydrolysis agent	→	Separated enantiomers
(+)-R-COO-NH_3-R'-(+) + HCl	→	(+)-R-COOH+(+)-R'-NH_3Cl
(-)-R-COO-NH_3-R'-(+) + HCl	→	(-)-R-COOH+(+)-R'-NH_3Cl

The diastereomer cleavage must be simple, selective, take place in quantitative yield, must not racemize the resolved compound, and must leave the resolving agent in a form that is easily recovered. The recovery is usually brought about by precipitation or extraction of the resolving agent. One important criterion in the choice of a resolving agent is the ease by which it may be dissociated and removed from the compound being resolved (50–54).

Computational methods are increasingly being brought to bear on the crystallographic aspects of chiral crystals. The possibility that the crystal structures of a particular diastereomeric salt pair could be predicted with an error in calculated lattice energy of less than 4 kcal/mole has been demonstrated for the system formed by a chlorine-substituted cyclic phosphoric acid and ephedrine (65). In another study, it was demonstrated that the stability difference of the diastereomeric salt pairs of three derivatives of 1-phenylethylammonium-2-phenylacetate was related to their resolution efficiency, even though similarities existed in their molecular structure and hydrogen-bonding motifs (66).

SUMMARY

With more and more therapeutic agents being administered as resolved enantiomers, methods for isolation of the desired chiral species will be in great need during development. Melting point or solubility phase diagrams can be extremely useful, since this information can be used to identify the agent in question as being either a conglomerate or a racemate. Should the compound happen to crystallize as a conglomerate, then enantiomer separation by direct crystallization could possibly represent the cost-effective route. Compounds identified as racemates would necessarily require separation by some other means, such as the formation of dissociable diastereomers.

REFERENCES

1. Lowry TM. Optical Rotatory Power. London: Longmans, Green, & Co., 1935.
2. Djerassi C. Optical Rotatory Dispersion. New York: McGraw-Hill, 1960.
3. Snatzke G. Optical Rotatory Dispersion and Circular Dichroism in Organic Chemistry. London: Heyden and Son, Ltd., 1967.
4. Crabbé P. ORD and CD in Chemistry and Biochemistry. New York: Academic Press, 1972.
5. Ciardelli F, Salvadori P. Fundamental Aspects and Recent Developments in Optical Rotatory Dispersion and Circular Dichroism. London: Heyden and Son, Ltd., 1973.
6. Purdie N, Brittain HG. Analytical Applications of Circular Dichroism. Amsterdam: Elsevier, 1994.
7. Berova N, Nakanishi K, Woody RW. Circular Dichroism: Principles and Applications. 2nd ed. New York: Wiley-VCH, 2000.
8. Barron LD. Analytical Molecular Light Scattering and Optical Activity. 2nd ed. Cambridge: Cambridge University Press, 2004.
9. Richardson FS, Riehl JP. Circularly polarized luminescence spectroscopy. Chem Rev 1977; 77:773–792.
10. Brittain HG. Excited-state optical activity. In: Schulman SG, ed. Molecular Luminescence Spectroscopy: Methods and Applications, Part 1. New York: Wiley-Interscience, 1985:583–620.
11. Brittain HG, Grinberg N. Techniques of chiroptical spectroscopy. In: Cazes J, ed. Handbook of Analytical Instrumentation. 3rd ed. New York: Marcel Dekker, 2005:271–294.
12. Cahn RS, Ingold CK, Prelog V. Specification of molecular chirality. Angew Chem Int Ed Engl 1966; 5:385–415.
13. Jacques J, Collet A, Wilen SH. Enantiomers, Racemates, and Resolutions. New York: John Wiley & Sons, 1981.
14. Brittain HG. Crystallographic consequences of molecular dissymmetry. Pharm Res 1990; 7:683–690.
15. Jacques J. The Molecule and Its Double. New York: McGraw-Hill, 1993.
16. Gal J. The discovery of biological enantioselectivity: Louis Pasteur and the fermentation of tartaric acid, 1857—a review and analysis 150 years later. Chirality 2008; 20:5–19.
17. Rousseau JJ. Basic Crystallography. New York: John Wiley & Sons, 1998.
18. Cotton FA. Chemical Applications of Group Theory. 2nd ed. New York: Wiley-Interscience, 1971.

19. Zorkii PM, Razumaeva AE, Belsky VK. The systematization of molecular crystal structures. Acta Cryst 1977; A33:1001–1004.
20. Mason SF. Molecular Optical Activity and the Chiral Discriminations. Cambridge: Cambridge University Press, 1982:165–166.
21. Kitaigorodskii AI. Organic Chemical Crystallography. New York: Consultants Bureau, 1955.
22. Brock CP, Schweizer WB, Dunitz JD. On the validity of Wallach's rule: on the density and stability of racemic crystals compared with their chiral counterparts. J Am Chem Soc 1991; 113:9811–9820.
23. Lowry TM. Molecular dissymmetry. In: Lowry TM, ed. Optical Rotatory Power. London: Longmans, Green, & Co., 1935; 36:25–36.
24. Kuroda R, Mason SM. Crystal structures of dextrorotatory and racemic sodium ammonium tartrate. J Chem Soc Dalton Trans 1994; B50:59–68.
25. Collet A, Brienne MJ, Jacques J. Optical resolution by direct crystallization of enantiomer mixtures. Chem Rev 1980; 80:215–230.
26. Collet A, Brienne MJ, Jacques J. Dédoublements Spontanéset Conglomérats d'Énantiomères. Bull Soc Chim France 1972; 127–142.
27. Collet A, Brienne MJ, Jacques J. Étude des mélanges d'antipodesoptiques. XIII. Compléments à Dédoublements Spontanés. Bull Soc Chim France 1977:494–498.
28. Leclercq M, Collet A, Jacques J. Étude des Mélanges d'Antipodes Optiques. XII. Mesure de la Stabilite des Racemiques Vrais. Tetrahedron 1976; 32:821–828.
29. Gourlay MD, Kendrick J, Leusen FJJ. Rationalization of racemate resolution: predicting spontaneous resolution through crystal structure prediction. Cryst Growth Design 2007; 7:56–63.
30. Kondepudi DK, Crook KE. Theory of conglomerate crystallization in the presence of chiral impurities. Cryst Growth Design 2005; 5:2173–2179.
31. Findlay A, Campbell AN. The influence of constitution on the stability of racemates. J Chem Soc (London) 1928; 1768–1775.
32. Meyerhoffer W. Stereochemische Notizen. Ber Dtsch Chem Ges 1904; 37:2604–2610.
33. Jacques J, Gabard J. Étude des Mélanges d'Antipodes Optiques. III. Diagrammes de Solubilité pour les Divers Types de Racémiques. Bull Soc Chim France 1972; 342–350.
34. Secor RM. Resolution of optical isomers by crystallization procedures. Chem Rev 1963; 63:297–309.
35. Collet A. Optical Resolution by Crystallization Methods. In: Kristulovic AM, ed. Chiral Separations by HPLC. Chichester: Ellis Horwood, 1989:81–104.
36. Zaugg HE. A mechanical resolution of DL-methadone base. J Am Chem Soc 1955; 77:910.
37. Fieser FF. Experiments in Organic Chemistry. Lexington: D.C. Heath Pub, 1955:188–190.
38. Collet A, Jacques J. Étude des Mélanges d'Antipodes Optiques. V. acidesmandéliquessubstitués. Bull Soc Chim France 1973; 3330–3334.
39. Brienne NJ, Collet A, Jacques J. A convenient optical resolution of sec-phenethyl alcohol by preferential crystallization of its 3,5-dinitrobenzoate. Synthesis 1983; 1983 (9):704–705.
40. Jacques J, Leclercq M, BrienneMJ. La Formation de selsaugmente-t-elle la Fréquence des Dédoublements Spontanés. Tetrahedron 1981; 37:1727–1733.
41. Kozma D, Böcskei Z, Simon K, et al. Racemic compound formation-conglomerate formation. Part 1. Structural and thermoanalytical study of hydrogen manonate, hydrogen phthalate, and hydrogen succinate of α-phenylethylamine. J Chem Soc Perkin Trans 1994; 2:1883–1886.
42. Böcskei Z, Kassai C, Simon K, et al. Racemic compound formation-conglomerate formation. Part 3. Investigation of the acidic salts of α-phenylethylamine by achiral dicarboxylic acids. Optical resolution by preferential crystallization and a structural study of (R)-α-phenylethylammonium hydrogen itaconate. J Chem Soc Perkin Trans 1996; 2:1511–1515.
43. Freer AA, Bunyan JM, Shankland N, et al. Structure of (S)-(+)-ibuprofen. Acta Cryst 1993; C49:1378–1380.

44. Shankland N, Wilson CC, Florence AJ, et al. Refinement of ibuprofen at 100 K by single-crystal pulsed neutron diffraction. Acta Cryst 1997; C53:951–954.
45. Brittain HG. Unpublished results.
46. Saigo K, Kimoto H, Nohira H, et al. Molecular recognition in the formation of conglomerate crystal: the role of cinnamic acid in the conglomerate crystals of 1-phenylethylamine and 1-(4-Isopropylphenyl)ethylamine salts. Bull Chem Soc Japan 1987; 60:3655–3658.
47. Kimoto H, Saigo K, Ohashi Y, et al. Molecular recognition in the formation of conglomerate crystal. 2. the role of arenesulfonic acid in the conglomerate crystals of amino acid salts. Bull Chem Soc Japan 1989; 62:2189–2195.
48. Dwivedi SK, Sattari S, Jamali F, et al. Ibuprofen racemate and enantiomers: phase diagram, solubility, and thermodynamic studies. Int J Pharm 1992; 87:95–104.
49. Ebbers EJ, Plum BJM, Ariaans GJA, et al. New resolving bases for ibuprofen and mandelic acid: qualification by binary phase diagrams. Tetrahedron Asymmetry 1997; 8:4047–4057.
50. Wilen SH. Resolving agents and resolutions in organic chemistry. In: Allinger NL, Elien EL, Wilen SL, eds. Topics in Stereochemistry. Vol. 6. New York: Wiley-Interscience, 1971:107–176.
51. Wilen SH. Strategies in optical resolutions. Tetrahedron 1977; 37:2725–2736.
52. Sheldon RA, Porskamp PA, ten Hoeve W. Advantages and limitations of chemical optical resolution. In: Tramper J, van der Plas HC, Linko P, eds. Biocatalysts in Organic Synthesis. Amsterdam: Elsevier, 1985:59–80.
53. Collet A. Separation and purification of enantiomers by crystallization methods. Enantiomer 1999; 4:157–172.
54. Toda F. Enantiomer Separation: Fundamentals and Practical Methods. Dordrecht: Kluwer Academic, 2004.
55. Wilen SH. Tables of Resolving Agents and Optical Resolutions. South Bend, IN: University of Notre Dame Press, 1972.
56. Newman P. Optical Resolution Procedures for Chemical Compounds. Vols 1–3. New York: Manhattan College Press, 1984.
57. Ugi I. Use of diastereomers avoids the ambiguities of the threo, erythro nomenclature. Z Naturforsch 1965; 20B:405–421.
58. Lorenz H, Seidel-Morgenstern A. Binary and ternary phase diagrams of two enantiomers in solvent systems. Thermochim Acta 2002; 382:129–142.
59. Shiraiwa T, Sado Y, Fujii S, et al. Optical resolution of (±)-phenylsuccinicacid by using (-)-proline as resolving agent. Bull Chem Soc Japan 1987; 60:824–826.
60. Valente EJ, Zubkowski J, Egglestron DS. Discrimination in resolving systems I: ephedrine-mandelic acid. Chirality 1992; 4:494–504.
61. Valente EJ, Miller CW, Zubkowski J, et al. Discrimination in resolving systems II: ephedrine-substituted mandelic acids. Chirality 1995; 7:652–676.
62. Hiramatsu H, Okamura K, Tsujioka I, et al. Crystal structure–solubility relationships in optical resolution by diastereomeric salt formation of DL-phenylglycine with (1S)-(+)-camphor-10-sulfonic Acid. J Chem Soc Perkin Trans 2000; 2:2121–2128.
63. Brianso MC. Structures Atomiques etMoléculaires des Sels Diastéréoisomères des α-Phényl-α-Méthylacétates d'α-Phényl-Éthylammonium p et n. Acta Cryst 1976; B32:3040–3045.
64. Lopez de Diego H. Crystal structure of (S)-1-phenylethylammonium (R)-mandelate and a comparison of diastereomericmandelate salts of 1-phenylethylamine. Acta Chem Scand 1994; 48:306–311.
65. Leusen FJJ. Crystal structure prediction of diastereomeric salts: a step toward rationalization of racemate resolution. Cryst Growth Design 2003; 3:189–192.
66. Karamertzanis PG, Anandamanoharan PR, Fernandes P, et al. Toward the computational design of diastereomeric resolving agents: an experimental and computational study of 1-Phenylethylammonium-2-Phenylacetate derivatives. J Phys Chem B 2007; 111:5326–5336.

5 Indirect methods for the chromatographic resolution of drug enantiomers

Wladyslaw Golkiewicz and Beata Polak

INTRODUCTION

The chiral resolution by chromatographic methods such as high-performance liquid chromatography (HPLC) or gas chromatography (GC) can be divided into two categories, direct and indirect separations. In the direct method, a racemic mixture is directly separated on a column that contains a chemically immobilized chiral stationary phase (CSP) or by the addition of a chiral selector to the mobile phase.

The main principle of the indirect method has been well described by Drayer (1):

> The physical properties of enantiomers are identical in an achiral environment. However, *chemical reaction that adds another asymmetric center creates a diastereomeric pair, each of which has physical properties that are not completely the same. Therefore, although an enantiomeric pair cannot be separated by ordinary chromatographic means or fractional recrystalization, the diastereomeric pair can often be separated easily by these means.*

An example of such a reaction can be that between two enantiomers of naproxen (R)- and (S)-naproxen, a nonsteroidal anti-inflammatory drug) and one enantiomer of 1-phenylethylamine. Naproxen represents a sample of a drug whose composition is to be determined and 1-phenylethylamine is the unichiral derivatizing reagent.

The stereochemical course of this reaction is shown in Figure 5.1.

It is obvious that, if two racemic mixtures were to be used in the reaction, that is, (R,S)-naproxen and (R,S)-1-phenylethylamine, four different diastereomers would be obtained. Unfortunately, although four different diastereomers would be synthesized, there would be two cases of racemic mixtures being formed, with the result that two racemic pairs of diastereomers would be obtained. Such mixtures can be separated in an achiral chromatographic system, but only into two peaks; each consisting of two diastereomers. This is why, according to the reaction shown in Figure 5.1, in indirect chromatographic resolution of enantiomers, only one enantiomer of high enantiomeric purity is used to convert enantiomers into diastereomers.

In the literature, an enantiomer that after reaction with the two enantiomers converts the racemic mixture to the diastereomeric mixture may be referred to as the chiral derivatizing reagent (CDR) or unichiral reagent (sometimes also called chiral derivatizing agent, CDA). The widespread interest in the analysis and quantification of the enantiomers of chiral drugs has led to the development and use of a variety of CSP in the direct methods and CDR in the indirect methods. There are various CDR that are characterized by high molecular absorptivity (UV or VIS) or high fluorescence quantum yield (φ). High racemic purity of the CDR is an important requirement in this analysis.

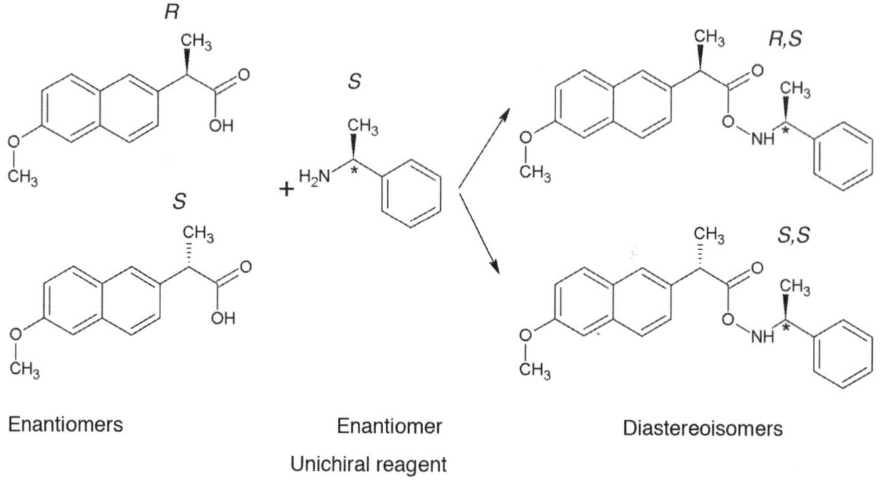

FIGURE 5.1 Course of the reaction for the resolution of naproxen enantiomers by the indirect method. [Note the 1-phenylethylamine has the same configuration about the chiral center (*marked by asterisks*) in both diastereomers.]

When a CDR is not enantiomerically pure and contains even a small amount of the second enantiomer, the racemic resolution of the target compound will not be determined accurately. For example, trace R-CDR in S-CDR reacting with S-drug containing a trace of R-drug would give a large S-CDR–S-drug, which would be relatively unaffected by the diminishingly small amount of the coeluting enantiomer R-CDR–R-drug derived by the reaction of the two trace impurities. However, the presence of R-CDR–S-drug coeluting with S-CDR–R-drug would have a significant adverse effect on the estimation of the trace R-drug in S-drug.

Enantiomeric analysis of chiral drugs can be accomplished by a variety of analytical techniques, including liquid and gas chromatography, capillary electrophoresis (CE), and supercritical fluid chromatography (SFC). HPLC, which is the most commonly used, has several attractions, for example, operation at ambient temperature, so that the risk of the chiral drug racemization on-column is reduced, possibility of using different detector systems including circular dichroism, and possibility of using different type of CSP or different precolumn derivatization using different types of CDR. Moreover, HPLC has also been used successfully at the preparative scale.

Conversion of a mixture of enantiomers by chiral derivatization into a volatile diastereoisomeric species allows the combination of GC and the indirect method. There are many reviews and publications on separation of enantiomers by this means, for example, Srinivas et al. (2) or Segura et al. (3). Also, Liu and Lin (4) dedicated part of their review to the indirect method as used in GC. However, some authors (2,5) argue that higher possibility of racemization in experimental temperature makes the GC indirect resolution not a method of choice for the enantiomeric separation of drugs.

INDIRECT CHROMATOGRAPHIC RESOLUTION OF DRUG ENANTIOMERS

There have been a number of reviews dealing with chiral liquid chromatography, both in direct and indirect format published in the past decade. Properties of CDR can be found in reviews written by Ilisz et al. (6), Sun et al. (7), and Toyo'oka (8). The chiral resolution of different classes of drugs was reviewed by Sun et al. (7), amine drugs of abuse by Lin and Liu (4), and nonsteroidal anti-inflammatory drugs by Davies (9). Pharmaceutical and biomedical applications of chiral analysis of pharmaceutical compounds in biological fluids were reviewed by Ducharme et al. (10) and Haginaka (11). Toyo'oka (12) divided all CDR into two groups: the first group that absorb in the UV or VIS regions and second group that can be analyzed using fluorimetric detection. Srinivas compared the stereoselective data for 17 racemic drugs obtained using direct and indirect methods (13) and evaluated experimental strategies for the development of indirect methods (14,15). Lindner wrote a review article on this topic in 1988 (16), which still can be a very good practical guide to understand principles of indirect method. The progress in chromatographic enantioseparations of drugs was described by Bojarski et al. (17).

The above-mentioned reviews are good sources for literature data, which can be a quick step to solve many separation problems. In this chapter, the information contained in a chapter written by J. Gal (18) is supplemented and updated and the latest achievements in synthesis of a new CDR and selected applications using indirect method are reported. Only derivatization for liquid and gas chromatography will be discussed, with the omission of analogous procedures for Capillary Electrophoresis (CE) or Nuclear Magnetic Resonance (NMR).

GENERAL CONSIDERATIONS
Chromatographic General Considerations

In considering derivatization, it is useful to use two chromatographic parameters used for characterization of the chromatographic separation, namely, resolution R_S and separation factor α.

Resolution is a quantitative measure of the distance between adjacent chromatographic peaks and their widths

$$R_S = \frac{2(t_2 - t_1)}{(w_1 + w_2)} \qquad (1)$$

where t_2 and w_2 are the retention time and width for second eluting peak, and t_1 and w_1 are the retention time and width of first eluting peak. Theoretically, if $R_S = 1.0$, complete separation of two compounds is obtained but in practice two adjacent peaks overlap each other and R_S should be >1.5.

The second chromatographic parameter to bear in mind is the separation factor, α, which is the ratio of the capacity factors, k_2 and k_1, for two adjacent peaks, where k_2 is the later eluting peak:

$$\alpha = \frac{k_2}{k_1} \qquad (2)$$

In very general terms, if efficiency of the column used is high (large number of the theoretical plates) and peaks are very narrow at the base, $\alpha = 1.05$ is enough to obtain satisfactory separation, but for more tailing peaks the separation factor should have the value, $\alpha = 1.1$ or more.

The Derivatization Reaction

The derivatizing reagent is used to react with the enantiomers in order to change the chemical and physical properties of these compounds. Therefore, it is necessary to ascertain that the expected diastereomers are obtained. This can be done by using typical analytical techniques as mass spectrometry (MS), UV, IR, etc. This procedure is especially important in the case of derivatization of compounds that possess more than one functional group.

The derivatization reaction of the enantiomer can be carried out in aqueous or nonaqueous solvents, depending on properties of the enantiomers and CDR, for example, in case of the acid halides, it is necessary to use a nonaqueous solvent such as chloroform, since the reagent readily undergoes hydrolysis with water and therefore water contamination should be avoided. However, the derivatization of analytes well dissolved in water can be carried out in water or aqueous media, what can be a critical advantage of analysis.

Another requirement for the reaction conditions must be minimization of the possibility of any racemization of either the enantiomers or the CDR. Vakily et al. (19) convincingly demonstrated that the use of high pH or high ionic strength during derivatization can cause significant racemization so that the obtained results may not accurately reflect the true enantiomeric disposition in human tissue. Although the possibility of racemization during preparation of enantiomeric drug derivates may not be ruled out, employing mild to moderate conditions usually minimize such an effect. Therefore, reaction conditions, mainly type of buffer, ionic strength, temperature, and pH should be selected with caution, to ensure that they do not promote racemization.

It has been stated (13) that a minor contamination of the CDR with its second enantiomer does not significantly alter the results of analysis. More information on racemization of the CDR, during synthesis, storage, and the derivatization reaction can be found in a paper written by Srinivas (13).

Both derivatized enantiomers may have an unequal reaction rate with the CDR. To overcome such an effect, the CDR must be added in excess and the reaction should go to completion with both enantiomers having reacted totally with the CDR. However, it is worth noting that using an excess of the CDR aggravates any problems associated with the presence of CDR excess and by-product after the reaction.

The Chiral Derivatizing Reagent

The choice of a derivatization reagent is of crucial importance. It is essential to consider several important aspects before choosing the most appropriate CDR. Toyo'oka (8) pointed out several of these:

1. The optical and chemical purity of the CDR should be very high. Commercially available CDR are not 100% enantiomerically pure and the second enantiomer is often present at a level of 0.5% to 1%. As already suggested, precise trace analysis of one enantiomer in the presence of large amount of the second enantiomer requires an exceptionally high purity of the CDR.
2. Degree of racemization during storage and synthesis of diastereomers should be as low as possible. The factors that may facilitate racemization are temperature, type of solvent used, interactions between the enantiomer and solvents molecules, pH, presence of chemical impurities, etc.

3. The resulting diastereomers have to show high chemical stability (at least 1 day) (12).
4. The reactivity of the CDR should be high in order that the derivatization may be carried out under mild conditions.
5. The CDR should possess specificity for the target functional group(s) and quantitatively react with enantiomers. This is especially important when the molecules of enantiomers contain more than one functional group.
6. The solubility of CDR in water or water-organic solvents is also very important because many bioactive chiral compounds are soluble in aqueous solutions.
7. The reagent should be commercially available—from the practical point of view both enantiomers of the CDR should be commercially available so that the elution order of the diastereomers may be switched if this is helpful.
8. Elution order of the diastereomers is generally performed by comparing the retention times of the resolved peaks of the racemic drug with those obtained in separate derivatization, from each pure enantiomer of the CDR. This is especially necessary when a trace enantiomer is determined in the presence of a large amount of the second enantiomer.

The selection of the CDR is also largely dependent on the type of functionalities of the enantiomers to be derivatized and the preferred detection system planned to use, for example, UV, VIS, or fluorescence.

Most of the popular CDR contain chemical structures that give rise to absorbance in the UV or VIS region. The quantitation of the enantiomers at trace level demands the use of a more sensitive detection system in either HPLC (UV, fluorescence, mass spectrometry) or GC (nitrogen-phosphorus detection, electron-capture, mass spectrometry). Usually in such a case a CDR with fluorophoric properties is used (20).

The structures of most popular chiral derivatized reagents are presented in Table 5.1.

Lindner (16) suggests that sufficient chromatographic separation factors (α) can be obtained when the asymmetric centers in the diastereomers, contributed by the derivatized enantiomers and by the CDR, are close to each other, and when rigid and relatively bulky and/or planar substituents (e.g., phenyl or naphthyl groups) on both the enantiomer and the CDR "are contributing to an overall molecular structural and conformational rigidity."

It should be borne in mind that the enantioselectivity connected with the CDR used is always related to a selected chromatographic system, whether a normal-phase or reversed-phase chromatography. Reversed-phase chromatography is normally preferred for the analysis of biological samples.

Additionally, the cost and ease of synthesis of the CDR may also dictate the selection of the reagent. Although a large number of different CDR have been synthesized for enantiomeric analysis in HPLC and GC (3,4), only some of them have been applied in analytical practice so far.

New Chiral Derivatizing Reagents

In some cases, a number of CDR that provide UV-VIS absorbance have been applied to the derivatization of various functional groups, but the sensitivity of the derivative subsequently proved to be not enough when dealing with real

TABLE 5.1 Chemical Structure of CDR Used for Derivatization

Structures and acronyms	Structures and acronyms
TAGIT	GITC
DDTIC	(S,S)-PDITC
(R,R)-DANI	NAP-IT
DBD-PyNCS	FLEC

INDIRECT CHROMATOGRAPHIC RESOLUTION OF DRUG ENANTIOMERS 75

TABLE 5.1 Chemical Structure of CDR Used for Derivatization (*Continued*)

Structures and acronyms	Structures and acronyms
MTPA-Cl	TPC
DTTAAN	NEA

samples. To overcome this problem, various types of the CDR with fluorescence properties have been developed (12–15).

More detailed information on the reaction conditions and properties of CDR can be found in original papers cited below and in various reviews and monographs: derivatization of amine group (21–30), carboxyl group (31–34), and thiol group (35–37). These new CDR received little attention until recently.

TYPES OF APPLICATION

The chromatographic separation of racemic mixtures is under constant development taking into account new chiral stationary phases (CSP in the direct method), new CDR, well-established conditions of the derivatization reaction and column separations (indirect method). The most important applications are in pharmaceutical, clinical, and environmental analysis. Investigations of the enantiomeric purity or composition of drugs using the chiral derivatization method is done in laboratory research, quality control, or dosage forms. Another application for the indirect method is explaining of enantioselective fate of drugs in living organisms, which have important implications in human health and effective therapy.

Sometimes it is necessary to detect traces of the inactive enantiomer, which can exhibit undesirable effects, in the presence of high excess of the active enantiomer. In such cases, the small peak may be overlapped by the tailing of the larger peak of the active enantiomer. Accordingly, in almost all cases the small peak should appear as the first peak. As already mentioned, reversal of the enantiomeric migration order is possible by changing the stereoconfiguration of the CDR. Now it is possible as commercial suppliers will usually offer both enantiomers of a given CDR.

The application of the indirect method for supplying of pure enantiomers obtained from a racemic mixture on semipreparative scale is now very rare, due to easy access to simpler and more efficient methods as preparative scale chiral chromatography or salt crystallization.

HPLC and GC are still the methods of choice because there are large numbers of commercially available CDR of high enantiomeric purity and well-established reactions leading to diastereomeric derivatives with excellent separation and detection possibilities.

EXAMPLES OF SEPARATION

Typically, amines, carboxylic acids, and hydroxyl groups are the functional groups that are derivatized in many drugs and their metabolites. The pair of enantiomers that make up the racemic mixture and the CDR must possess easily derivatizable and compatible functional groups. The reaction should be quick to reduce the chance of racemization of enantiomers and variation in formation rate of derivatives.

It is convenient to describe CDR applications according to the derivatizing functional group in the drug (20).

DERIVATIZATION OF AMINO GROUP

Given that many drugs and biochemically important compounds have at least one amino group in their structure, derivatization reactions of primary and secondary amines have been extensively investigated.

According to Gal (18), the formation of an amide bond, between amines and carboxylic acids or acid chlorides, appears to be the most common reaction used in chiral derivatization of primary and secondary amino groups. In fact amide-forming reactions proceed readily and many chiral carboxylic acids of high enantiomeric purity are available. The statement that the reaction of amide formation is the most popular in derivatization of amines is certainly true in GC. However, in HPLC, the most frequently used CDR in derivatization of amines would now (Fig. 5.2) seem to be isothiocyanates and chloroformates.

Formation of Thioureas

Isothiocyanate reagents have following advantage over isocyanate reagents:

- The improved hydrolytic stability facilitates the derivatization of hydrophilic compounds and the analysis of drugs from biological matrices.
- Reaction with amino group is more selective than with hydroxyl group and no other free functional group needs to be protected.
- Under the same reaction conditions the hydroxyl group is not derivatized.
- Thiourea derivatives allow very sensitive UV detection.

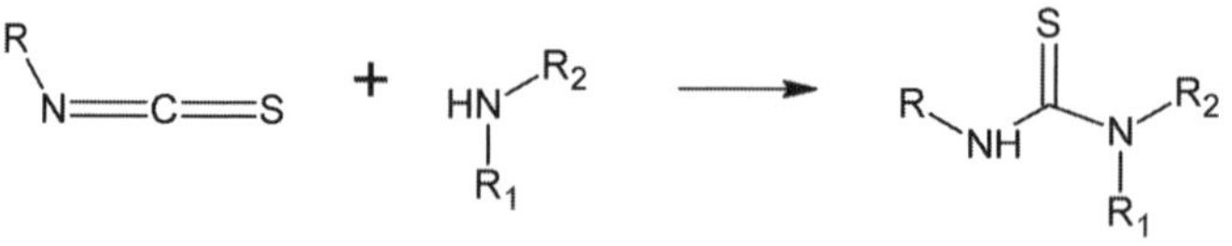

FIGURE 5.2 Formation of thioureas.

These properties are particularly advantageous in the case of derivatization of amino alcohols, amino acids, catecholamines, and other highly polar compounds.

Wu et al. (38) developed a stereoselective reversed-phase high-performance liquid chromatography (RP-HPLC) assay to separate and quantify enantiomers of propranolol and 4-hydroxypropranolol after derivatization with 2,3,4,6-tetra-O-acetyl-α-D-glucopyranosyl-isothiocyanate (TAGIT; for chemical structure, see Table 5.1). Zeng et al. (39) developed a similar assay, which was applied to determine atenolol enantiomers using 2,3,4,6-tetra-O-acetyl-β-D-glucopyranosyl isothiocyanate (GITC) in rat hepatic microsome. The same group (40) used the earlier described method for the determination of propranolol enantiomers in transgenic Chinese hamster lung (CHL) cell lines. A baseline separation of propranolol derivatives with GITC was achieved using RP-HPLC. The assay was applied to study of the depletion of (S)-(-) and (R)-(+)-propranolol in transgenic Chinese hamster cell lines. Zeng et al. (41) developed an RP-HPLC assay that allowed the separation of the enantiomers of esmolol together with its acid metabolites (Fig. 5.3). The hydrolysis of racemic esmolol in human plasma (Fig. 5.3B) resulted in the formation of an equivalent amount of the acid metabolite (Fig. 5.3E).

Kleidernigg and Lindner (21) synthesized (S,S)–N–3,5-dinitrobenzoyl-trans-diaminocyclohexane-isothiocyanate (DDITC*; see Table 5.1), which can serve as a highly selective, stable, and enantiomerically pure CDR for the indirect resolution of chiral primary and secondary amines, especially amino alcohols. The derivatization of amino alcohols with DDITC takes place under mild conditions and the resulting diastereomers are well separable by RP-HPLC.

The same authors separated many amino alcohols derivatives, the separation factor (α) of the thioureas ranging between 1.05 and 2.00, and the peak resolution (R_q) ranging from 0.67 to 6.67. Comparison of three DDITC and GITC derivatives of amino alcohols showed that the separation factor α and resolution R_S were usually higher for the DDITC derivatives than those for the GITC derivatives. A disadvantage of DDITC is that derivatization of secondary amines containing a tertiary butyl group at amino function (e.g., timolol) is impossible.

Kleidernigg and Lindner (22) synthesized a new CDR, structurally related to obtained earlier DDITC, namely, (1S,2S)-N-[(2-isothiocyanato)-(cyclohexyl)-pivalinoyl amide] (PDITC; see Table 5.1). This new reagent can serve as a highly selective stable reagent for the indirect resolution of chiral primary and secondary amines and thiol compounds.

It was shown that the resulting diastereomeric thioureas can be separated by RP-HPLC as demonstrated with a number of amino alcohol type of drugs and other examples. Chromatographic data for selected amino alcohols derivatized with (S,S)-PDITC and GITC were compared. The separation factors, α-values of the PDITC derivatives, ranging from 1.03 to 2.08, the resolution R_S ranging from 0.5 to 12.9 were in most cases superior to the GITC derivatives (Fig. 5.4).

Peter et al. (23,42) developed new CDR, (1R,2R)-1,3-diacetoxy-1-(4-nitrophenyl)-2-propyl isothiocyanate [(R,R)-DANI; see Table.5.1] that broadened the number of CDR available for the enantioseparation of a series of β-blockers. The application of DANI for the enantioseparation of a series of amino alcohols diastereomers showed that several β-blocking agents could be baseline separated with separation factors, α, ranging from 1.13 to 1.22 and

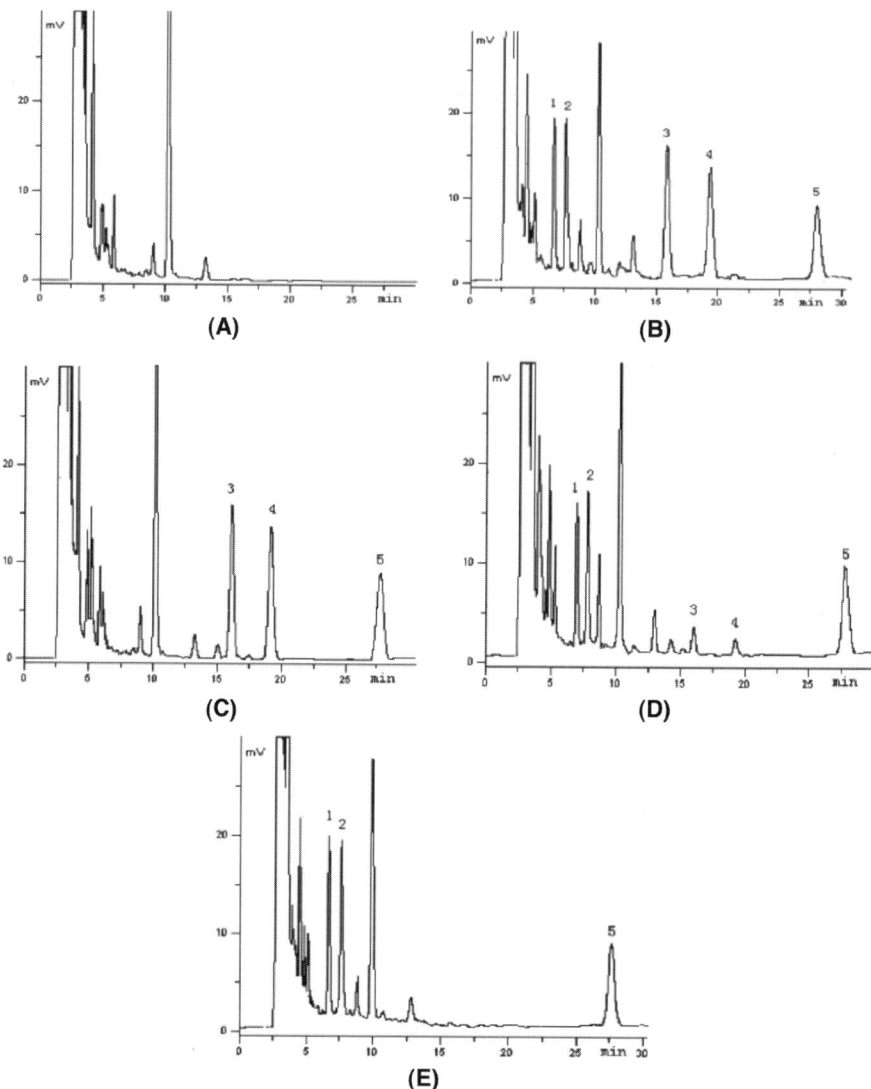

FIGURE 5.3 Comparison of chromatograms of esmolol analysis in plasma. (**A**) Blank plasma; (**B**) plasma spiked with IS and 1.7 µg/mL of esmolol and the acidic metabolite; (**C**) plasma samples after hydrolysis for 0 hour; (**D**) 4.8 hours; (**E**) 8 hours; (1) (−)-*S*-acid metabolite; (2) (+)-*R*-acid metabolite; (3) (−)-(*S*)-esmolol; (4) (+)-(*R*)-esmolol; (5) (−)-(*S*)-propranolol. *Source*: From Ref. 41.

resolution R_S from 0.78 to 2.72. The derivatization could be carried out quantitatively under mild conditions, at room temperature for 5 hours when the molar ratio of the CDR to the β-blocker was 2:1. In the case of a higher excess of the CDR, for example, 4:1, the reaction time shortened to 1.5 hours (43).

DANI as the CDR was also used for the separation of twelve synthetically and pharmaceutically interesting racemic amino alcohols (44) with two adjacent

INDIRECT CHROMATOGRAPHIC RESOLUTION OF DRUG ENANTIOMERS

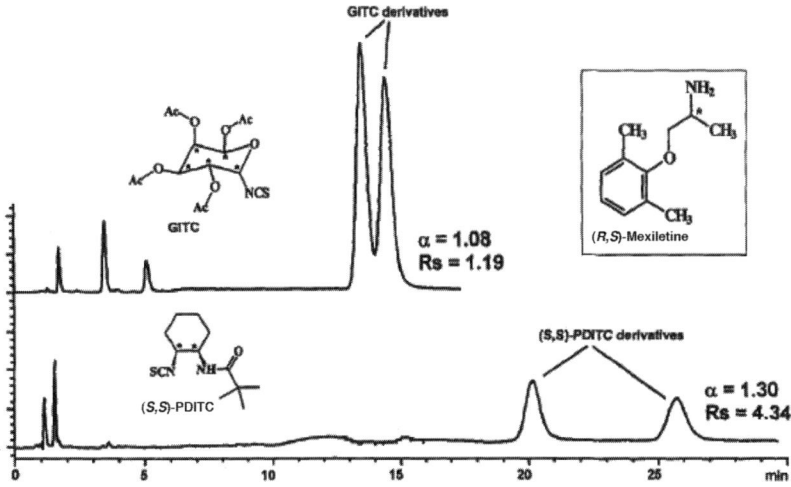

FIGURE 5.4 Resolution of (R,S)-mexiletine derivatized as (S,S)-PDITC (*bottom*) and GITC (*top*) thioureas. *Source*: From Ref. 22.

chiral centers. Since the investigated amino alcohols contain two chiral centers, four enantiomers exist, and after separation four peaks on chromatogram should appear. It was not a case for most investigated amino alcohols; they could not be separated into four peaks (some only for two). This is a limitation of DANI as compared with TAGIT (the GITC isomer). However, the enantiomers of 2-amino-1-hydroxymethylcyclohexane, 2-amino-1-hydroxymethyl-4-cyclohexane, and 2-aminomethylcyclohexanol were well resolved with DANI whereas they could not be separated at all with GITC.

Büschges et al. (45) developed from the 2-arylpropionic acid (S)-(+)-naproxen, namely, 1-(6-methoxy-2-naphthyl)ethyl isothiocyanate (NAP-IT; see Table.5.1), structurally closely related to 1-(naphthyl)ethyl isocyanate (NEIC). The new CDR, NAP-IT was tested utilizing chiral amino compounds such as β-adrenoreceptor antagonists and antiarrhythmic agents as model compounds for drugs with primary or secondary amino moieties. The thiourea derivatives were well resolved on a reversed-phase column but not under normal-phase conditions on a silica gel column.

Toyo'oka et al. (46) synthesized a fluorescent chiral tagging reagent, 4-(3-isothiocyanatopyrrolidin-1-yl)-7-(N,N-dimethylaminosulfonyl)-2,1,3-benzoxadiazole [R(−)-DBD-PyNCS; see Table 5.1] and used for the separation of racemic pairs of β-blockers. The derivatives with DBD-PyNCS were completely separated, except timolol. The separation factors α were in the range 1.04 to 1.32.

The DBD-PyNCS was applied to the determination of the concentration of the enantiomers of propranolol in rat plasma and saliva after oral administration. Thioureas of β-blockers fluoresce in the long-wavelength region so the detection limits are at femtomole levels for those having the isopropylamino structure and picomole levels for those having the *tert*-butylamino structure.

Jin et al. (47) developed a method for the determination of thyroxine (the thyroid hormones) enantiomers, used in a pharmaceutical formulation designed to treat thyroidism, by RP-HPLC–MS with precolumn derivatization by DBD-Py-NCS. Retention times and separation factors, α, were strongly dependent on the mobile phase components and their concentrations. Water-acetonitrile containing 0.1% of formic acid was selected as the mobile phase. It was found that decreasing the acetonitrile concentration caused an increase in the separation factor, α, up to 1.61, and improvement of the resolution, R_S, with the highest value being 2.96.

Formation of Carbamates
One of the most popular chloroformate chiral reagents is FLEC [1-(9-fluorenyl) ethyl chloroformate] (see Table 5.1). Both enantiomers are available commercially in enantiomerically pure form, and diastereomers are easily detectable due to the formation of a highly fluorescent product upon derivatization (48). The sensitivity of fluorescence detection for the methamphetamine derivative was almost 200 times higher than that of UV detection. The reaction proceeds at room temperature in basic solutions and in mild conditions. Good separation in RP-HPLC is obtained with increasing hydrophobicity of the diastereomers.

Gunaranta and Kissinger (49) investigated the metabolism of R-, S-, and racemic amphetamine in rat liver microsomes by microdialysis and liquid chromatography with precolumn derivatization of amphetamines with FLEC. The reaction of the FLEC reagent with amphetamine occurs at room temperature under basic conditions as shown in Figure 5.5.

An inconvenience of this reaction was a slight difference in reaction rates for (R)- and (S)-amphetamine and the long time of derivatization, about 7 hours. Another disadvantage was that baseline separation of the enantiomers was not achieved for both amphetamine and its metabolite, 4-hydroxyamphetamine. The study showed that the (S)-amphetamine was metabolized to a larger extent than the (R)-enantiomer.

Two reports have described the use of FLEC for the separation of reboxetine diastereomers by RP-HPLC (50) or NP-HPLC hyphenated with a CSP column (chiral OD-H) (51). It was found that a chiral column separated diastereomers better than achiral reversed-phase or normal-phase columns.

Novel, nonfluorinated enantiomers of a chiral quinolone that exhibits good activity against gram-positive, gram-negative, and atypical bacteria were

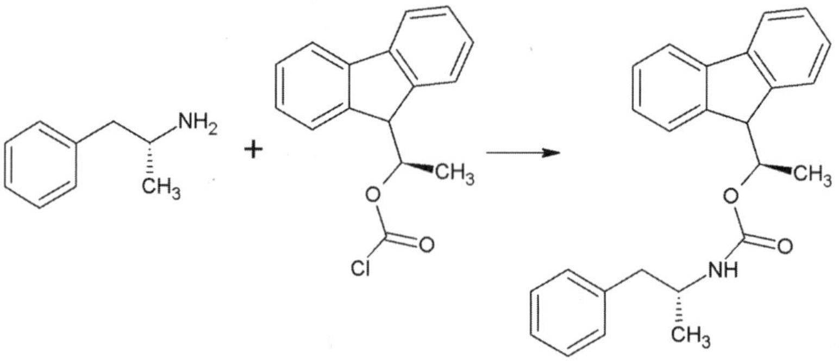

FIGURE 5.5 Reaction between FLEC chiral reagent and (R)-amphetamine.

INDIRECT CHROMATOGRAPHIC RESOLUTION OF DRUG ENANTIOMERS

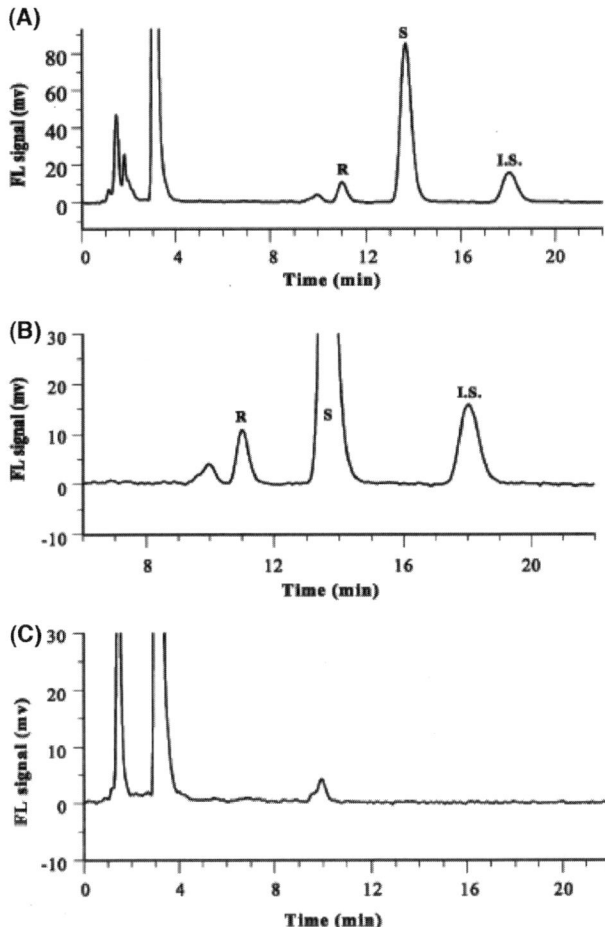

FIGURE 5.6 Chromatograms of plasma samples spikes with (*R*) (0.8 μg/mL) and (*S*) (1.0 μg/mL) enantiomers of PGE-9509924. (**A**) Full-scale chromatogram, (**B**) expanded scale, (**C**) blank plasma sample. Detection with fluorescence detector. *Source*: From Ref. 52.

determined in dog plasma by Zontendam et al. (52). The enantiomers were derivatized with FLEC and separated by RP-HPLC (Fig. 5.6).

The reaction was carried out at room temperature and was quantitatively completed after 30 minutes. Plasma samples were analyzed by electrospray LC-MS/MS (liquid chromatography–quadrupole mass spectrometer) or by HPLC with fluorescence detection. Results obtained with LC-MS/MS detection and fluorescence detection were essentially identical.

The FLEC has been evaluated by Campins-Falco et al. (53) for the enantioselective analysis of amphetamines by RP-HPLC. The chromatographic conditions affording the best resolution of diastereomers and sensitivity were determined for amphetamine, methamphetamine, ephedrine, pseudoephedrine, 3,4-methylenedioxyamphetamine (MDA), 3,4-methylenedioxymethamphetamine (MDMA), and 3,4-methylenedioxyethylamphetamine (MDEA). Low resolution was obtained for separation of amphetamine (α =1.06; R_S = 0.97) and methamphetamine (α =1.04; R_S = 0.91), but for the other investigated substances chromatographic parameters were very good (separation factor α in the range of 1.09–1.23).

The possibility of using FLEC for the enantioselective analysis of amphetamine and related substances in biological fluids was evaluated by analyzing urine spiked with MDA.

Formation of Amides

Amide-forming reactions between amines and carboxylic acids or acids chlorides proceed readily, according to the scheme shown in Figure 5.7. Separation of the amide diastereomers is now less popular in HPLC but still exploited in GC. Reaction proceeds under relative mild conditions, at ambient temperature or a little elevated temperature (e.g., 60 °C for 5 minutes).

Stereoselective metabolism of famprofazone (analgesic) in humans was studied by Shin (54). The enantiomers were derivatized with α-methoxy-α-(trifluoromethyl)-phenylacetyl chloride (MTPA-Cl; see Table 5.1) and next protected with N-methyl-N-triethylsilyl trifluoroacetamide (MSTFA) at the hydroxyl groups. The derivatives of famprofazone and its metabolites were well resolved by capillary gas-liquid chromatography except for two pairs of N-MTPA, O-silyl derivatives: (–)-norpseudoephedrine and (–)-norephedrine; (+)-p-hydroxymethamphetamine and (–)-p-hydroxynorephedrine, which overlapped. The extract from human urine was separated using GC-MS and the stereochemical identities of metabolites were confirmed by comparison of the EI (electron impact) mass spectra and retention times of metabolites and the authentic standards.

Kim et al. (55) developed a stereoselective method for the quantitative determination of metoprolol enantiomers and its metabolites in human urine using GC-MS. In GC, it is not uncommon to use a second derivatizing, usually nonchiral, reagent to derivatize additional functional group(s) (in this case the hydroxyl group) on the enantiomeric analyte. For example, first metoprolol enantiomers were derivatized with MSTFA to get O-TMS (trimethylsilyl) derivatives and next by chiral MTPA-Cl to form diastereomeric amides.

Typical GC-MS (SIM) chromatograms for separation of metoprolol enantiomers and metoprolol metabolites are shown in Figure 5.8.

The reagent peaks were well separated from those from the diastereomers and did not interfere with detection of the diastereomers. The detection limit for (R)-(+)- and (S)-(–)-metoprolol was 0.5 mg/mL.

Derivatization of amphetamines with one of the frequently (4) used CDR, (S)-(–)-trifluoroacetylprolyl chloride (TPC; see Table 5.1) has inherent limitation. Racemization of this CDR produces diastereomers even from pure (S)-(+)-methamphetamine. Paul et al. (56) instead of TPC utilized MTPA-Cl to prepare

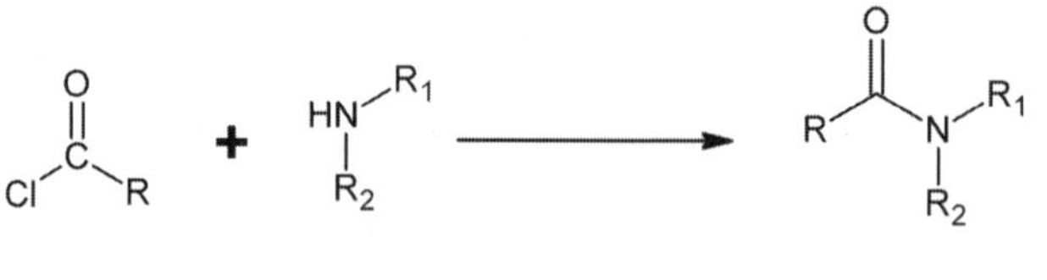

FIGURE 5.7 Formation of amides.

INDIRECT CHROMATOGRAPHIC RESOLUTION OF DRUG ENANTIOMERS 83

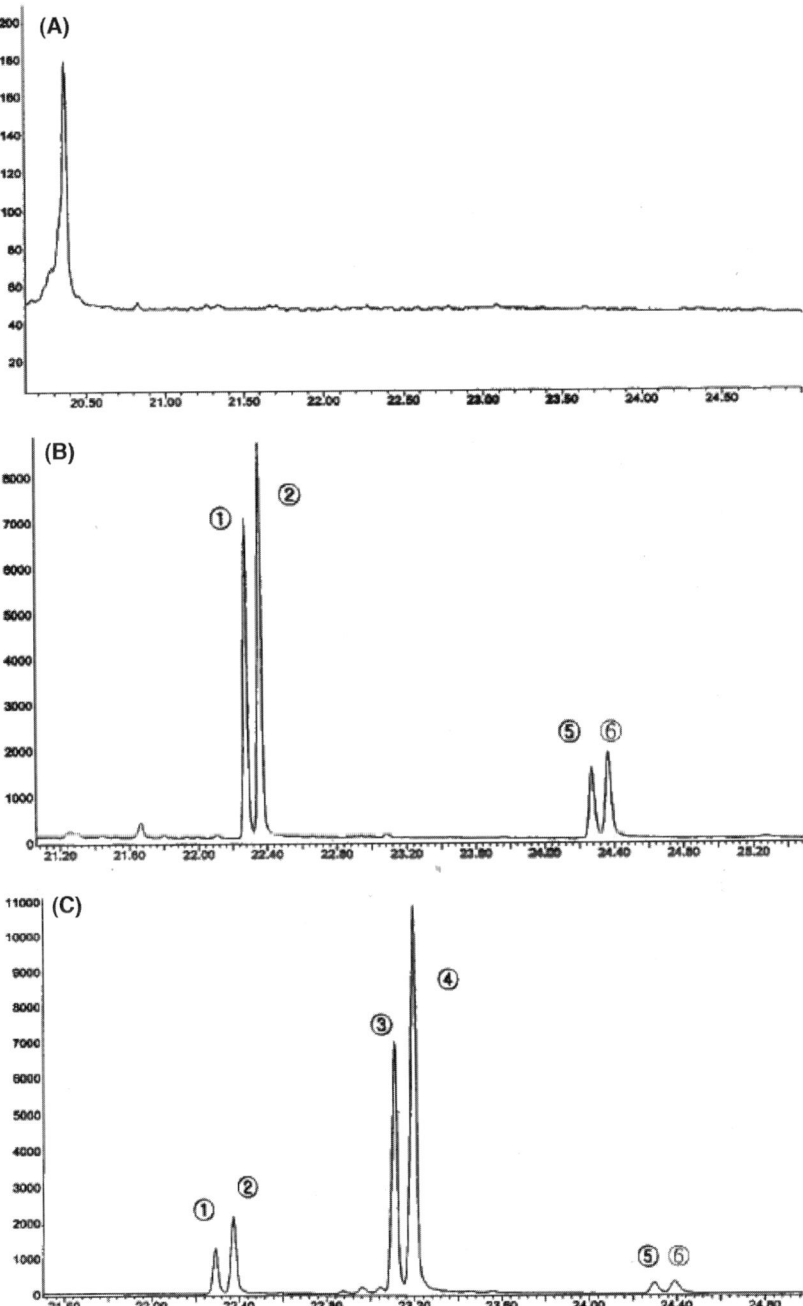

FIGURE 5.8 GC-MS chromatograms of selected-ion monitored at m/z 404. (**A**) Blank urine sample, (**B**) 1 µg/mL of metoprolol tartrate spiked urine sample, (**C**) 5.25-hour urine sample after oral administration of 100-mg metoprolol tartrate. (1) (R)-(+)-metoprolol, (2) (S)+(−)-metoprolol, (3, 4) metoprolol metabolites, (5, 6) bisoprolol (internal standards). *Source*: From Ref. 55.

the amides diastereomers of racemic amphetamines, which were separated and quantified in urine specimens by GC-EI-MS. No racemization was observed with this reagent. The procedure was applied to test 43 urine specimens from the U.S. Navy Laboratory for the presence of amphetamine, methamphetamine, MDA, MDMA, and MDEA.

Tao and Zeng (57) described a method of the determination of several chiral phenylethylamine drugs in urine or rat hepatic microsomes. Drugs such as fenfluramine, amphetamine, and methamphetamine were derivatized with TPC and analyzed by GC-MS. The method allowed study of the metabolic depletion of amphetamine and fenfluramine in rat hepatic microsomes. It was stated that metabolism of the (R)-enantiomers of amphetamine and fenfluramine was more rapid than (S)-enantiomers of these drugs. The authors (57) developed two methods of derivatization for investigated drugs: aqueous phase and organic phase derivatization.

Wang et al. (58) tried to solve the problems in which forensic toxicologists have been involved for years: (i) enantioselective data and (ii) the differentiation between illegal consumption of abused amphetamine-generating drugs and legitimate administration of prescribed amphetamine-generating drugs. They elaborated a simple, rapid, reliable analytical method involving the reaction of amphetamine and methamphetamine with CDR (TPC) and GC-isotope dilution mass spectrometry. According to the authors, the developed method should meet the requirements of most of the workplace urine drug-testing programs. Using TPC as CDR, a resolution of 2.2 and 2.8 for amphetamine and methamphetamine was achieved.

Hair analysis is a useful tool for detecting and monitoring drugs of abuse over a long time period. The enantioselective separation of amphetamine and related drugs has been found to be useful for correct interpretation of amphetamine-type stimulants (ATS). Since the metabolism of ATS is stereospecific, the use of the proportions of the metabolite enantiomers is a powerful tool for the interpretation of the drug administered.

Martins et al. (24) synthesized a novel chiral reagent (2S,4R)-N-heptafluorobutyryl-4-heptafluoro-butyroxy-prolyl chloride (HFBOPCl) and applied it for the enantiomeric quantification of amphetamines and methylene-dioxylated amphetamines in hair. Drug enantiomers were converted to their diastereomeric derivatives, in 15 minutes at room temperature in carbonate buffer. All 10 resulting diastereomers were baseline separated in 12 minutes retention times.

The developed method was applied to analyzing hair specimens from 24 suspected ATS abusers. Both (R) and (S) amphetamine and methamphetamine enantiomers were detected in all cases and one hair specimen was tested positive for MDMA and MDA. In most cases the concentrations of (S)-amphetamine and (S)-methamphetamine exceeded those corresponding to the (R)-enantiomers. No additional derivatization with achiral reagent was necessary.

Besides MTPA-Cl, also (R)-(+)-α-methoxy-α-(trifluoromethyl) phenylacetic acid (MTPA) was used for chiral identification and determination of the enantiomers of methamphetamine, ephedrine, pseudoephedrine, and methcathinone by GC-MS and compared with the results obtained by NMR (59). The substances were derivatized with MTPA to give diastereomeric derivatives, in the presence of dicyclohexylcarbodiimide (DCC). Baseline resolution of eight diastereomeric MTPA derivatives was obtained with resolution of at least $R_S = 1.6$. The accuracy of the determination of the enantiomeric ratios was determined by comparison of the theoretical (known concentration of solution) and experimentally determined results from GC-MS and NMR. Excellent agreement was obtained with the theoretical (known) values and obtained by both methods.

Ephedrine and pseudoephedrine are common over-the-counter pharmaceuticals, sometimes used for conversion to methamphetamine in illicit manufacturing processes, so the identification of ephedrine and pseudoephedrine and their metabolites is of great importance.

Wang et al. (60) used MTPA for the derivatization of ephedrine, norephedrine, pseudoephedrine, and cathinone following GC-MS separation. The authors used 5 CDR (TPC, GITC, TAGIT, FLEC, and MPTA) and found that MTPA was the most effective CDR for the derivatization of ephedrines, allowing complete baseline resolution of 10 structurally closely related compounds.

Formation of Ureas

Primary and secondary amines react with isocyanates to give corresponding ureas (Fig. 5.9).

Derivatization of amines with isocyanates is less popular than with isothiocyanates, probably because more severe reaction conditions are required, for example, a longer reaction time and a higher temperature.

There are two commercially available chiral isocyanates: 1-phenylethyl isocyanate (PEIC) and 1-(1-naphthyl)ethylisocyanate (NEIC).

A stereoselective HPLC method determination of mefloquine, a chiral drug administered for prophylaxis and treatment of malaria, was developed by Souri et al. (61). The method was applied for the determination of mefloquine enantiomers in plasma, urine, and blood. Mefloquine was derivatized with NEIC and the resulting diastereomers were separated on a silica normal-phase column with the separation factors $\alpha > 1.3$.

Szymura-Oleksiak et al. (62) described a stereoselective direct chromatographic method for assay of enantiomers of acebutolol and its active metabolite diacetolol in human serum (CSP Chiralpak AD). The results obtained by the direct method were compared with results obtained by the indirect method. The NEIC derivatives of acebutolol, diacetolol, and pindolol (internal standard) were separated in reversed-phase mode. It is interesting that the separation factors (α) for acebutolol and diacetolol enantiomers had the same values for both the direct and indirect methods. The serum concentrations of acebutolol and diacetolol enantiomers determined in five patients on long-term acebutolol treatment showed that data obtained by both methods were very similar.

PEIC (also known as α-methylbenzyl isocyanate, MBIC) was used by Beal and Telt (63) for the derivatization of pindolol enantiomers in human plasma and urine. The PEIC derivatives of pindolol and metoprolol (internal standard) were separated by RP-HPLC and detected using fluorescence. The elaborated

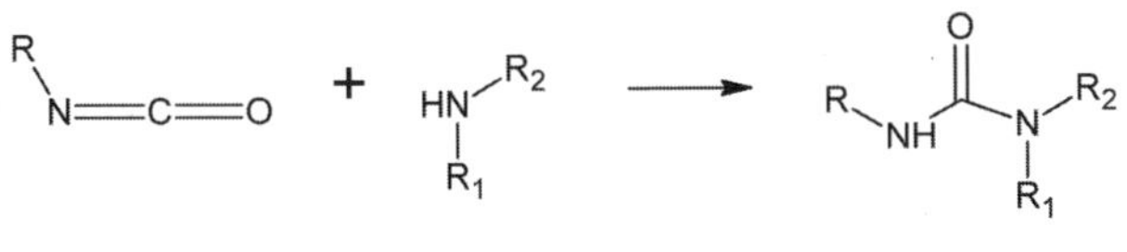

FIGURE 5.9 Formation of ureas.

assay allowed the measurement of clinically relevant concentrations of each pindolol enantiomer.

DERIVATIZATION OF HYDROXYL GROUPS

A variety of CDR have been employed in the derivatization of hydroxyl groups: FLEC, PEIC (MBIC), MTPA, 2-methoxy-2-(1-naphthyl)propionic acid (M-α-NPA), or (R,R)-O,O'-di-p-toluoyl tartaric acid anhydride (DTTAAN). Compounds with a hydroxyl group are usually difficult to derivatize due to the limited reactivity and competitive reaction with water in the reaction medium. Therefore, reactions with acid chloride CDR are mainly performed in anhydrous solvents such as chloroform and benzene under protection from moisture (13).

The carbonation of hydroxyl compounds with chloroformate reagents seems to be a suitable derivatization reaction because of the good reactivity of the –COCl group in the reagent. Fransson and Ragnarsson (64) described the separation of six α-hydroxy acids with the separation factors 1.24 or higher; the only exception was malic acid where separation factor, α, was in the range 1.10 to 1.15. Retention of the acidic diastereomeric carbonate derivatives was conveniently regulated by pH of the mobile phase (RP-HPLC).

L-carnitine is a vitamin-like amino acids derivative, highly therapeutically effective, used for various nutritional and pharmaceutical applications. In contrast, D-carnitine displays serious side-effect so that the D-carnitine concentration must be precisely determined. Freimüller and Altorfer (65) developed a method of separation of carnitine isomers after derivatization with FLEC using RP-HPLC. Reliability and validity for the method to quantify D-carnitine in L-carnitine samples in the range 0.1% to 1% was ensured.

Isocyanate CDR such as PEIC can react with alcoholic hydroxyl groups to produce carbamate diastereomers. Kim et al. (66) described the determination of (S)-terbutaline in enantiomeric excess of (R)-terbutaline. The enantiomers were converted to diastereomeric derivatives using PEIC and separated by RP-HPLC and ^{1}H-NMR. The obtained results were compared and NMR results agreed with those obtained by the RP-HPLC method.

Also carboxylic acids, mainly as acid chlorides, were used for derivatization of hydroxyl groups. The esters derived from the reaction between M-α-NPA and various chiral alcohols were separated by normal-phase HPLC (67).

Taji et al. (68) described the resolution of racemic alcohols as esters of M-α-NPA, using normal-phase HPLC. The separation factor, α, ranged from 1.15 for 2-butanol esters to 1.93 for 2-hexadecanol esters. The large values of α allowed the isolation of pure enantiomers by semipreparative scale column chromatography and reduction of diastereomers with LiAlH$_4$ or hydrolysis with KOH/EtOH.

Delmopinol is a new chiral substance active against plaque. Egginger et al. (69) established an enantioselective method to elucidate the pharmacokinetic behavior of this drug. The delmopinol enantiomers were extracted from plasma by solid-phase extraction and derivatized with (R,R)-O,O'-di-p-toluoyl tartaric acid anhydride [(R,R)-DTTAAN; see Table 5.1] yielding diastereomeric derivatives, which were separated by RP-HPLC followed by highly sensitive electrochemical detection. The limit of quantitation in plasma was approximately 3 pmol (1 ng) per enantiomer per 0.5 mL plasma.

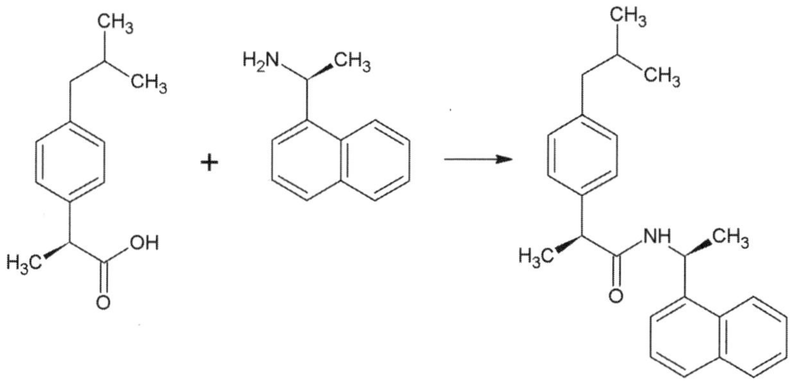

FIGURE 5.10 Reaction between (S)-ibuprofen enantiomer and (R)-NEA.

DERIVATIZATION OF CARBOXYLIC GROUPS

Derivatization of carboxyl compounds is usually carried out by amide or ester formation. The esterification reaction with chiral alcohols requires drastic conditions, so derivatization with chiral amines is carried out more frequently. The reaction with chiral amines, in the presence of activation reagents, usually proceeds under relative mild conditions at room temperature.

An enantiospecific assay for determination of ibuprofen enantiomers in serum and urine has been developed by Tan et al. (70). Enantiomers of ibuprofen were derivatized with (R)-1-(1-naphthen-1-yl)ethylamine (NEA; see Fig. 5.10) using 1-(3-dimethylaminopropyl)-3-ethyl-carbodiimide and 1-hydroxybenzotriazole as coupling reagent, to yield diastereomeric amides. Chromatographic resolution was achieved using RP-HPLC.

Some new CDR were applied for separation of ibuprofen and ketoprofen enantiomers. Hayamizu et al. (31) developed a new labeling reagent, methylated Nε-Dansyl-L-lysine, for compounds with a carboxylic moieties. The derivatization reaction proceeds under mild condition with this coupling reagent. The assay was employed for quantitation of ibuprofen enantiomers in human urine.

Santa et al. (32) synthesized four new fluorescent CDR with benzofurazan structure, similar to the reagents developed earlier by Toyo'oka (12). These reagents with amino functional groups were tested by derivatization of ketoprofen followed by RP-HPLC and fluorescence or MS detection.

Trans-4-hydroxy-2-nonenoic acid (HNEA), a marker of lipid peroxidation, enantiomers were separated (33) using direct (Chiralpak AD-RH as CSP) and indirect methods (1S,2S-)(+)-2-amino-1-(4-nitrophenyl)-1,3-propanediol (ANPAD as CDR). The direct method provided better precision of determination but not so low a limit of quantification. Accordingly, the indirect separation method was applied for the determination of the enantiomeric ratio of HNEA enantiomers in rat brain mitochondria lysate.

DETERMINATION OF AMINO ACIDS

The development of chromatographic methods to determine the enantiomeric composition of amino acids (AA) is an important subject in many fields, namely, medicines, pharmacy (production of a new peptides), food, and natural products. Unusual D-AA have been found in plants and mammalian tissues, either in the free form or as components of peptides and proteins. The information on occurrence of D-AA in plants was presented in reviews by Robinson (71), Gamburg and Rekoslavskaya (72), and Friedman (73).

CDR, earlier proposed for the determination of amino and carboxyl groups, can also be used for the indirect separation of AA enantiomers, but some of new CDR were also developed.

Brukner and Wachsman (74) developed a series of CDR in which one chlorine in cyanuric chloride was substituted by an alkoxy or aryloxy group, whereas the second chlorine was replaced by (S)-amino acid derivative (e.g., amide). These CDR, possessing a reactive chlorine, were tested for their capability of derivatization of amino acids enantiomers. Derivatization reactions were conducted for up to 6 hours, but after 2 hours no increase of peak areas occurred. Separation factor for tested AA and best CDR used were in the range of 1.14 to 2.11.

Peter et al. (42) developed new CDR for chiral derivatization of compounds possessing an amino group. They utilized (S,S)-2-amino-1-(4-nitrophenyl)-1,3-propanediol (for the general structure of DANI see Table 5.1), which is a side-product of chloramphenicol synthesis.

Applicability of DANI was demonstrated through the example of resolution of 18 proteinogenic α-AA. The diastereomeric thioureas were separated by RP-HPLC. The separation factors were in the range 1.09 to 1.54 for methanol as mobile phase modifier and 1.06 to 1.24 for acetonitrile. The reaction was completed within 2 hours at 60°C.

Nimura et al. (75) investigated D-AA in mammalian tissues to determine localization, dynamics, and physiological function, particularly free D-aspartate. For these studies, they synthesized N-(tert-butylthiocarbamoyl)-L-cysteine ethyl ester (BTCC), which together with o-phthalaldehyde react with AA enantiomers to produce fluorescent diastereomers that are well separated using RP-HPLC. Separation factors, α, for investigated AA were in the range 1.31 to 1.56. What more, reaction time was only 1.5 minutes!

Peter et al. applied commercially available compound (S)-N-(4-nitrophenoxylcarbanoyl) phenylalanine methoxyethyl ester, (S)-NIFE, for enantioseparation of ring- and α-methyl-substituted phenylalanines and phenylalanine amides (76), 18 unnatural β-amino acids (77), and proteinogenic AA (78). The effect of pH and kind of the modifier of a mobile phase, reaction time on separation were investigated.

Brukner and Westhauser (79) used commercially available N-isobutyryl-L-cysteine together with o-phthaldialdehyde to convert AA enantiomers occurring in plants into diastereomers. The resulting fluorescent isoindole derivatives were separated by RP-HPLC and GC-MS. The data obtained by authors demonstrated that D-enantiomers of certain AA occur in all investigated plants. Notably, D-Ala, D-Asp, and D-Glu were found in many plants, for example, apple, water melon, mango, coconut milk, D-Lys in few plants, and D-Leu only in coconut milk.

SUMMARY

Significant pharmacodynamic and pharmacokinetic differences between enantiomers cause great interest in stereochemical aspects of these drugs. Thus, it is very important to develop reliable methods of enantioseparation and quantitative determination. One of the methods is derivatization of enantiomers leading to formation of diastereoisomers described in this chapter. Separation techniques employed for this goal vary depending on the purpose sought and the type of stationary phase to be used. Indirect methods allow reasonable separations without the need of using expensive CSP.

Derivatizing reagents change the chemical and physical properties of investigated compounds.

There are several requirements for derivatization reagents:

- Enantiomeric purity
- Stability
- Good sensitivity and selectivity of reaction
- Stability of resulting derivatives
- Negligible risk of racemization
- Possibility of using the derivatization reaction for bioanalytical applications
- Availability of an established protocol (volume of chiral reagent, duration of reaction, temperature reaction, reaction media, pH, and other conditions)

The most popular groups as the target for derivatization are amino, hydroxy, and carboxylic groups. These tagging groups may be derivatized in various ways: amino solutes could be transformed into ureas, thioureas, carbamates, amides; hydroxyl compounds could be transformed to esters, carbamates; acids could be transformed to amides or esters.

However, there are many groups for which derivatization is difficult (tertiary amines, lactones, alkenes, and alkynes) due to lack of effective reagents. So development of new selective and sensitive derivatization reagents is still necessary.

There is no instant way to find good conditions for separation of diastereomers using the indirect method. In the order of 90% of separations of diastereomers is carried out using RP-HPLC and C_{18} column, and occasionally trial and error is still used in the development of the separation. Nowadays normal-phase chromatography on a silica gel column with a nonaqueous mobile phase is rarely used. It was found that sometimes the type of the organic modifiers strikingly influences the separation. In case of RP-HPLC, the effect of the type of modifier (methanol, acetonitrile, tetrahydrofuran) was investigated and most authors stated that better separation of diastereomers can be obtained when methanol is used (42,75). However, it was also shown that using a mobile phase composition of methanol-water or methanol-aqueous buffer, it was not possible to separate some diastereomers, whereas acetonitrile containing eluents did allow appropriate resolution. Many times decreasing of the polar modifier concentration in a mobile phase leads to improving of separability but, however, this leads to longer retention times.

Some researchers believe that the impetus for chiral separation advancement and enhancement has remained very high over the past decade. In this regard, both direct and indirect separation methods have been very valuable. According to the opinions Gübitz and Schmid in (80), that "A certain disadvantage of this (indirect) approach is the additional step (derivatization)" but "on the other hand many problems cannot be solved by direct separation approaches." With the advances that have been made in analytical science since 2004, the number of problems that cannot be solved will be ever diminishing. This is especially the case because of the use of MS detection becoming more commonplace and poor detectability being the problem that the indirect method is often used to solve. However, not all laboratories routinely use MS detection and not all chiral LC separations are readily compatible with MS detection (81). It is not surprising then that there is still evidence from recent literature in interest in the development of new chiral derivatizing agents (82) and in the application of the indirect method to fascinating new applications where very low limits of detection are required (83). In fact, the indirect method is still of value even when MS detection is being used (84).

REFERENCES
1. Drayer DE. The early history of stereochemistry: from the discovery of molecular asymmetry and the first resolution of a racemate by Pasteur to the asymmetrical chiral carbon of van't Hoff and Le Bel. In: Wainer IW, ed. Drug Stereochemistry, Analytical Methods and Pharmacology. 2nd ed. New York, Basel, Hong Kong: Marcel Dekker, Inc., 1993:1–24.
2. Srinivas NR, Shyn WC, Barbhaiya RH. Gas chromatographic determination of enantiomers as diastereomers following pre-column derivatization and applications to pharmacokinetic studies: a review. Biomed Chromatogr 1995; 9:1–9.
3. Segura J, Ventura R, Jurado C. Derivatization procedures for gas chromatographic–mass spectrometric determination of xenobiotics in biological samples, with special attention to drugs of abuse and doping agents. J Chromatogr B 1998; 713:61–90.
4. Liu JT, Lin RH. Enantiomeric composition of abused amine drugs: chromatographic methods of analysis and data interpretation. J Biochem Biophys Methods 2002; 54:115–146.
5. Aboul-Enein HY, Ali I. Introduction. In: Aboul-Einen HY, Ali I, eds. Chiral Separations by Liquid Chromatography and Related Technologies. New York, Basel: M. Dekker, Inc., 2003:1–20.
6. Ilisz I, Berkecz R, Peter A. Application of chiral derivatizing agents in the high-performance liquid chromatographic separation of amino acids enantiomers: a review. J Pharm Biomed Anal 2008; 47:1–15.
7. Sun XX, Sun LZ, Aboul–Einen HY. Chiral derivatization reagents for drug enantio-separation by high-performance liquid chromatography based upon pre-column derivatization and formation of diastereomers: enantioselectivity and related structure. Biomed Chromatogr 2001; 15:116–132.
8. Toyo'oka T. Indirect enantioseparation by HPLC using chiral benzofurazan-bearing reagents. In: Gübitz G, Schmidt MG, eds. Methods in Molecular Biology. Vol.43. Chiral Separations: Methods and Protocols. Totowa: Humana Press Inc., 2004:231–246.
9. Davies NM. Methods of analysis of chiral non-steroidal anti-inflammatory drugs. J Chromatogr B 1997; 691:229–261.
10. Ducharme J, Fernandez C, Gimenez F, et al. Critical issues in chiral drug analysis in biological fluids by high-performance liquid chromatography. J Chromatogr B 1996; 686:65–75.
11. Haginaka J. Pharmaceutical and biomedical applications of enantioseparations using liquid chromatographic techniques. J Pharm Biomed Anal 2002; 27:357–372.
12. Toyo'oka T. Resolution of chiral drugs by liquid chromatography based upon diastereomer formation with chiral derivatization reagents. J Biochem Biophys Methods 2002; 54:25–56.
13. Srinivas NM. Clinical pharmacokinetic data of racemic drug obtained by the indirect method following precolumn diastereomer formation: is the influence of racemization during chiral derivatization significant? Biomed Chromatogr 2004; 18:343–349.
14. Srinivas NM. Evaluation of experimental strategies for the development of chiral chromatographic methods based on diastereomer formation. Biomed Chromatogr 2004; 18:207–233.

15. Srinivas NM. Simultaneous chiral analyses of multiple analytes: case studies, implication and method development considerations. Biomed Chromatogr 2004; 18:759–784.
16. Lindner W. Indirect separation of enantiomers by liquid chromatography. In: Zief M, Crane L, eds. Chromatographic Chiral Separations. Chromatographic Science Series. Vol. 40. New York: M. Dekker, 1988:91–130.
17. Bojarski J, Aboul–Einen HY, Ghanem A. What's new in chromatographic enantioseparations. Curr Anal Chem 2005; 1:59–77.
18. Gal J. Indirect methods for the chromatographic resolution of drug enantiomers: synthesis and separation of diastereomeric derivatives. In: Wainer IW, ed. Drug Stereochemistry, Analytical Methods and Pharmacology. 2nd ed. New York, Basel, Hong Kong: Marcel Dekker Inc., 1993:65–106.
19. Vakily M, Corrigan B, Jamail F. The problem of racemization in the stereospecific assay and pharmacokinetic evaluation of ketorolac in human and rats. Pharmaceutical Research 1995; 12:1652–1657.
20. Toyo'oka T. Recent progress in liquid chromatographic enantioseparation based upon diastereomer formation with fluorescent chiral derivatization reagents. Biomed Chromatogr 1996; 10:265–277.
21. Kleidernigg OP, Posch K, Lindner W. Synthesis and application of a new isothiocyanate as a chiral derivatizing agent for the indirect resolution of chiral amino alcohols and amines. J Chromatogr A 1996; 729:33–42.
22. Kleidernigg OP, Lindner W. Synthesis of new stable aliphatic isothiocyanate-based chiral derivatizing agent and application to indirect separation of chiral amino and thiol compounds. Chromatographia 1997; 44:465–472.
23. Peter M, Peter A, Fűlőp F. Development of new isothiocyanate-based chiral derivatizing agent for amino acids. Chromatographia 1999; 50:373–375.
24. Martins LF, Yegles M, Chung H, et al. Sensitive, rapid and validated gas chromatography/negative ion chemical ionization-mass spectrometry assay including derivatisation with a novel chiral agent for the enantioselective quantification of amphetamine-type stimulants in hair. J Chromatogr B 2006; 842:98–105.
25. Toyo'oka T, Liu Y-M, Development of optically active fluorescent Edman-type reagents. Analyst 1995; 120:385–390.
26. Toyo'oka T, Jin D, Tomoi N, et al. R-(–)-4-(3-isothiocyanatopyrrilidin-1-yl)-7-(N,N-dimethylaminosulfonyl)-2,1,3-benzoxadiazole, a fluorescent chiral tagging reagent: sensitive resolution of chiral amines and amino acids by reversed–phase liquid chromatography. Biomed Chromatogr 2001; 15:56–67.
27. Kleidernigg OP, Lindner W. Indirect separation of chiral proteinogenic α- amino acids using the fluorescence active (1R,2R)-N-[(2-isothiocyanato)-cyclohexyl]-6-methoxy-4-quinolinylamide) as chiral derivatising agent. A comparison. J Chromatogr A 1998; 795:251–261.
28. Peter M, Fűlőp F. Comparison of isothiocyanate chiral derivatizing reagent for high-performance liquid chromatography. Chromatographia 2002; 56:631–636.
29. Peter M, Peter A, Fűlőp F. Development of new isothiocyanate-based chiral derivatizing agent for amino acids. Chromatographia 1999; 50:373–375.
30. Al-Kindy S, Santa T, Fukushima T, et al. Enantiomeric determination of amines by high-performance liquid chromatography using chiral fluorescent derivatization reagents. Biomed Chromatogr 1998; 12:276–280.
31. Hayamizu T, Kudoh S, Nakamura H. Methylated N-ε-dansyl-L-lysine as a fluorogenic reagent for the chiral separation of carboxylic acids. J Chromatogr B 1998; 710:211–218.
32. Santa T, Luo J, Lim Ch-K, et al. Enantiomeric separation and detection by high-performance liquid chromatography-mass spectrometry of 2-arylpropionic acids derivatized with benzofurazan fluorescent reagents. Biomed Chromatogr 1998; 12:73–77.
33. Brichae J, Honzatko A, Picklo MJ. Direct and indirect high-performance liquid chromatography enantioseparation of *trans*-4-hydroxy-2-nonenoic acid. J Chromatogr A 2007; 1149:305–311.

34. Yasaka Y, Ono Y, Tanaka M. (S)-(+)-1-methyl-2-(6,7-dimethoxy-2,3-naphthalimido) ethyl trifluoromethane sulfonate as a fluorescence chiral derivatising reagent for carboxylic acid enantiomers in high-performance liquid chromatography. J Chromatogr A 1998; 810:221–225.
35. Jin D, Takehana K, Toyo'oka T. Chiral separation on racemic thiols based on diastereomer formation with a fluorescent chiral tagging reagent by reversed-phase liquid chromatography. Anal Sci 1997; 13:113–115.
36. Muramatsu N, Toyo'oka T, Yamaguchi K, et al. High-performance liquid chromatographic determination of erdosteine and its optical active metabolite utilizing a fluorescent chiral tagging agent, R-(–)-4-(N,N-dimethylaminosulfonyl)-7-(3-aminopyrrolidyn-1-yl)-2,1,3-benzoxadiazole. J Chromatogr B 1998; 719:177–189.
37. Jin D, Toyo'oka T. Indirect resolution of thiol enantiomers by high-performance liquid chromatography with a fluorescent chiral tagging reagent. Analyst 1998; 123:1271–1277.
38. Wu ST, Ping-Chang Y, Gee WL, et al. Stereoselective high-performance liquid chromatography determination of propranolol and 4-hydroxypropranolol in human plasma after pre-column derivatization. J Chromatogr B 1997; 692:133–140.
39. Li X, Yao TW, Zeng S. Reversed-phase high-performance liquid chromatographic analysis of atenolol enantiomers in rat hepatic microsome after chiral derivatization with 2, 3,4,6 –tetra-O-acetyl-β-D-glycopyranosylisothiocyanate. J Chromatogr B 2000; 742:433–439.
40. Zhou Q, Yao TW, Zeng S. Chiral reversed phase high-performance liquid chromatography for determining propranolol enantiomers in transgenic Chinese hamster CHL cell lines expressing human cytochrome P450. J Biochem Biophys Methods 2002; 54:369–376.
41. Tang VH, He Y, Yao TW, et al. Simultaneous determination of the enantiomers of esmolol and its acids metabolite in human plasma by reversed–phase liquid chromatography with solid phase extraction. J Chromatogr B 2004; 805:249–254.
42. Peter M, Peter A, Fűlőp F. Application of (1S,2S)- and (1R,2R)-1,3-diacetoxy-1-(4-nitrophenyl)-2-propylisothiocyanate to the indirect enantioseparation of racemic proteinogenic amino acids. J Chromatogr A 2000; 871:115–126.
43. Peter M, Gyeresi A, Fűlőp F. Liquid chromatographic enantioseparation of beta-blocking agents with (1R,2R)-1,3-diacetoxy-1-(4-nitrophenyl)-2-propyl isothiocyanate as chiral derivatizing agent. J Chromatogr A 2001; 910:247–253.
44. Peter M, Fűlőp F. Indirect high performance liquid chromatographic enantioseparation of racemic amino alcohols with 1,3-diacetoxy-1-(4-nitrophenyl)-2-propyl isothiocyanate as derivatizing agent. J Liq Chrom Rel Technol 2000; 23:2459–2473.
45. Bűschges R, Linde H, Mutschler E, et al. Chloroformates and isothiocyanates derived from 2-arylpropionic acids as chiral reagents: synthetic routes and chromatographic behaviour of the derivatives. J Chromatogr A 1996; 725:323–334.
46. Toyo'oka T, Toriumi M, Ishii Y. Enantioseparation of β-blockers labelled with a chiral fluorescent reagent, R (–)-DBD-PyNCS, by reversed-phase liquid chromatography. J Pharm Biomed Anal 1997; 15:1467–1476.
47. Jin D, Kumar AP, Song G-C, et al. Determination of thyroxine enantiomers in pharmaceutical formulation by high-performance liquid chromatography-mass spectrometry with precolumn derivatization. Microchem J 2008; 88:62–66.
48. Chen YP, Hsu MC, Chien CS. Analysis of forensic samples using precolumn derivatization with D-(9-fluorenyl)ethyl chloroformate and liquid chromatographic with fluorimetric detection. J Chromatogr A 1994; 672:135–140.
49. Gunaranta C, Kissinger PT. Investigation of stereoselective metabolism of amphetamine in rat liver microsomes by microdialysis and liquid chromatography with precolumn chiral derivatization. J Chromatogr A 1998; 828:95–103.
50. Frigerio E, Pianezzola E, Strolin Benedetti M. Sensitive procedure for the determination of reboxetine enantiomers in human plasma by reversed-phase high-performance liquid chromatography with fluorimetric detection after chiral derivatization with (+)-1-(9-fluorenyl)ethyl chloroformate. J Chromatogr A 1994; 660:351–358.

51. Walters R, Buits S. Improved enantioselective method for the determination of the enantiomers of reboxetine in plasma by solid-phase extraction, chiral derivatization, and column-switching high-performance liquid chromatography with fluorescence detection. J Chromatogr A 1998; 828:167–176.
52. Zontendam PH, Canty JF, Martin MJ, et al. Development of a chiral assay for a novel, nonfluorinated quinolone, PGE-9509924, in dog plasma using high performance liquid chromatography with electroscopy tandem mass spectrometry or fluorescence detector. J Pharm Biomed Anal 2002; 30:1–11.
53. Campins-Falco P, Verdu-Andres J, Herraez-Hernandez R. Separation of the enantiomers of primary and secondary amphetamines by liquid chromatography after derivatization with (–)-1-(9-fluorenyl) ethyl chloroformate. Chromatographia 2003; 57:309–315.
54. Shin HS. Stereoselective metabolism of famprofazone in humans: N-dealkylation and β- and p-hydroxylation. Chirality 1997; 9:52–58.
55. Kim KH, Lee JH, Ko M, et al. Determination of metoprolol enantiomers in human urine by GC-MS using (–)-α-methoxy-α-(trifluoromethyl)phenylacetyl chloride as a chiral derivatizing agent. Chromatographia 2002; 55(1/2):81–85.
56. Paul BD, Jemionek J, Lesser D, et al. Enantiomeric separation and quantitation of (+/–)-amphetamine, (+/–)-methamphetamine, (+/–)-MDA, (+/–)- MDMA, and (+/–)-MDEA in urine specimens by GC-EI-MS after derivatization with (R)-(–)- or (S)-(+)-alpha-methoxy-alpha-(trifluoromethy)phenylacetyl chloride (MTPA). J Anal Toxicol 2004; 28:449–455.
57. Tao QF, Zeng S. Analysis of enantiomers of chiral phenethylamine drugs by capillary gas chromatography/mass spectrometry/flame-ionization detection and precolumn chiral derivatization. J Biochem Biophys Methods 2002; 4:103–113.
58. Wang S-M, Wang T-Ch, Giang Y-S. Simultaneous determination of amphetamine and methamphetamine enantiomers in urine by simultaneous liquid–liquid extraction and diastereomeric derivatization followed by gas chromatographic–isotope dilution mass spectrometry. J Chromatogr B 2005; 816:131–143.
59. LeBelle MJ, Savard C, Dawson BA, et al. Chiral identification and determination of ephedrine, pseudoephedrine, methamphetamine and methcathione by gas chromatography and nuclear magnetic resonance. Forensic Sci Int 1995; 71:215–223.
60. Wang S-M, Lewis RJ, Canfield D, et al. Enantiomeric determination of ephedrines and norephedrines by chiral derivatization gas chromatography–mass spectrometry approaches. J Chromatogr B 2005; 825:88–95.
61. Souri E, Farsam H, Jamali F. Stereospecific determination of mefloquine in biological fluids by high-performance liquid chromatography. J Chromatogr B 1997; 700:215–222.
62. Szymura-Oleksiak J, Walczak M, Bojarski J, et al. Enantioselective high-performance liquid chromatographic assay of acebutolol and its active metabolite diacetolol in human serum. Chirality 1999; 11:267–271.
63. Beal JL, Tett SE, Determination of pindolol enantiomers in human plasma and urine by simple liquid-liquid extraction and high-performance liquid chromatography. J Chromatogr B 1998; 715:409–415.
64. Fransson B, Ragnarsson V. Separation of enantiomers of α-hydroxy acids by reversed–phase liquid chromatography after derivatization with 1-(9-fluorenyl)-ethyl chloroformate. J Chromatogr A 1998; 827:31–36.
65. Freimüller S, Altorfer H. A chiral HPLC method for the determination of low amounts of d-carnitine in l-carnitine after derivatization with (+)-FLEC. J Pharm Biomed Anal 2002; 30:209–218.
66. Kim KH, Kim HJ, Kim J-H, et al. Determination of the optical purity of (R)-terbutaline by ^{1}H-NMR and RP-LC using chiral derivatizing agent, (S)-(–)-α-methylbenzyl isocyanate. J Pharm Biomed Anal 2001; 25:947–956.
67. Harada N, Watanabe M, Kuwahara S, et al. 2-Methoxy-2-(1-naphthyl)propionic acid, a powerful chiral auxiliary for enantioresolution of alcohols and determination of their absolute configurations by the ^{1}H NMR anisotropy method. Tetrahedron Assymetry 2000; 11:1249–1253.

68. Taji H, Kasai Y, Rugio A, et al. Practical enantioresolution of alcohols with 2-methoxy-2- (1-naphthyl)propionic acid and determination of their absolute configurations by the ^{1}H NMR anisotropy method. Chirality 2002; 14:81–84.
69. Egginger G, Blaschke E, Olsson AM, et al. Stereoselective high-performance liquid chromatographic assay of (+/−)-delmopinol in plasma using solid-phase extraction, a chiral derivatizing agent and electrochemical detection. J Chromatogr A 1994; 666:275–282.
70. Tan SC, Jackson SHD, Swift CG, et al. Enantiospecific analysis of ibuprofen by high performance liquid chromatography: determination of free and total drug enantiomer concentrations in serum and urine. Chromatographia 1997; 46:23–32.
71. Robinson T. D-amino acids in higher plants. Life Sci 1976; 19:1097–1102.
72. Gamburg KZ, Rekoslavskaya NI. Formation and functions of D-amino acids in plants. Soviet Plant Physiol 1992; 38:904–912.
73. Friedman M. Chemistry, nutrition and microbiology of D-amino acids. J Agric Food Chem 1999; 47:3457–3479.
74. Brukner H, Wachsman M. Design of chiral monochloro-s-triazine reagents for the liquid chromatographic separation of amino acid enantiomers. J Chromatogr A 2003; 998:73–82.
75. Nimura N, Fujiwara T, Watanabe A, et al. A novel chiral thiol reagent for automated precolumn derivatization and high-performance liquid chromatographic enantioseparation of amino acids and its application to the aspartate racemase assay. Anal Biochem 2003; 315:262–269.
76. Olajos E, Peter A, Casimir R, et al. HPLC enantioseparation of phenylalanine analogs by application of (S)-N -(4-nitrophenoxycarbonyl)phenylalanine methoxyethyl ester as a new chiral derivatizing agent. Chromatographia 2001; 54:77–82.
77. Peter A, Arki A, Vekes E, et al. Direct and indirect high-performance liquid chromatographic enantioseparation of β- amino acids. J Chromatogr A 2004; 1031:171–178.
78. Peter A, Vekes E, Torok G. Application of (S)-N-(4-nitrophenoxycarbonyl) phenylalanine methoxyethyl ester as a new chiral derivatizing agent for proteinogenic amino acid analysis by high-performance liquid chromatography. Chromatographia 2000; 52:821–826.
79. Brukner H, Westhauser T. Chromatographic determination of L- and D- amino acids in plants. Amino acids 2003; 24:43–55.
80. Gűbitz G, Schmid MG. Chiral separation principles: an introduction. In: Gűbitz G and Schmid MG, eds. Methods in Molecular Biology, Chiral Separations: Methods and Protocols. Totowa, NJ: Humana Press, 2004:1–28.
81. Imrie GA, Noctor TAG, Lough WJ. Drug bioanalysis by LC-MS: some pragmatic solutions to commonly occurring problems. Chromatography Today 2009; 2(2):27–30.
82. Bhushan R, Dubey R. Synthesis of (S)-naproxen-benzotriazole and its application as chiral derivatizing reagent for microwave-assisted synthesis and indirect high performance liquid chromatographic separation of diastereomers of penicillamine, cysteine and homocysteine. J Chromatogr A 2011; 1218:3648–3653.
83. Hashim N, Khan, SJ. Enantioselective analysis of ibuprofen, ketoprofen and naproxen in wastewater and environmental water samples. J Chromatogr A 2011; 1218:4746–4754.
84. Wang H, Ma C, Zhou J, et al. Stereoselective analysis of tiopronin enantiomers in rat plasma using high-Performance liquid chromatography-electrospray ionization mass spectrometry after chiral derivatization. Chirality 2009; 21(5):531–538.

6 HPLC chiral stationary phases for the stereochemical resolution of enantiomeric compounds: The current state of the art

W. John Lough

TRENDS IN THE DEVELOPMENT OF HPLC CHIRAL STATIONARY PHASES

In any account of drug stereochemistry, the development of commercially available chiral stationary phases (CSP) warrants some attention. This area of scientific endeavor has made a significant impact on the study of drug stereochemistry and indeed on the entire pharmaceutical industry. It has also been a major success story over the past quarter century, reaching maturity very quickly but also seeing nonstop activity and useful developments in the ensuing period. Not surprisingly then, it has been extensively chronicled already (1–6). However, such has been the volume of activity that it is still necessary to pick out the important trends from the wealth of detail available.

In the latter half of the 1980s in particular, commercially available CSP made a highly significant contribution to transforming the way chiral drugs were developed. Despite concern over the links between chirality and the thalidomide tragedy (7), synthetic chiral drugs were still generally being developed as their racemates, a situation not entirely unconnected to the fact that in the early 1980s it was still difficult to resolve and determine individual enantiomers without the use of indirect methods (see chap. 5) such as diastereomer formation. While there was interest among regulators in the analysis of individual enantiomers, as illustrated by the work of Wainer and Doyle in the laboratories of the Food and Drug Administration (FDA) in the United States on Pirkle-type CSP (Fig. 6.1) (see Ref. 8, and references therein), and pharmaceutical companies were aware of this, the technology to develop single enantiomer drugs was simply not in place. This situation changed in 1985 with the commercial introduction of Hermansson's Enantiopac® CSP (9), based on the immobilization of the plasma protein α_1-acid glycoprotein (AGP) (Fig. 6.2), by LKB of Sweden. While the Pirkle-type CSP had been on the market and had been highly acclaimed for their "breadth of spectrum," the phases of this type available at the time only gave enantioresolution for a limited class of compounds. The Pirkle-type 1A containing an electron-deficient dinitrobenzoyl group only really gave enantioseparations for compounds containing electron-rich aromatic systems, typically a naphthalene ring, and tended to be used primarily for neutral compounds. However, with the advent of the commercial AGP phase, suddenly it was possible to obtain direct chiral separations for a wide range of different classes of racemic drugs. AGP is the main plasma-binding protein for basic drugs, and accordingly, even though the tertiary structure would have been affected by the immobilization, the AGP CSP was especially useful for enantioseparations of basic drugs (10). However, it was also used for quite a few acidic (11) and neutral drugs. Also, another immobilized

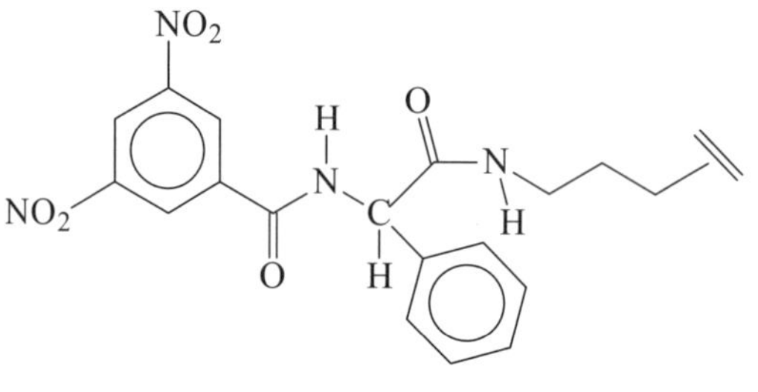

FIGURE 6.1 Synthetic multiple-interaction (Pirkle-type) chiral stationary phase featuring a modified natural amino acid chiral selector, in this case N-(3,5-dinitrobenzoyl)phenylglycine. Using normal phase solvents, this CSP gave good enantiomer separations for a wide range of neutral compounds, for example, containing a naphthalene ring system, where there is scope for π-π interactions with the dinitrobenzoyl group.

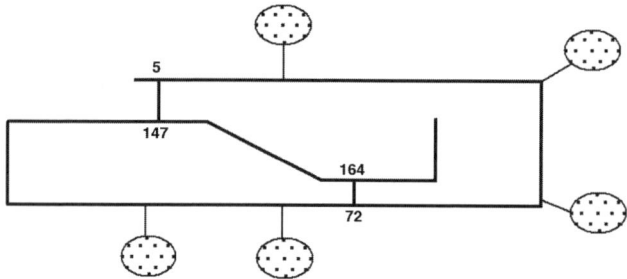

FIGURE 6.2 Schematic representation of α_1-acid glycoprotein showing the peptide backbone, two disulfide bridges, and five carbohydrate units. This chiral selector when immobilized to form a CSP gives rise to enantiomer separations for a "broad spectrum" of chiral analytes, being particularly successful for basic compounds.

protein CSP, that based on bovine serum albumin, was used primarily for acidic drugs (12). Other classes of CSP such as those based on cyclodextrins and derivatized polysaccharides followed so that by the late 1980s the situation had been transformed to the point that the separation of the enantiomers of synthetic chiral drugs had become quite commonplace. Such was the interest in the subject area that several books on chiral liquid chromatography were published (1,2), and on the back of the advances that had been made, the FDA was soon able to start formulating guidelines related to single enantiomer drugs (see chap. 14). Of particular interest was the J. T. Baker guide to CSP authored by Wainer (3). Building upon previous publications (13), Wainer subdivided CSP into five different types (Wainer types I–V) and provided advice on how to set about developing a chiral separation. As such, this publication gained some prominence at the time but it could be said that it has gained even more prominence in recent times. In patent cases involving challenges to patents for single enantiomer drugs generally held by the innovator pharmaceutical company that had produced the original racemic version (see chap. 15), the Wainer guide is often central to discussions, being held up as evidence not only that chiral high-performance

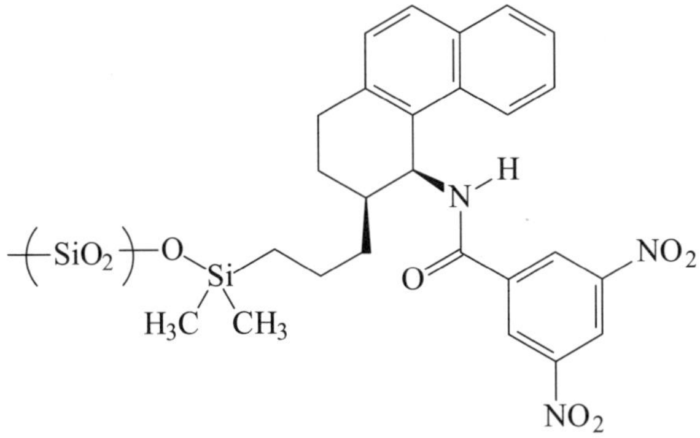

FIGURE 6.3 (*S,S*) Whelk O1 CSP (Regis Technologies, Inc.) is a higher-generation Pirkle-type synthetic multiple-interaction CSP. Compared to earlier Pirkle-type CSP, it contains two chiral centers, a π-base group as well as a π-acid group, and has greater structural rigidity.

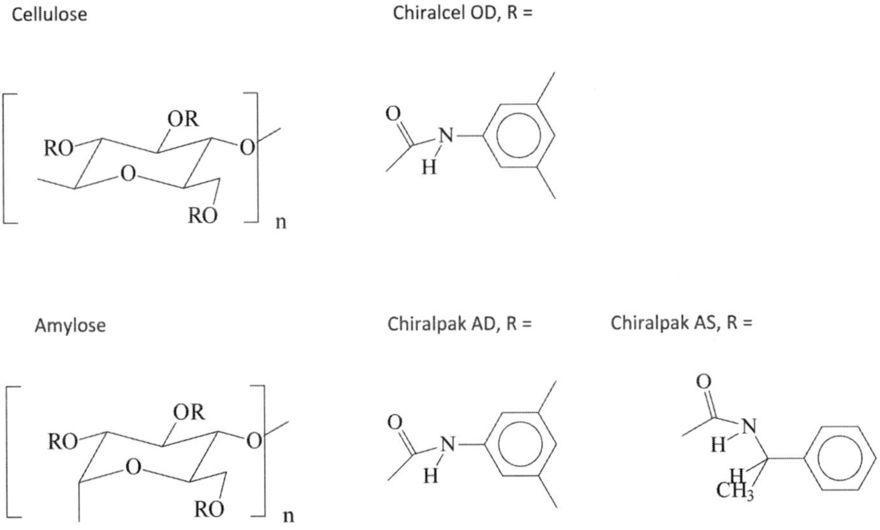

FIGURE 6.4 Derivatized polysaccharide CSP showing groups commonly used to functionalize polysaccharides such as cellulose or amylose.

liquid chromatography (HPLC) was an obvious option in the late 1980s but also that innovation was not needed in setting about developing a chiral separation.

Even given the remarkable early success of commercial CSP, by the time the second edition of the Wainer Drug Stereochemistry book (4) was published in 1993, there had been significant additional progress. Higher-generation Pirkle CSP had been developed (Fig. 6.3) and derivatized polysaccharide CSP based on carbamate rather than ester derivatization were now well established, with derivatized amylose CSP being used as well as derivatized cellulose CSP (Fig. 6.4). Such was the breadth of spectrum of these newer CSP that it was possible to contemplate adopting a chiral LC screening approach to method development rather than a carefully mapped out sequential strategy that might have been adopted in the late 1980s. From the mid-1990s onward, major pharmaceutical companies adopted these screening approaches involving working through parallel columns with different mobile phases and then optimizing the best "hits." This approach to method development is still prevalent today in major companies but, of course, the screens have been much refined and are generally more sophisticated and efficient.

With chiral separations seemingly no longer presenting too much of a challenge [personal communication, Keith Truman (GlaxoWellcome, Ware, U.K.) 1996, "Chiral separation is no longer an issue; we just do it" (!)], a major focus for chiral LC over the next decade and on to today has been doing chiral separations on a preparative scale, whether it be in early discovery to carry out pharmacological testing on individual enantiomers or in production to do such separations on a process scale. This is why a major chapter (see chap. 7) on this topic has been included in this volume. The next really significant development in chiral LC was not entirely unrelated to the focus on preparative chiral LC. Immobilized derivatized polysaccharides can obviously be used with a wider range of mobile phases than the previously existing derivatized CSP, which were prepared by coating the chiral selector onto the stationary phase. Equally obviously this meant that on the very large preparative scale there was much less chance of a drug being contaminated with traces of chiral selector. The advent of immobilized derivatized polysaccharide CSP was also related to the coated equivalents coming off patent. Once this happened, most companies in the commercial CSP market brought out their own clones, mostly of Chiralcel OD and then, slightly later, Chiralpak AD. Developments in commercial CSP have continued but there have been no new products that have made a really major impact to transform the field. However, the chiral ion-exchange CSP introduced by Wolfgang Lindner's group are worth noting as they represent the first new genuinely useful class of commercially available CSP for quite some time.

CHIRAL HPLC SCREENING

As already suggested, by the mid-1990s, there were several successful broad-spectrum CSP commercially available and, since the technology for automation was also available or under development in-house (for a typical example, see chap. 7, Fig. 7.6), it was possible for major pharmaceutical companies to develop unattended chiral LC screening systems. Such systems were developed specifically to meet business needs and as such details of many of these systems, while often described at symposia without giving too much information on the compounds involved, have not appeared as often in the scientific literature as they might have done. It is difficult to generalize, but the direction seemed to move from systems based on popular/successful CSP such as Chiralcel OD, Chiralcel OJ, Chiralpak AD, and Chiralpak AS (four products that Chiral Technologies (Cedex, France) and parent company Daicel Industries (Tokyo, Japan) were to proclaim as "Gold Medal" CSP) and Whelk O1 to systems with additional complementary CSP in order to push the success rate up toward 100%. This effort to achieve the ultimate success rate in obtaining enantioresolution through a screening approach could also be made by adopting a strategy that involves both NP-HPLC and RP-HPLC and also other separative techniques. In keeping with realism though, efforts were also made to keep screens as simple and practical as possible by eliminating redundancies (e.g., only including one of two CSP that give very similar results, using neural networks to omit experiments with, based on data from previous work, very little chance of success) and thereby unnecessary experiments.

Although information about screening systems used in pharmaceutical companies does not all get published, there are still useful illustrative examples in the scientific literature. Collaborating with Sanofi-Synthelabo, the group of Vander Heyden in Belgium described a normal-phase chiral LC screen based on Chiralcel OD [immobilized cellulose tris-(3,5-dimethylphenylcarbamate)], Chiralcel OJ [immobilized cellulose tris-(4-methylbenzoate)], and Chiralcel AD [immobilized amylose tris-(3,5-dimethylphenylcarbamate)], which had been used on a diverse set of 38 chiral drugs (14). It was reported that the degree of enantioresolution that could be obtained was highly dependent on the type of organic modifier used in the mobile phase and that enantiomeric separation was observed for 89% of cases. In a follow-up paper also published in 2002 (15), a similar success rate (this time for 37 diverse chiral drugs) was reported for a reversed-phase screen, also using three derivatized polysaccharide CSP. In the same way that organic modifier had been important under normal-phase conditions, control of drug ionization by the mobile phase pH was important in reversed phase, relatively extreme pH values of 2.0 and 9.0 being used. The screens described by AstraZeneca workers in 2003 (16) were perhaps more typical of the state of the art in industry at the time involving compounds (55 compounds, 57 enantiomeric pairs in total) encountered as starting materials, intermediates, or final APIs in industry, and two screens, one using four derivatized polysaccharide CSP and five mobile phases and the second using three macrocyclic antibiotic CSP and two mobile phases. The former showed enantioselectivity for 87% of the test compounds, the latter 65%, and, importantly, the combined screen showed 96%.

Between 2007 and 2009, a series of articles (17–20) appeared in the journal Current Pharmaceutical Analysis in which screening strategies were compared. HPLC, supercritical fluid chromatography (SFC), capillary electrophoresis (CE), and gas chromatography (GC) screens were considered orthogonal (17), enantioselectivity differences between normal-phase LC and SFC screens were attributed to the mobile phase modifiers used (18), with a success rate of 60% to 95%, depending on the compound sets applied reversed-phase screens were claimed to be viable alternatives to normal-phase LC and SFC screens (19) and Pirkle-type phases were compared with derivatized polysaccharides phases (20). In some more recent publications, advantages are highlighted or comparisons made in studies on screening featuring newer CSP unique to particular vendors. In a paper by workers from Chiral Technologies Europe (21), which have greater mobile phase versatility than their coated equivalents, a study on the suitability of immobilized derivatized polysaccharides for screening under reversed-phase LC conditions was reported. Also, the groups of Vander Heyden and Chankvetadze have both carried out extensive studies involving new coated derivatized polysaccharides [e.g., using the chiral selectors cellulose tris(3-chloro-4-methylphenylcarbamate) and amylose tris(2-chloro-5-methylphenylcarbamate)] that are commercially available from Phenomenex (Torrance, California, USA) (22–25).

RECENT DEVELOPMENTS IN COMMERCIALLY AVAILABLE CHIRAL STATIONARY PHASES

Overview

Practitioners carrying out chiral separations by LC use commercially available CSP and so the state of the art in chiral LC is best represented by the recent developments from CSP vendors/manufacturers. As might already be apparent from the discussion of chiral LC screening, much of the scientific literature in chiral LC is concerned with the study and application of commercial CSP rather than with research into new CSP that will be the commercial CSP of tomorrow.

CSP were subdivided into classes according to the nature/structure of the chiral selector (2) or into types (3) according to the likely chiral recognition mechanism. There was a time when CSP could even almost be subdivided very simply according to the manufacturer [Pirkle-type (Regis Technologies, Inc., Illinois, U.S.); cyclodextrins (Astec, Whippany, New Jersey, U.S.); crown ethers, derivatized polysaccharides (Daicel Industries, Tokyo, Japan); Chiral-AGP (Chrom Tech Ltd., Congleton, U.K.)]. However, especially since adsorbed/coated derivatized polysaccharide CSP came off patent, companies have been taken over and more vendor companies have entered the market, most companies try to sell as wide a range of CSP as possible. Some of the newer CSP might not fit so obviously into one of the original five types (3), and so, in looking at recent developments in commercial CSP, it is still probably most appropriate to look at classes according to chiral selector despite the increase in such classes.

Derivatized Polysaccharides

As derivatized polysaccharide LC stationary phases (Fig. 6.4) are historically important CSP and have formed the basis of most chiral LC screening method development strategies, they have already received several mentions in this chapter. In early versions of these, hydroxyl groups on cellulose had been derivatized to form tris-esters. However, this class of CSP, products of Daicel Industries from the work of the research group of Yoshio Okamato (26) really began to make its impact from mid-1988 onward with the introduction of carbamate derivatives and the use of amylose as well as cellulose. These CSP have had a very strong presence in the market since their introduction. However, as already mentioned, there were changes from round about 2006 onward. Polysaccharide CSP immobilized by covalent bonding (27) offered greater physical and chemical robustness with respect to mobile phases that could be used. This is very useful in preparative and production-scale separations (no leakage of chiral selector into isolated API, easier to choose good solvent for sample solubility), which had become an increasingly important market as well as giving greater flexibility in method development. Another, almost parallel, development was the patent situation allowing competitors of Daicel (or Chiral Technologies in the United States and Europe) to develop their own "clones" of Chiralcel OD and Chiralpak AD. For example, Regis, Macherey Nagel Inc. (KG, Düren, Germany), and Eka Nobel Inc. (Bohus, Sweden) developed RegisCell™, Nucleocel Delta, and Cellucoat™, respectively. All of these were very similar in retentivity and selectivity to Chiralcel OD. This is illustrated for RegisCell in Figure 6.5. Cellcoat, produced by AECS-QuikPrep Ltd. (Bridgend, U.K.), was quite different in its properties from Chiralcel OD though still an adsorbed/coated CSP with the same chiral selector. From a nonscientific

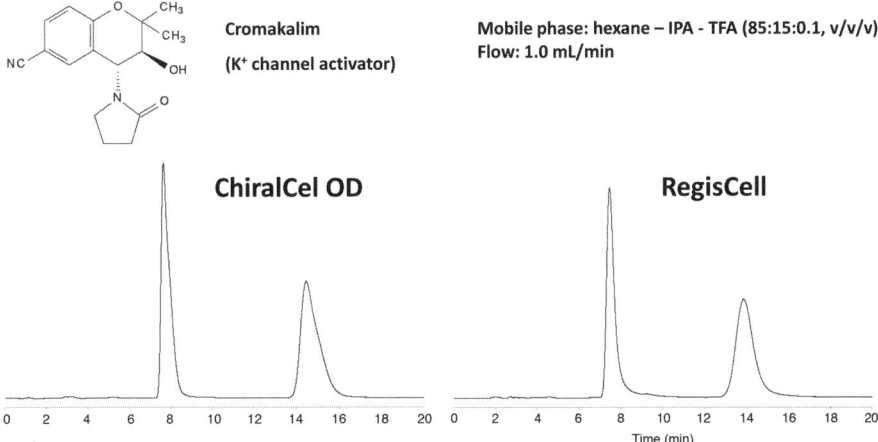

FIGURE 6.5 Chromatograms illustrating the similarity between the original Daicel derivatized polysaccharide CSP and clones thereof. Taking into account minor differences in temperature, mobile phase composition, and column history, there are no significant differences in separations obtained (experience from panel of 50 racemic drugs with two mobile phases). *Source*: Courtesy of R. Wimal H. Perera, University of Sunderland.

perspective, it was interesting to note that in general these companies producing clones elected not to undercut the price of the original Daicel CSP by a significant amount. A different approach was taken by Phenomenex. In introducing the Lux range of CSP, they not only offered a Chiralcel OD-type CSP [Lux Cellulose-1 using cellulose tris(3, 5-dimethylphenylcarbamate) as the chiral selector] but also offered, with the ubiquitous claim of "complementary selectivity," CSP based on different derivatized polysaccharide selectors such as Lux Cellulose-2 using cellulose tris(3-chloro-4-methylphenylcarbamate). This range was based on technology that Phenomenex acquired along with the company Sepaserve, which was founded by Bezhan Chankvetadze, the distinguished Georgian separation scientist and erstwhile coworker of Okamato (28,29).

A less high-profile recent derivatized polysaccharide CSP is ZirChrom®-CelluloZe (chiral selector is modified 3,5-dimethylphenyl-carbamoyl cellulose) in which the support material is zirconia rather than a siliceous material. While this might not seem like a dramatic difference, it is potentially useful in allowing access to high pH mobile phases when using reversed-phase mobile phase conditions.

Higher-Generation "Pirkle-Type" CSP

The previously mentioned Pirkle-type 1A (Fig. 6.1) and the Whelk O1 CSP (Fig. 6.3) are perhaps the best known of the Pirkle-type CSP, more helpfully named by Doyle (10) as synthetic multiple-interaction CSP. Pirkle famously developed his first CSP using the principle of microscopic reversibility (30) and paid due attention to the "three-point interaction rule" (31) in his phase design. Pirkle's group developed several generations of these CSP with the most successful being commercialized by Regis Technologies, Inc. Many of this type of CSP were

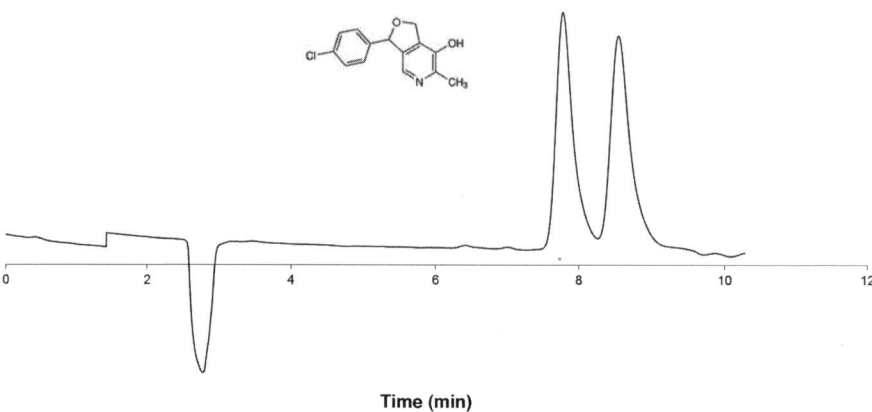

FIGURE 6.6 Chromatogram showing the operation of the Whelk O1 CSP under reversed-phase conditions. *Source*: Courtesy of R. Wimal H. Perera, University of Sunderland.

made not only by Pirkle but by quite a few other groups. The DACH-DNB CSP of Gasparinni and the ULMO CSP of Lindner were also commercialized by Regis (both described in Ref. 6, along with other Regis Pirkle-type CSP). However, the Whelk O1 exhibited a breadth of spectrum that ranked up with the best of the derivatized polysaccharide CSP and therefore is much more frequently used than other Pirkle-type CSP. Of course, being covalently bonded, it had over a decade start on the covalently bonded derivatized polysaccharides in having mobile phase versatility suitable for preparative work and being readily able to operate under reversed-phase conditions (Fig. 6.6), having in fact been designed originally by molecular modeling to achieve the RP-HPLC separation of naproxen enantiomers (32). An improved version of the Whelk O1 CSP using ultrapure 5-μm Kromasil silica particles has been available for 3 to 4 years now. Perhaps this is a belated advance, but it is a sign of things to come. 3-μm and sub-2-μm versions of DACH-DNB have already been prepared (33) and indications from recent conference abstracts is that the same device has been applied to Whelk O1. This is the direction in which commercial CSP will inevitably be going and this will lead to higher efficiency or, more likely, much faster separations. However, there have been verbal reports that the improvements in going to very small particles observed with CSP are not as much as that for standard C18 RP-HPLC phases. This would not be surprising as improvements for CSP might be limited by the slower mass transfer that often accompanies the complex, multisite interactions that can be involved in chiral recognition.

Macrocyclic Antibiotics

As Pirkle was to Regis and Okamoto to Daicel, so was Dan Armstrong to Astec, but to an even greater degree. Astec has been bought up by Sigma Aldrich, but

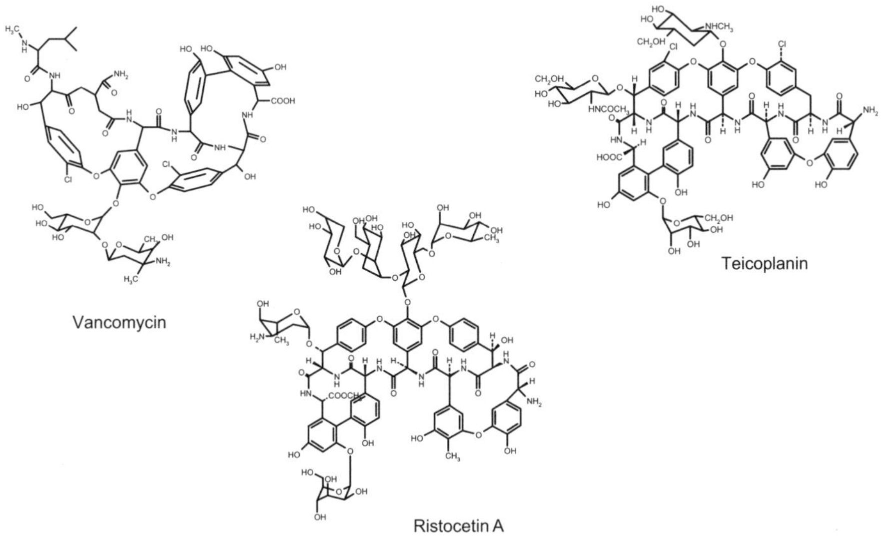

FIGURE 6.7 Structures of macrocyclic antibiotics used in the Chirobiotic™ range of CSP.

there is still a very strong link with CSP emanating from Professor Armstrong's laboratories. Prominent among these CSP are those belonging to the class of macrocyclic antibiotics. As may be seen from the structures of the principal macrocyclic antibiotics used in CSP (Fig. 6.7), they contain amide/peptide linkages and sugar residues. They also possess an element of tertiary structure almost as if they were miniature proteins and, as such, they show a good breadth of spectrum for achieving enantiomer separations for a wide range of compound classes without the slow mass transfer and other issues associated with the use of very large molecules like proteins. Also, like protein CSP they may be used under reversed-phase conditions with low proportions of polar organic solvent to aqueous buffer in the mobile phase. However, they may also be used under normal-phase conditions and, have been used extensively in polar organic (34) and polar ionic (35) modes, the latter being exploited frequently for its suitability with MS detection (36). Aglycone versions of these CSP (i.e., with the sugar residues removed from the chiral selector) were also produced to give CSP with complementary enantioselectivity. Another modification, in the wake of the aforementioned upsurge in interest in preparative chiral separations was the development of V2, T2, and R2 CSP with greater stability and loadability (37). In keeping with the method development practices of the pharmaceutical industry, Sigma Aldrich also sell these so-called Chirobiotic® CSP together along with a guidebook with recommendations for mobile phases to be used for method-development screening.

Oligosaccharide CSP
Oligosaccharide CSP also belong to the Armstrong "stable," cyclodextrin CSP (38) being the first major CSP products from Astec. CSP based on native cyclodextrins (α-, β-, and γ-cyclic toroidal rings containing 6-, 7-, and

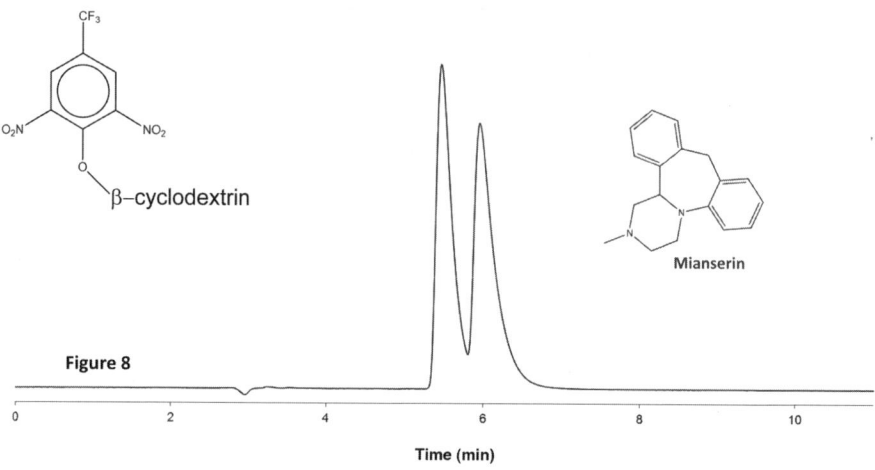

FIGURE 6.8 Chromatogram showing the separation of mianserin enantiomers on one of the more recently introduced members of the Cyclobond™ series of CSP, that is, Cyclobond I 2000 DNP. With the mono-aryl functionalization, retention and separation may be obtained with proportions of methanol in the mobile phase as high as 40%. *Source*: Courtesy of R. Wimal H. Perera, University of Sunderland.

8-D-glucose units, respectively) were followed by a range of derivatized cyclodextrin CSP with complementary enantioselectivity. This Cyclobond® range of CSP was improved in the late 1990s to give Cyclobond 2000 products. The most recent Cyclobond CSP to be introduced is the Cyclobond™ I 2000 DNP (39) (Fig. 6.8) product. This followed some of Armstrong's earlier work being revisited to find an improved method for the derivatization. In a similar fashion to macrocyclic antibiotic CSP, cyclodextrin CSP were originally used in the reversed phase mode but have now been used in normal-phase mode and extensively in polar organic and polar ionic modes. With so many complementary cyclodextrin CSP having been produced, some have become less attractive in that there are fewer unique separations that cannot be done by other cyclodextrin CSP. For example, the Cyclobond I SN CSP, featuring a naphthylethylcarbamate derivative of β-cyclodextrin, has now been withdrawn from the market.

Very recent new oligosaccharide products from Sigma Aldrich are the cyclofructan CSP (40) (Fig. 6.9). These CSP are especially useful for primary amines. While this might not seem especially exciting as primary amine enantioseparations may easily be achieved, for example, by crown ether CSP (addressed later), at least these separations may be had using mobile phases different from the highly aqueous, acidic mobile phases needed for the crown ether CSP. Also, it must be remembered that the enantioseparation of primary amines is an important application area since, in a synthetic route to a homochiral drug, the chiral resolution may often be carried out on a primary amine intermediate. Further, in some of his most recent work, Armstrong has

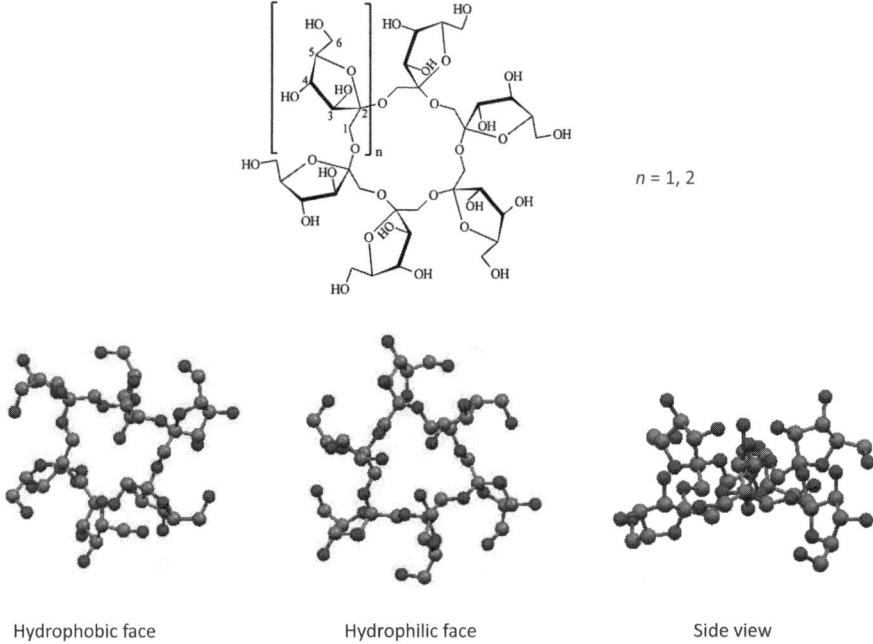

FIGURE 6.9 Structure of the cyclofructan chiral selector. *Source*: Courtesy of Daniel W. Armstrong, University of Texas Arlington.

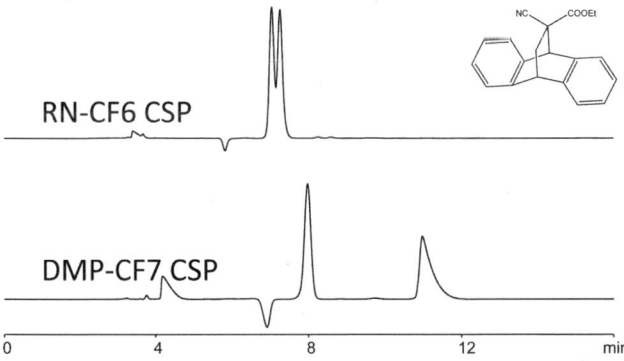

FIGURE 6.10 Complementary selectivity of (*R*)-naphthylethyl-functionalized CF6 and dimethylphenyl-functionalized CF7 CSP using normal-phase conditions (heptane—ethanol—TFA (95:5:0.1, v/v/v). *Source*: Courtesy of Daniel W. Armstrong, University of Texas Arlington.

produced aromatic functionalized cyclofructan CSP (41), which give separations for a wider range of compound classes including those with secondary amine, tertiary amine, sulfonamide, or acid groups (Fig. 6.10). These CSP exhibit best performance under normal-phase conditions with an (*R*)-naphthylethyl CSP showing complementary selectivity to a dimethylphenyl CSP.

Protein CSP

While protein CSP, in particular the Enantiopac and subsequent second-generation Chiral-AGP products based on the immobilization of the plasma protein, AGP (Fig. 6.2), played a major role in the early days of commercially available CSP, they had been overshadowed by derivatized polysaccharide CSP by the early to mid-1990s. However, given the broad spectrum of classes of chiral drugs for which they give a separation of the enantiomers under reversed-phase mobile phase conditions, albeit with low proportions of the polar organic component, they remained of interest and still do today. While CSP based on proteins will be less stable and have slower mass transfer than small-molecule CSP, this has long since ceased to be a significant issue. Given the timing of preparation of pharmacopoeial monographs (after a drug comes off patent) and the source of methods therein (often from the Development labs of the drug's innovator company), pharmacopoeial assays for trace enantiomeric impurity in single enantiomer drugs often employ conditions based on an AGP CSP. Despite much activity on other proteins and the commercialization of some such as Chiral-HSA [based on human serum albumin; useful for drug-protein binding studies (42)] and Chiral-CBH (based on the stable enzyme, cellobiohydrolase) as well as the longstanding BSA CSP (Macherey Nagel's Resolvosil), Chiral-AGP remains the most popular commercial protein CSP. This is despite it being compared less favorably to an ovomucoid CSP (OVM CSP) in 1991 in a somewhat contentious report (43) and even in the more circumspect conclusions of a more extensive study reported in 2008 (44) as not separating large chiral molecules better than an OVM CSP.

Over the years and particularly in recent years, much of the reported work on Chiral-AGP has involved drug bioanalysis no doubt because of its reversed-phase mode of operation. Because of this, there is an interest in MS detection with LC using Chiral-AGP. Sensitivity can be a problem when dealing with mobile phases when they contain less than 5% of the polar organic component, but Imrie et al. found that this could be easily circumvented by the postcolumn addition of acetonitrile (45). Also, Chiral Technologies, Inc., responsible for the Chiral-AGP product in Europe/United States, have been involved in the development of a rational screening strategy for work on Chiral-AGP in which liquid chromatography–mass spectrometry (LC-MS) compatibility is a major priority (46).

Ion-Exchange CSP

Chiralpak® QN-AX and Chiralpak® QD-AX (Fig. 6.11) are enantioselective weak anion-exchange (AX) CSP from Chiral Technologies, Inc., which are based on O-9 (*tert*-butylcarbamoyl) quinine and quinidine, respectively. They were developed by Wolfgang Lindner's group in Vienna (47) for the enantioseparation of chiral acids such as compounds containing carboxylic, phosphonic, phosphinic, phosphoric, or sulfonic groups. Quinine and quinidine being enantiomers of one another, the use of Chiralpak QD-AX and Chiralpak QN-AX gives rise to a reversal of elution order (Fig. 6.12). Many of the reports of the use of these CSP involve derivatized amino acids, occasionally to spectacular effect (Fig. 6.13). However, they also work well for nonsteroidal anti-inflammatory drugs (Fig. 6.14) and mandelic acids.

Chiral Technologies Europe have now entered into an exclusive worldwide licensing agreement with Lindner's group to manufacture and market

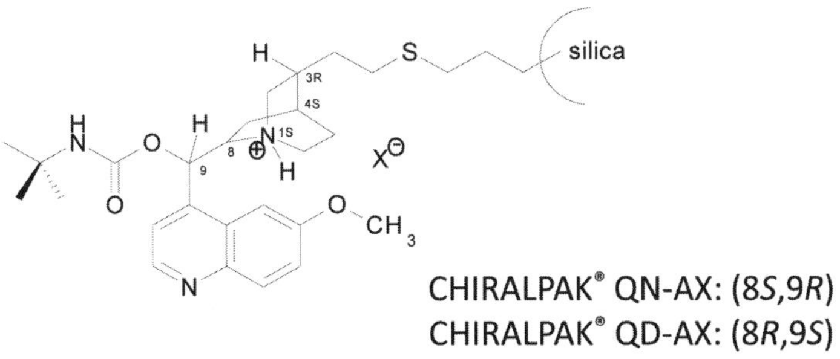

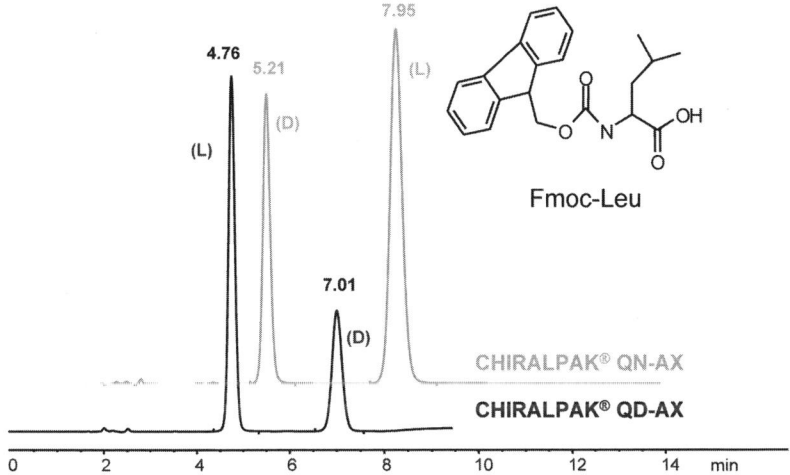

FIGURE 6.11 Structure of chiral anion-exchangers from Chiral Technologies Inc. *Source*: Courtesy of Brian Freer, Chiral Technologies Europe.

FIGURE 6.12 Method development for acidic compounds in the polar organic mode showing reversal of elution order between quinine and quinidine CSP [using mobile phase: methanol—acetic acid—ammonium acetate (98:2:0.5, v/v/w)]. *Source*: Courtesy of Brian Freer, Chiral Technologies Europe.

"novel Cinchona alkaloid-based zwitterion-exchange type chiral stationary phases." With a zwitterionic CSP recently introduced, it would not be surprising if Lindner's cation-exchange CSP (48) also made it to the market.

Other Classes of CSP

While not attempting to provide an exhaustive list of CSP classes, it is worth noting others not mentioned thus far. Ligand exchange and crown ether CSP have been around since the 1980s and early 1990s (49,50). They still serve a useful purpose in providing good chiral separations for specific classes of compound, primary amines for crown ether CSP, and amino acids and α-hydroxy-acids for ligand exchange CSP. As well as the Crownpak® CR(+)

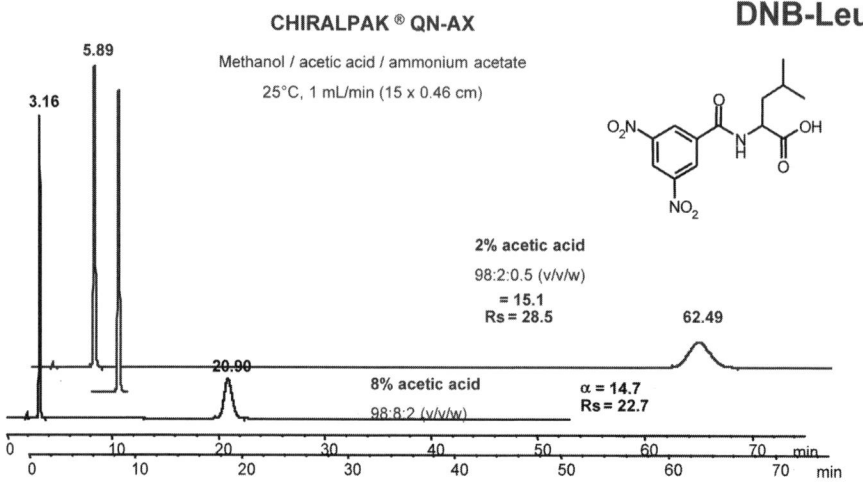

FIGURE 6.13 Method development on Chiralpak™ QN-AX in polar organic mode for acidic compounds: increasing counter ion strength reduces retention; increasing pH will increase retention. *Source*: Courtesy of Brian Freer, Chiral Technologies Europe.

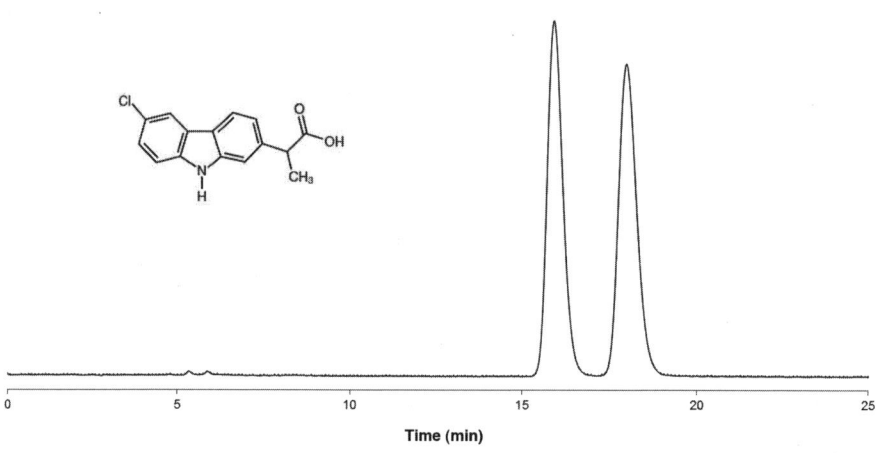

FIGURE 6.14 Chromatogram showing the separation of the enantiomers of the nonsteroidal anti-inflammatory drug, carprofen on Chiralpak™ QD-AX. *Source*: Courtesy of R. Wimal H. Perera, University of Sunderland.

and Crownpak® CR(−) CSP from Daicel, the crown ether CSP based on (+)-(18-crown-6)-2,3,11,12-tetracarboxylic acid, Chirosil RCA(+) and Chirosil SCA(−), from RS Tech, Daejeon, Korea, are now marketed by Regis Technologies, Inc. There have been reports of secondary amines being resolved on such tartaric acid–derived CSP (51), but this is very much the exception rather than the rule. Regarding ligand exchange CSP, the Daicel products are Chiralpak WH and Chiralpak MA(+). Regis now sell not only a Davankov CSP but also a Davankov reagent for use as a mobile phase additive for use with achiral alkyl-silica phases. Sigma Aldrich's ligand exchange CSP are named CLC phases.

Polymeric CSP have also been around since the 1980s (52), but in this case there have been recent developments, specifically the introduction of polycyclic amine CSP from Sigma Aldrich. The PCAP CSP, available in R,R and S,S forms, are based on the reproducible polymerization of a cyclic diamine from the surface of silica. In the more recent P-CAP-DP^{TM} aromatic groups are introduced. These polycyclic amine CSP have been used in normal phase and polar organic modes. The key selling points for these CSP seem to be that they are stable covalently bound materials, with high stability, no memory effect, high sample loadability, and easy scale-up.

RESEARCH AND POSSIBLE FUTURE DIRECTIONS IN HPLC CHIRAL STATIONARY PHASES

Much of the scientific research literature on chiral LC over the past 2 to 3 years is on applications, some quite elegant, on existing commercial CSP, or, more rarely, heralds a new commercial CSP (53). Inevitably, new research will involve the use of CSP based on sub-2-µm particles or on fused core silica particles, and applications will involve the use of a chiral dimension in 2D-LC. There seems to be little on the horizon with respect to new classes of CSP that will transform chiral LC. This should come as no surprise as much has already been done in this area and there is no longer any real need for new types of CSP. Nonetheless, it is still interesting to note the ideas of Welch on microscale chiral LC (54). With a strong background in synthetic organic chemistry, one of Welch's thoughts is that microscale LC is now readily accessible, and chiral LC columns requiring only minute amounts of immobilized chiral selector may be used. Therefore, using very expensive chiral selectors or chiral selectors prepared by lengthy syntheses now becomes economically viable, thus opening up some very exciting possibilities. No doubt some of these possibilities will be realized. Also, it is worth noting that had microscale chiral LC, though available (55), been more commonplace in the late 1980s and early 1990s, then the possibilities would really have been exciting at that time.

REFERENCES

1. Wainer IW, Drayer DE, eds. Drug Stereochemistry. New York: Marcel Dekker, Inc., 1988.
2. Lough WJ, ed. Chiral Liquid Chromatography. Glasgow: Blackie, 1989.
3. Wainer IW. A Practical Guide to the Selection and Use of HPLC Chiral Stationary Phases. Phillipsburg, USA: J.T. Baker Inc., 1988.
4. Wainer IW, ed. Drug Stereochemistry, Analytical Methods and Pharmacology. 2nd ed. New York, Basel, Hong Kong: Marcel Dekker, Inc., 1993.
5. Lough WJ, Wainer IW, eds. Chirality in the Natural & Applied Sciences. Oxford: Blackwell Publishing Ltd., 2002.
6. Lough WJ. Chiral analysis of pharmaceuticals. In: Lee DC, Webb M, eds. Pharmaceutical Analysis. Oxford: Blackwell Publishing Ltd., 2003:74–104.
7. De Camp WH. Importance of enantiomer separations. In: Lough WJ, ed. Chiral Liquid Chromatography. Glasgow: Blackie, 1989.

8. Doyle TD. Synthetic multiple-interaction chiral bonded phases. In: Lough WJ, ed. Chiral Liquid Chromatography. Glasgow: Blackie, 1989.
9. Hermansson J. Liquid chromatographic resolution of racemic drugs using a chiral α_1-acid glycoprotein column. J Chromatogr 1984; 298:67–78.
10. Schill G, Wainer IW, Barkan SA. Chiral separation of cationic drugs on an α_1-acid glycoprotein bonded stationary phase. J Liquid Chromatogr 1986; 9:641–666.
11. Hermansson J, Eriksson E. Direct liquid chromatographic resolution of acidic drugs using a chiral α_1-acid glycoprotein column (enantiopac). J Liquid Chromatogr 1986; 9:621–639.
12. Allenmark, S. Optical resolution by liquid chromatography on immobilised bovum-serum albumin. J Liquid Chromatogr 1986; 9:425–442.
13. Armstrong DW. Optical isomer separation by liquid chromatography. Anal Chem 1987; 59:93–100.
14. Perrin C, Vu VA, Matthijs N, et al. Screening approach for chiral separation of pharmaceuticals Part I. Normal-phase liquid chromatography. J Chromatogr A 2002; 947(1):69–83.
15. Perrin C, Matthijs N, Mangelings D, et al. Screening approach for chiral separation of pharmaceuticals Part II. Reversed-phase liquid chromatography. J Chromatogr A 2002; 966(1–2):119–134.
16. Anderson ME, Aslan D, Clarke A, et al. Evaluation of generic chiral liquid chromatography screens for pharmaceutical analysis. J Chromatogr A 2003; 1005 (1–2):83–101.
17. Akin A, Antosz F, Ausec JL, et al. An orthogonal approach to chiral method development screening. Current Pharm Anal 2007; 3(1):53–70.
18. Wong MM, Holzheuer WB, Webster GK. A comparison of HPLC and SFC chiral method development screening approaches for compounds of pharmaceutical interest. Current Pharm Anal 2008; 4(2):101–105.
19. Holzheuer WB, Wong MM, Webster GK. Reversed phase chiral method development screening for compounds of pharmaceutical interest. Current Pharm Anal 2009; 5(4):346–357.
20. Holzheuer WB, Wong MM, Webster GK. Evaluation of Pirklestationary phases in chiral method development screening for compounds of pharmaceutical interest. Current Pharm Anal 2009; 5(1):10–20.
21. Tong Z, Dung N, Pilar F. Reversed-phase screening strategies for liquid chromatography on polysaccharide-derived chiral stationary phases. J Chromatogr A 2010; 1217(7):1048–1055.
22. Dossou KSS, Chiap P, Chankyetadze B, et al. Optimization of the LC enantioseparation of chiral pharmaceuticals using cellulose tris(4-chloro-3-methylphenylcarbamate) as chiral selector and polar non-aqueous mobile phases. J Separation Sci 2010; 33(12):1699–1707.
23. Peng L, Jayapalan S, Chankvetadze B, et al. Reversed-phase chiral HPLC and LC/MS analysis with tris(chloromethylphenylcarbamate) derivatives of cellulose and amylose as chiral stationary phases. J Chromatogr A 2010; 1217(44):6942–6955.
24. Younes AA, Mangelings D, Heyden YV. Chiral separations in normal phase liquid chromatography: Enantioselectivity of recently commercialized polysaccharide-based selectors. Part I: Enantioselectivity under generic screening conditions. J Pharm Biomed Anal 2011; 55(3):414–423.
25. Younes AA, Mangelings D, Heyden YV. Chiral separations in normal-phase liquid chromatography: Enantioselectivity of recently commercialized polysaccharide-based selectors. Part II. Optimization of enantioselectivity. J Pharm Biomed Anal 2011; 56(3):521–537.
26. Okamoto Y, Kawashima M, Aburatani R. Chromatographic Resolution 12. Optical resolution of beta-blockers by HPLC on cellulose triphenylcarbamate derivatives. Chem Letts 1986; 7:1237–1240.
27. Ikai T, Yamamoto C, Kamigaito M, et al. Immobilization of polysaccharide derivatives onto silica gel-facile synthesis of chiral packing materials by means of intermolecular polycondensation of triethoxysilyl groups. J Chromatogr A 2007; 1157(1–2):151–158.

28. Chankvetadze B, Yashima E, Okamoto Y. Tris(chloro-disubstituted and methyl-disubstituted phenylcarbamate)s of cellulose as chiral stationary phases for chromatographic enantioseparation. Chem Letts 1993; 4:617–620.
29. Chankvetadze B, Yashima E, Okamoto Y. Chloromethylphenylcarbamate derivatives of cellulose as chiral stationary phases for high-performance liquid-chromatography. J Chromatogr A 1994; 670(1–2):39–49.
30. Pirkle WH, House DW, Finn JM. Broad-spectrum resolution of optical isomers using chiral high-performance liquid-chromatographic bonded phases. J Chromatogr 1980; 192(1):143–158.
31. Pirkle WH, Hyun MH, Tsipouras A, et al. A rational approach to the design of highly effective chiral stationary phases for the liquid chromatographic separation of enantiomers. J Pharm Biomed Anal 1984; 2(2):173–181.
32. Pirkle WH, Welch CJ, Lamm B. Design, synthesis, and evaluation of an improved enantioselective naproxen selector. J Org Chem 1992; 57(14):3854–3860.
33. Cancelliere G, Ciogli A, D'Acquarica I, et al. Transition from enantioselective high performance to ultra-high performance liquid chromatography: A case study of a brush-type chiral stationary phase based on sub-5-micron to sub-2-micron silica particles. J Chromatogr A 2010; 1217(7):990–999.
34. Armstrong DW, Liu YB, Ekborgott KH. Covalently bonded teicoplanin chiral stationary-phase for HPLC enantioseparations. Chirality 1995; 7(6):474–497.
35. Hashem H, Tründelberg C, Attef O, et al. Effect of chromatographic conditions on liquid chromatographic chiral separation of terbutaline and salbutamol on Chirobiotic V column. J Chromatogr A 2011; 1218(38):6727–6731.
36. Desai MJ, Armstrong DW. Transforming chiral liquid chromatography methodologies into more sensitive liquid chromatography-electrospray ionization mass spectrometry without losing enantioselectivity. J Chromatogr A 2004; 1035(2):203–210.
37. Beesley TE. Preparative purification of basic chiral racemates. LC GC North America 2004; S:31.
38. Coventry L. Cyclodextrin inclusion complexation. In: Lough WJ, ed. Chiral Liquid Chromatography. Glasgow: Blackie, 1989:148–165.
39. He LF, Beesley TE. A new high performance cyclodextrin derivative LC phase for chiral separations-Cyclobond I 2000 DNP. LC GC North America 2006; S:30.
40. Sun P, Wang C, Breitbach ZS, et al. Development of new HPLC chiral stationary phases based on native and derivatized cyclofructans. Anal Chem 2009; 81(24):10215–10226.
41. Kalikova K, Janeckova L, Armstrong DW, et al. Characterization of new R-naphthylethylcyclofructan 6 chiral stationary phase and its comparison with R-naphthylethyl beta-cyclodextrin-based column. J Chromatogr A 2011; 1218(10):1393–1398.
42. Hollosy F, Valko K, Hersey A, et al. Estimation of volume of distribution in humans from high throughput HPLC-based measurements of human serum albumin binding and immobilized artificial membrane partitioning. J Med Chem 2006; 49(24):6958–6971.
43. Kirkland KM, Neilson KL, McCombs DA. Comparison of a new ovomucoid and a 2nd-generation alpha-1-acid glycoprotein-based chiral column for the direct high-performance liquid-chromatography resolution of drug enantiomers. J Chromatogr 1991; 545(1):43–58.
44. Zhou L, Mao B, Ge Z. Comparative study of immobilized alpha 1 acid glycoprotein and ovomucoid protein stationary phases for the enantiomeric separation of pharmaceutical compounds. J Pharm Biomed Anal 2008; 46(5):898–906.
45. Imrie GA, Noctor TAG, Lough WJ. Drug bioanalysis by LC-MS: some pragmatic solutions to commonly occurring problems. Chromatography Today 2009; 2(2):27–30.
46. Michishita T, Franco P, Zhang T. New approaches of LC-MS compatible method development on alpha(1)-acid glycoprotein-based stationary phase for resolution of enantiomers by HPLC. J Separation Sci 2010; 33(23–24):3627–3637.
47. Gyimesi-Forras K, Akasaka K, Lammerhofer M, et al. Enantiomer separation of a powerful chiral auxiliary, 2-methoxy-2-(1-naphthyl)propionic acid by liquid chromatography using chiral anion exchanger-type stationary phases in polar-organic

mode; investigation of molecular recognition aspects. Chirality 2005; 17(S):S134–S142.
48. Hoffmann CV, Lammerhofer M, Lindner W. Novel strong cation-exchange type chiral stationary phase for the enantiomer separation of chiral amines by high-performance liquid chromatography. J Chromatogr A 2007; 1161(1–2):242–251.
49. Lam S. Chiral ligand exchange chromatography. In: Lough WJ, ed. Chiral Liquid Chromatography. Glasgow: Blackie, 1989:83–101.
50. Hilton M, Armstrong DW. Evaluation of a chiral crown-ether lc column for the separation of racemic amines. J Liquid Chromatogr 1991; 14(1):9–28.
51. Steffeck RJ, Zelechonok Y, Gahm KH. Enantioselective separation of racemic secondary amines on a chiral crown ether-based liquid chromatography stationary phase. J Chromatogr A 2002; 947(2):301–305.
52. Johns DM. Binding to synthetic polymers. Chiral ligand exchange chromatography. In: Lough WJ, ed. Chiral Liquid Chromatography. Glasgow: Blackie, 1989:177–184.
53. Han X, He L, Zhong Q, et al. Synthesis and evaluation of a synthetic polymeric chiral stationary phase for LC based on the N,N '-[(1R,2R)-1,2-diphenyl-1,2-ethanediyl]bis-2-propenamide monomer. Chromatographia 2006; 63(1–2):13–23.
54. Welch CJ. Microscalechiral HPLC in support of pharmaceutical process research. Chirality 2009; 21(1):114–118.
55. Matlin SA, Stacey VE, Lough WJ. Hexahelicine chiral stationary phase: part 1. Phase synthesis and use in HPLC resolution of enantiomers. J Chromatogr 1988; 450:157–162.

7 Preparative and production scale chromatography in enantiomer separations

Geoffrey B. Cox

INTRODUCTION

At its origins, chromatography was a preparative technique, developed by Tswett in the early 1900s for the isolation of plant pigments. Given the ubiquity of analytical high-performance liquid chromatography (HPLC) today, it is hard to realize that it was not until the late 1960s that the analytical use of liquid chromatography was made possible by the development of high-efficiency packing media and instrumental chromatographic systems. Up to that point, most liquid chromatography was carried out with the aim of isolating useful quantities of materials from mixtures. Almost all of these separations were carried out at the laboratory scale; most of the procedures used were certainly not appropriate for larger-scale separations. In 1964, however, UOP introduced a countercurrent chromatographic process, which is used widely for the production of ultrapure *p*-xylene and in the production of high-fructose corn syrup. By 1984 it was estimated (1) that the installed capacity for the *p*-xylene process was 600,000 MTA. The scale of high-fructose corn syrup production can be gauged by WHO figures for its 2007 per capita consumption in the United States—around 40 lb (2). Until the development of compression technology—radial compression by Waters Associates and dynamic axial compression introduced by Prochrom SA (Champigneulles, France)—larger-scale column liquid chromatography was used only at low pressures and with rather low productivity, mainly because of the problems involved in packing and maintaining the chromatographic beds. These developments allowed the use of smaller particle sizes and higher efficiency columns for the more difficult separations.

Preparative chromatography in the pharmaceutical industry has for the main part been restricted to early stages in the discovery and development processes. This is not to say that the technique is not used in production. One of the first reports of the large-scale use of HPLC as a production process was in the production of biosynthetic human insulin (3). Other reports of the use of HPLC in the production of high-cost, high-potency pharmaceutical products such as steroids and prostaglandins have appeared (4). More recently, the combination of the techniques—of countercurrent chromatography and axial compression columns—has resulted in economic high-performance industrial-scale enantiomer separations within the pharmaceutical industry (5).

The first attempts at the separation of enantiomeric species involved the use of natural materials as adsorbents. Lactose (6) was used with limited success for the separation of the enantiomers of a camphor derivative. The major breakthrough in enantioseparation techniques occurred with the use of cellulose (7) and more particularly cellulose triacetate (8). The latter material became the first chiral stationary phase (CSP) that had some generality of application and was used extensively for preparative separations of enantiomers. As described elsewhere in this publication (9), modification of this phase by replacing the acetate ester groups with other derivatives and the support of these chiral polymers on silica has resulted in a family of highly selective and successful CSPs for enantiomer separations by chromatography.

Since the release of the FDA guidelines (10) for the development of chiral drug molecules and the resulting need for the production of enantiomerically pure products, the use of chiral chromatography in the pharmaceutical industry has increased dramatically, both for analytical and preparative purposes. It is increasingly recognized that enantioselective chromatography is the fastest and most convenient way to resolve racemic mixtures, especially in the early stages of discovery when it is necessary to test both enantiomers for activity and toxicity. As the products move through chemical development, chemists and engineers tend to focus on finding alternative routes to the enantiomerically pure products. This has been helped by the downsizing of the simulated moving bed (SMB) systems originally developed for the petroleum and food industries, together with modifications that allow the use of high-performance chromatographic packing materials. Such systems are more suited to the cGMP environment of pharmaceutical production (5) and have led to a steady increase in the number of processes where chromatography is compared with the more traditional processes of crystallization, bioprocessing, and enantioselective synthesis in order to optimize the overall economics of manufacture. In several cases, enantioselective chromatography using SMB has turned out to be the method of choice (11).

In recent years, another technique, that of supercritical fluid chromatography, has been of increasing interest in the separation of gram to kilogram quantities of materials. This technique has the advantages of being fast and of using less solvent than is required for an equivalent HPLC separation. Currently, it has rarely been used at large scale, or for separations above a few kilograms in the pharmaceutical industry, although some systems have been employed in the purification of fish oils (12).

BASIC PRINCIPLES OF PREPARATIVE AND PRODUCTION CHROMATOGRAPHY

Although much chromatography can be carried out without knowledge of the underlying theory, an understanding of what is going on inside the column is important in designing separations and in troubleshooting when things do not proceed as expected. This is especially important in preparative chromatography where the behavior of solutes under mass overload conditions is not always intuitive.

Some Theory

Solutes in a chromatographic process are distributed between a moving liquid (the mobile phase) and a solid (stationary phase). Different solutes are distributed between the phases in different proportions. Because the distribution process is very much faster than the translation of the species through the column by the mobile phase, the solutes move through the column as coherent bands at a speed inversely related to the extent of the distribution into the stationary phase. This rapid distribution between the phases means that the process is, to a good approximation, at thermodynamic equilibrium. In analytical chromatography, the overall equilibrium between the phases is not disturbed by the low concentrations of the solute and the injected material travels through the column as a narrow band, broadened only by kinetic factors related to the particle size in the column and the mobile phase flow rate. The situation changes when larger concentrations of a solute are present in the column. At this point, other factors become important. Adsorption of the solute on the stationary phase results in an effective change in the surface available for adsorption. Where there is a limited surface area, for example, there is a lower concentration of adsorption sites available for adsorption of each successive molecule. In other cases, an adsorbed solute molecule can facilitate the adsorption of a second by favorable solute-solute interactions (or by unfavorable solute–mobile phase interactions when solvophobic effects are important), which lead to stacking of the solute molecules on the surface and increasing the probability of adsorption as solute concentrations on the surface increase. In both of these situations, the effective distribution between the phases is changed, which in turn changes the speed of that section of the chromatographic band through the column. These effects lead to a further broadening of the chromatographic band, but due to thermodynamics and not kinetics.

To a very good approximation, one can separate out the thermodynamic from the kinetic effects by assuming that the two processes are independent. In this case the variance of the two processes are additive.

$$\sigma_{tot}^2 = \sigma_{thermo}^2 + \sigma_{kinetic}^2 \tag{1}$$

Thus, we can study the band spreading processes due to the thermodynamic effects independently from those due to the kinetic processes, greatly simplifying the situation.

In analytical chromatography, the distribution between the two phases is taken to be a constant; the retention times do not change with sample load. This is only true for very small concentrations of the material in the column. The distribution coefficient (the Henry constant, a) is related to the retention factor and the phase ratio in the column by the equation:

$$\frac{C_s}{C_m} = a = k' \frac{V_m}{V_s} = \frac{k'}{\Phi} \tag{2}$$

where C_s and C_m are the concentrations of the solute in the stationary and mobile phases respectively, a is the Henry constant, k' is the retention factor, V_m and V_s are the mobile and stationary phase volumes respectively and Φ is the phase ratio.

The situation changes as the concentration of the solute in the column increases. In the simplest case, the adsorption of a solute molecule reduces the available concentration of surface sites for the adsorption of the next, and the stationary phase concentration is given by the Langmuir isotherm

$$C_s = \frac{aC_m}{1 + bC_m} \tag{3}$$

The ratio a/b gives the saturation capacity of the packing material, which is the stationary phase concentration corresponding to a monolayer of the solute on the surface of the packing material. This is an important parameter in preparative chromatography as the ratio of sample concentration to the saturation capacity determines the chromatographic properties of the solute at any point in the column. From Eq. 2, it is seen that the ratio of the stationary phase concentration to that in the mobile phase is directly related to the retention factor. Therefore, as the solute concentration in the injected band increases and the ratio of the stationary phase to mobile phase concentrations decreases according to Eq. 3, the retention also decreases. This results in a distortion of the chromatographic band in the column since the high-concentration parts move more quickly than the lower concentration zones. Figure 7.1A shows the in-column band shape of a solute, which follows Eq. 3 together with the corresponding eluting peak shape in Figure 7.1B. In such cases, increasing sample load results in changes to the peak elution only at the band front while the remainder of the peak follows the same profile as for the smaller samples. This results in a series of "nested" peaks as sample quantity increases, as shown in Figure 7.2. This figure is characteristic of the loading experiments that are conventionally carried out to determine the maximum load on the column, which permits separation of the components to the purity and recovery desired.

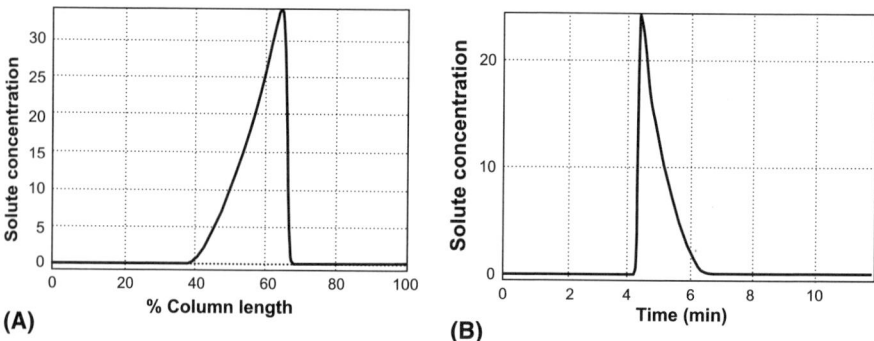

FIGURE 7.1 (**A**) In-column band shape of a solute, which follows Eq. 3 and (**B**) the corresponding eluting peak shape.

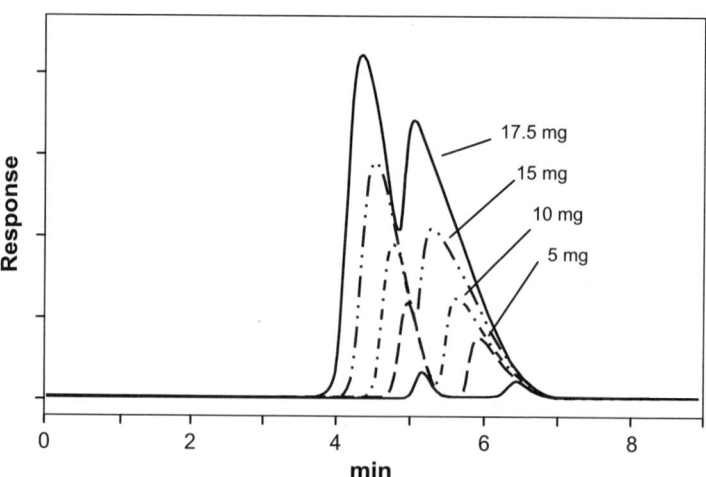

FIGURE 7.2 Loading study showing a series of "nested" peaks as sample quantity increases.

Of course, not all solutes follow the conveniently simple adsorption isotherm, but a sufficient number approximate to it, which highlight its usefulness as a model.

A relation for the band spreading due to the thermodynamic effects can be derived by assuming an infinite column efficiency (the "Ideal Model" of chromatography) (13). This equation, due to the group of Guiochon, may be used to estimate the retention factor for the peak maximum from the saturation capacity and the retention factor at low load.

$$k' = k'_0 \left(1 - L_f^{1/2}\right)^2 \qquad (4)$$

Here L_f is the loading factor, the ratio between the sample load and the column saturation capacity. This equation also allows the calculation of the column saturation capacity from an analytical injection (to determine k'_0) and an overloaded injection at a given load (to measure k'). Once this is obtained, it is possible to calculate the retention factor for the peak maximum at any other load, simplifying the decision of how much sample to load on the column. It should be noted that due to the assumption of infinite efficiency, the equation gives best estimates for high-efficiency columns and high sample loads where the kinetic band spreading is small relative to the thermodynamic effects and can safely be ignored. At low sample loads and with lower-efficiency columns, the equation gives values of saturation capacity that decrease with increased sample load; these trend toward a constant value at high load.

Non-Langmuir Isotherms

The only unusual peak shape that need concern us in enantiomer separations is that arising from what is known as an "S-shaped" isotherm. In this case, instead of the high concentration zone of the overloaded peak eluting more quickly than the lower concentration zones, it elutes more slowly. This has the effect of distorting the peak in the opposite sense to that seen in Figure 7.2. Such behavior often occurs when the adsorption of a solute molecule is enhanced rather than diminished by the prior adsorption of other molecules. This occurs up to a point; eventually the surface of the packing material becomes saturated with the solute and the retention of higher concentrations is reduced. Figure 7.3 shows a set of "nested" chromatograms arising from the injection of increasing quantities of α-methyl-α-phenylsuccinimide on a column packed with CHIRALPAK IA using Methyl t-butyl ether (MTBE) as mobile phase. The second eluting component is eluted under the influence of a predominantly "S-shaped" (quadratic) isotherm.

Multiple Solutes

The above picture is, of course, a simplified one. It does not take interactions between the solutes into account. As noted above, the high concentration of a solute in the column not only affects the distribution of that solute between the phases but can also affect the distribution of other solutes that overlap with the band as it travels through the column. The effect on other solutes is similar to the effects described above: species present at the high concentration zones are less retained than those in low concentration zones. This has the effect of causing earlier retention of bands eluting close to the high concentration band. Where the other component elutes more quickly than the main component, it is

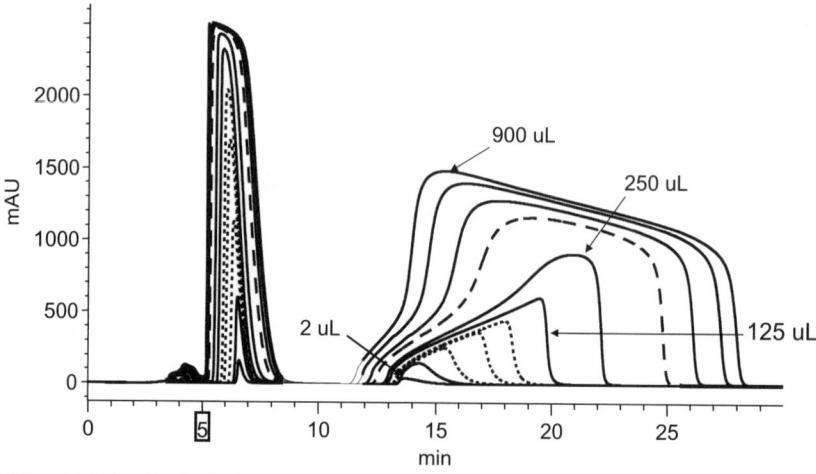

FIGURE 7.3 "Nested" chromatograms arising from the injection of increasing quantities of α-methyl-α-phenylsuccinimide on a column (250 × 4.6 mm) packed with CHIRALPAK IA using MTBE as mobile phase (1 mL/min). Solute concentration = 30 g/L. *Source*: Courtesy of R. W. Stringham and B. Lord, Chiral Technologies, Inc.

displaced from under the peak envelope, while in the case where it elutes more slowly, it can be pulled into the tail of the main peak. Of course, mathematical models can be used to describe this behavior. Where the isotherm describes the influence of other solutes in the system, it is known as a competitive isotherm. The simplest is based on the Langmuir isotherm.

$$C_s^i = \frac{a^i C_m^i}{1 + \Sigma_0^j b^j C_m^j} \tag{5}$$

The consequences of this relation are shown in Figure 7.4. A minor peak eluting at the front of the major one is displaced (Fig. 7.4A), while the same peak eluting afterward is pulled into the main envelope (the "tag-along" effect, Fig. 7.4B). Overlapping bands that are present at the same concentration as in Figure 7.4C show both effects; the earlier eluting band is displaced by the later one, while at the same time causing earlier elution of the later one, causing significant distortion of both peaks.

Column Efficiency

At first sight, given the broad peaks that are caused by mass overload in a column, there seems little point in using columns with high efficiency. However, the displacements seen as a result of the interaction of the solutes in the column are influenced by its efficiency. Thus, where strong displacement effects are seen, it is important to use high-performance columns. Unfortunately, the situation for most enantiomer separations is such that the displacement effects are rather small. This is in part due to the separation mechanism where the second eluting peak uses more of the CSP surface than the first eluting

PREPARATIVE ENANTIOSELECTIVE CHROMATOGRAPHY

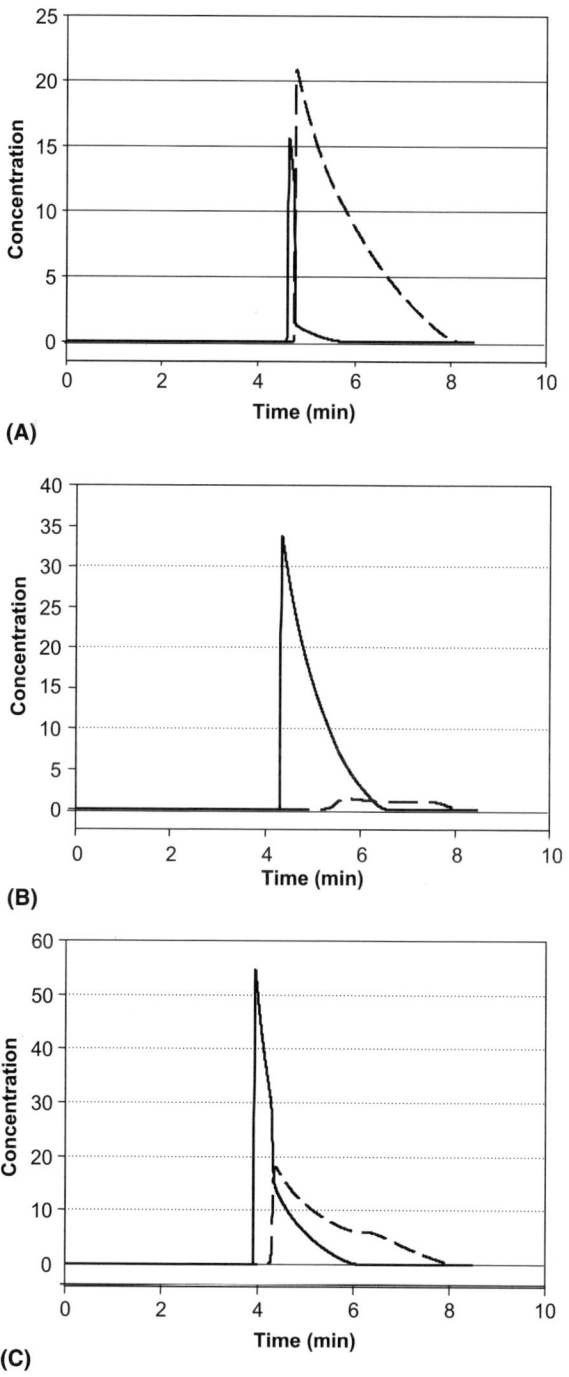

FIGURE 7.4 (**A**) Displacement of a minor peak eluting at the front of a major band. (**B**) Tag-along effect where the minor peak eluting afterward is pulled into the main envelope. (**C**) Overlapping bands that are present at the same concentration show both effects.

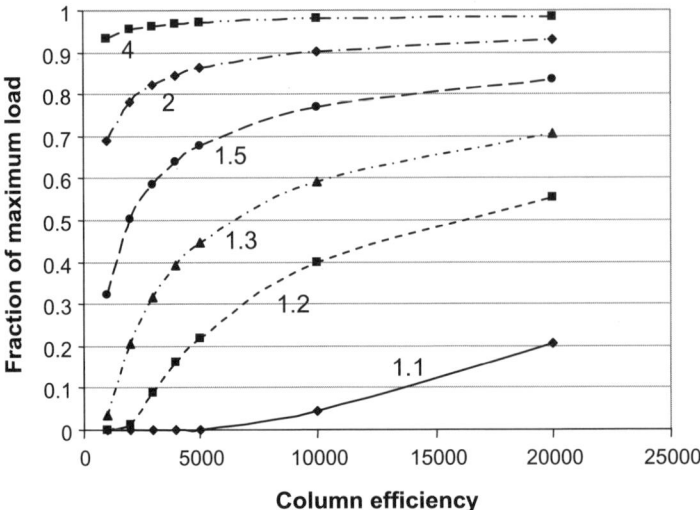

FIGURE 7.5 The influence of efficiency on the fraction of the maximum theoretical load as a function of the selectivity of the separation. Solid line: selectivity = 1.1; short dashed line: selectivity 1.2; short dashed dotted line: selectivity 1.3; long dashed line: selectivity 1.5; long dashed dotted line: selectivity 2.0; dashed double dotted line: selectivity 4.0.

enantiomer because of the chiral interactions. This results in different saturation capacities for the two components, moving the interactions between the solutes away from the classical Langmuir behavior. The end result is that most chiral separations are loaded to the point at which the bands just touch or overlap slightly. Under these circumstances, the column efficiency is less important, as long as the band spreading due to the column is small compared with that due to the mass overload. This is true for large values of selectivity. Figure 7.5 shows the results of calculations of the fraction of the maximum theoretical load (at infinite efficiency) that is possible for columns of different efficiency as a function of the selectivity of the separation. Several conclusions can be drawn from the figure. Where selectivity is low, the maximum load is small (18 mg in this example, assuming a column 250 × 50 mm and selectivity of 1.1) and even with 20,000 plates, the possible load is only 20% of the theoretical (3.6 mg). Spending a little time in increasing the selectivity can pay great dividends here: increasing selectivity to 1.2 allows an injection of 55% of the theoretical maximum load, in this case 34 mg can be loaded, a close to tenfold increase. At the other end of the selectivity scale, a 2000 plate column will allow at least 80% of the theoretical maximum load (700 mg and up) for selectivity of 2 and higher. Thus, a high-efficiency preparative column can perform all separations, whereas a lower-efficiency column may not have enough plates to allow resolution of the more difficult problems. Where only small samples are to be separated, use of the high-efficiency column can save time in that minimal selectivity optimization is required. For larger quantities, the time spent in optimizing the selectivity can be very well spent in terms of the run time and solvent use. At large scale, where pressure restrictions of equipment and columns become important, use of columns that have more efficiency than required for the separation can limit the production rate; ideally, column length and particle size should be chosen to maximize the production rate at the maximum operating pressure of the chromatographic system.

PREPARATIVE ENANTIOSELECTIVE CHROMATOGRAPHY

As noted earlier, the use of preparative chromatography for the isolation of individual enantiomers became practical with the adoption of cellulose triacetate. This was rapidly supplanted by other chiral phases, notably those developed by the group of Pirkle (14) and of Okamoto (15). These silica-based phases allowed much higher efficiency separations as well as bringing higher and alternative selectivity relative to the cellulose triacetate, which outweighed any reduction in capacity.

Method Development
Screening

A major difficulty in enantioselective chromatography is that there are no theories that adequately explain the selectivity of the various CSPs for specific solutes. This means that the only practical way to find a suitable phase is to screen all of the phases available for each separation. Given the number of commercial CSPs available, this is impractical and most workers in the field select a few phases that have been found to be most useful. Guidelines to this selection can be found in the literature (16–18); frequently, the phases chosen are selected from the polysaccharide-based CSPs together with one or two Pirkle-type phases and the macrocyclic antibiotic-based phases. There are two options for screening. Conventionally, a multipath switching valve is installed on an analytical liquid chromatograph and the columns are switched sequentially (see Fig. 7.6). Additionally, multichannel systems allowing eight

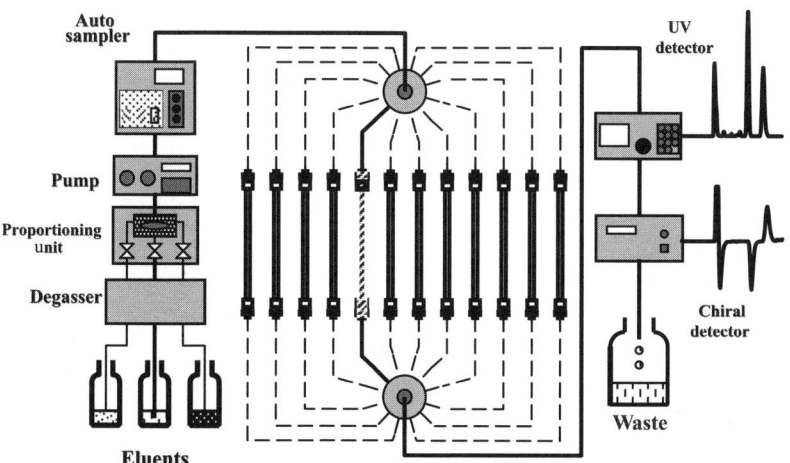

FIGURE 7.6 Schematic of a convention multicolumn screening system.

TABLE 7.1 Suggested Solvent Mixtures for Enantioselective Screening Experiments

Solvent-stable CSPs (e.g., immobilized polysaccharide-based media)
1. Hexane/2-propanol (80:20)
2. Hexane/ethanol (80:20)
3. Hexane/dichloromethane/alcohol (49:49:2)
4. MTBE/alcohol (98:2)

Polysaccharide-based CSPs
1. Hexane/ethanol (80:20)
2. Hexane/isopropanol (80:20)
3. Methanol/ethanol (50:50)
4. Acetonitrile

Abbreviation: CSP, chiral stationary phase.

channels in parallel are available. These allow testing of eight separation conditions at one time. The choice is one of convenience and economics, which is based on the number of samples to be run and the laboratory space available.

For analytical and small-scale preparative separations, the conditions used for the screens are generally chosen from a limited set, chosen as a function of the stationary phases available for the separation. For HPLC separations, the solvent choice is generally between ethanol and 2-propanol in mixtures with hexane or heptane and with pure methanol and acetonitrile as additional options. In SFC separations, the choice is often limited to one of methanol or 2-propanol as the mobile phase modifier. Where the stationary phases employed are solvent-resistant, other options are available. Solvent choices for initial screening are shown in Table 7.1. It should be noted that these recommendations will not cover all situations but will find a separation adequate for analysis or for small-scale separations between the components roughly 80% to 85% of the time. More specific aspects of solvent selection are discussed in the section that follows.

A typical set of chromatograms arising from a screen is shown in Figure 7.7. The separation most appropriate for the application (whether it be analytical or preparative) can often be selected directly from the results of the screen; in the event that the best separation is insufficient, the screening results are used as a basis for further optimization.

Selection of Solvents

The choice of solvents used for a screen depends upon the phases employed for the separation. Although the conventional polysaccharide-based phases generally give high success rates, they can be limited by the solvents available; as noted in chapter 6 they are soluble in many of the organic solvents of intermediate polarity such as dichloromethane and ethyl acetate and so the phases are not stable under such conditions. The recently introduced immobilized polysaccharide-based CSPs, the Pirkle phases and the macrocyclic antibiotic, are not subject to such limitations.

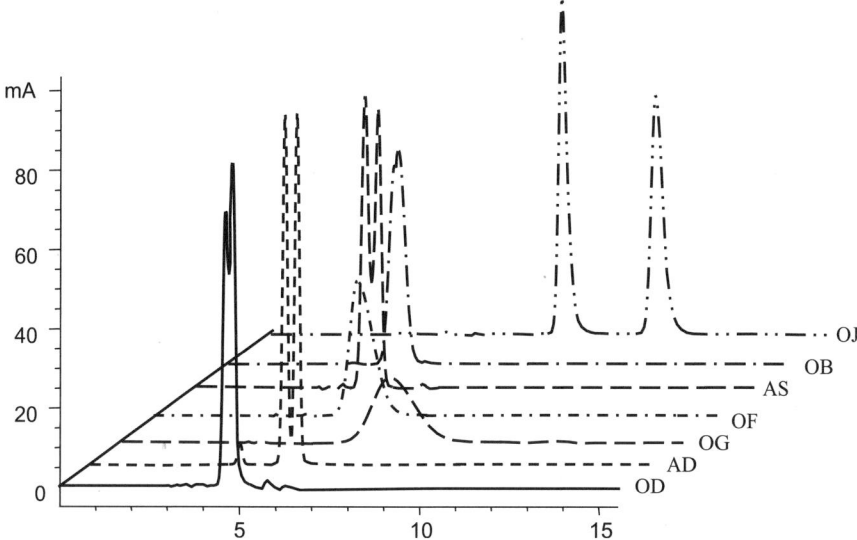

FIGURE 7.7 Typical screening results. Solute: Benzoin ethyl ether, mobile phase: 10% 2-propanol in hexane. Columns: solid line: CHIRALCEL OD; short dashed line: CHIRALPAK AD; long dashed line: CHIRALCEL OG; short dashed dotted line: CHIRALCEL OF; long dashed line: CHIRLAPAK AS; long dashed dotted line: CHIRALCEL OB; dashed double dotted line: CHIRALCEL OJ.

For preparative purposes, the optimal mobile phase is a compromise between several factors, including selectivity, solubility, viscosity, and isotherm type. If an optimal separation is required (e.g., for large-scale separations), it is worth evaluating several mobile phases to find which gives the best production rate (19). Again, the solvent stability of the immobilized or bonded phases allows a wider range of solvents when optimizing the solubility. The solvents used in preparative chromatography influence many parameters, including the selectivity, the column efficiency, the loading capacity for the critical solute(s), the production rate, and the ease and economics of product recovery. The choice of the best mobile phase for a separation is dictated by the physical and chemical properties of the solvents.

Viscosity. In laboratory-scale separations, the operating pressure is usually not restrictive and the usual limitation to flow velocity through the column is that of the capacity of the pump. At larger scale, however, the particle strength becomes the limiting factor since the columns become wide enough in diameter to eliminate the wall support for the particles in the column. This occurs for columns 10 cm or more in diameter. Without the benefits of the wall support, irregular particles start to fracture at around 40 bar pressure, whereas spherical silica particles generally start to fracture at around 70 bar (20) with even the strongest of them starting to create "fines" at pressures a little over 100 bar. This results in the requirement to limit the pressure drop in the system to that

TABLE 7.2 Viscosity (in cP at 20°C) of Common Solvents Used in Enantioselective Separations

	20°C[a]	25°C[b]
Acetonitrile	0.38	0.369
Dichloromethane	0.44	0.413
Ethanol	1.08	
Ethyl acetate	0.45	0.423
Heptane	0.42	
Hexane	0.31	0.300
Methanol	0.59	0.544
Methyl t-butyl ether	0.27	
Propan-2-ol	2.4	2.04
Tetrahydrofuran	0.55	0.456
Water	1.0	

[a] http://macro.lsu.edu/HowTo/solvents/viscosity.htm
[b] Sigma-Aldrich

permitted for the packing material. For this reason, large-scale chromatography is often carried out with particle sizes larger than used in analytical chromatography. There are, however, upper limits to the particle size chosen as increasing the particle size decreases the column efficiency and, as noted earlier, there is often a lower limit to the efficiency required for a separation. Once a particle size is chosen, the reduction in pressure drop in the system is usually accomplished by choosing low viscosity solvents where possible. Table 7.2 shows the viscosity of common solvents used in enantioselective separations.

Selectivity Often the choice of solvent is dictated by selectivity; if only one solvent gives any selectivity at all, this has to be used. There are, however, several options for solvent choice that may not be used in the initial screening and these become important when the selectivity achieved is insufficient for the purposes of the separation or where it is important to optimize the separation to maximize the production rate. For normal phase separations that typically use a mixture of hexane (or heptane) with an alcohol additive, the selectivity can be adjusted by the use of mixtures of alcohols (e.g., admixture of methanol with ethanol and hexane) or by use of other alcohols not normally used for screening. Examples of these latter alcohols are n-propanol and the butanol isomers (21). These and the more conventional alcohols are also often used together with acetonitrile in polar phase separations. The alcohol-acetonitrile systems are nonlinear and often a maximum in retention and selectivity is found between 5% and 15% alcohol in acetonitrile.

When solvent-stable stationary phases are used, there is the possibility to expand the solvent range used to the medium polarity solvents such as ethyl acetate, Tetrahydrofuran (THF), MTBE, and dichloromethane. These are usually used in admixture with hexane together with some concentration of an alcohol such as ethanol to adjust the solvent strength. Statistically, MTBE and THF or dichloromethane give the highest success rates for achieving enantioselectivity although once one starts to optimize the separations achieved, other solvents—even acetone—should be investigated, provided it

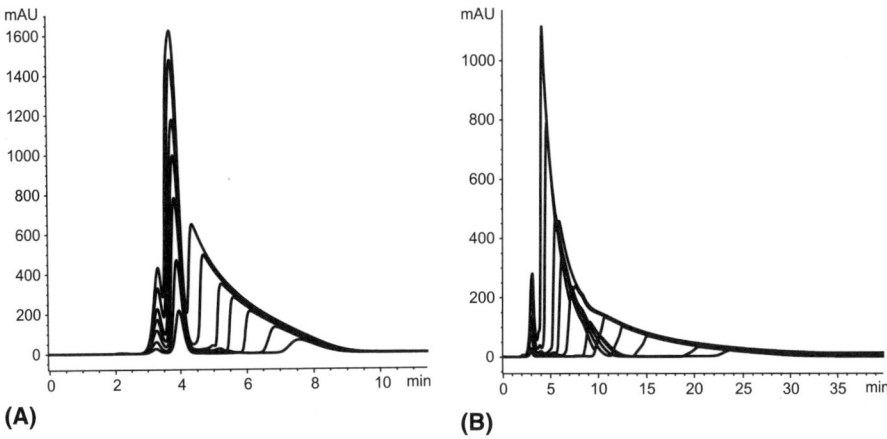

FIGURE 7.8 Loading studies of α-methyl-α-phenylsuccinimide using (**A**) ethyl acetate and (**B**) chloroform. C = 200 g/L, Vinj = 20, 50, 75, 100, 150, and 200 µL. Mobile phase: 100% chloroform, flow rate = 1 mL/min, P = 10 bar, T = 25°C as mobile phase. Other conditions as in Figure 7.3.

is possible to detect the product. With such solvents it is frequently convenient to use a different form of detection other than UV absorption, such as evaporative light scattering (and related techniques) or the often neglected refractive index detector. There is usually sufficient sample available when optimizing preparative separations for larger scale that use of a less sensitive detector is not a disadvantage. As noted elsewhere, use of such a detector can minimize unwanted signals due to minor impurities which have large UV extinction coefficients, the signals from which can obscure those due to the major components.

Effect of mobile phase on adsorption. The choice of mobile phase not only influences the retention, selectivity, and solubility of the solutes, but can also determine the shape of the adsorption isotherm with significant consequences for the production rate of the separation. A recent study has illustrated this (22). As noted earlier, Figure 7.3 shows the results of a loading study of α-methyl-α-phenylsuccinimide using MTBE as a solvent. Figure 7.8 shows the same solute separated on the same stationary phase using (*i*) ethyl acetate and (*ii*) chloroform as mobile phase. These solvents give markedly different effects; MTBE gives rise to essentially an "S"-shaped isotherm, while ethyl acetate is closer to a conventional Langmuir isotherm and chloroform gives a bi-Langmuir isotherm (one in which there are two sorption sites, one of which overloads very quickly, causing a rapid collapse of the separation during loading. Table 7.3 shows the results of calculation of production rate for both HPLC and SMB separations for this system. Related effects have been reported for SFC separations (23).

Sample solubility One way in which the solvent influences the loading capacity of the system is its influence on the solubility of the sample. The effects are easy to

TABLE 7.3 Production Rate for HPLC and SMB Separations of α-Methyl-α-Phenylsuccinimide

	MTBE	ACN85/IPA15	EtOAc	CHCl$_3$	Hex/THF
HPLC productivity (kg/kg/day)	0.92	1.09	2.66	0.18	0.49
SMB productivity (kg/kg/day)	2.16	0.98	4.97	0.69	2.96

Abbreviations: HPLC, high-performance liquid chromatography; SMB, simulated moving bed.

visualize. Where solubility is low, large injection volumes are necessary to introduce significant amounts of sample into the column. Unfortunately, large injection volumes cause "volume overload"—the band spreading due to the sample volume becomes large compared with that due to the column, giving wide peaks compared with a small injection volume. This limits the sample quantity that can be injected into the column as the band width due to the injection fills the space between the peaks. It is always better to introduce the sample in as small a volume as possible. One way to achieve this is often to dissolve the sample in a solvent rich in the more polar mobile phase component. While it is possible to succeed with this strategy, it has some pitfalls. One is that the sample is less soluble in the mobile phase and may precipitate out of solution as the injection band mixes with the mobile phase. This can, under the worst conditions, filter out on the inlet frit or can crystallize within the column itself. Both lead to high inlet pressures, sometimes blocking the column completely. This is particularly prone to occur in SFC separations where the sample is conventionally dissolved in the polar component of the mobile phase and may not be especially soluble in the supercritical mobile phase. Another pitfall is that by introducing the sample in a more polar solution, this can influence the separation adversely. The band of strong solvent usually moves through the column more quickly than the solutes, causing premature elution of part of the injected band. This effect decreases with transport through the column as the injected band mixes with the mobile phase but nonetheless can significantly distort the peaks.

Additives
An important feature of many of the CSPs used in preparative enantioselective separations is the need for additives to improve the selectivity and peak shapes for the separations of nonneutral solutes. This is because of the multiple functional groups present in chirally selective phases. For preparative purposes, it is useful to use additives that are volatile or that are otherwise easy to remove from the products. With polysaccharide phases, for example, it is necessary to use an acid additive (such as trifluoroacetic acid) when separating enantiomers of acidic compounds while it may be necessary to use a basic additive for basic compounds. Figure 7.9 shows the influence of added diethylamine on the separation of the enantiomers of chlorpheniramine. In this case, the diethylamine competes with the chlorpheniramine amino function for the active sites on the silica surface, dramatically improving the peak shapes. One possible disadvantage of the use of additives is the possibility of their catalyzing

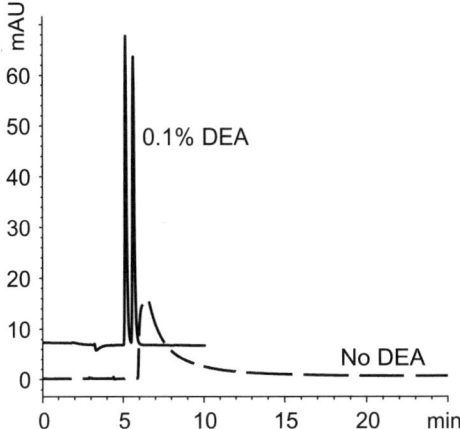

FIGURE 7.9 The influence of added diethylamine on the separation of the enantiomers of chlorpheniramine. CHIRALPAK AD-H 250 × 4.6 mm, 90/10 hexane/IPA, flow rate 1 mL/min. *Abbreviation*: DEA, diethylamine.

reactions; racemization or—in the case of acidic additives in the presence of alcoholic mobile phase modifiers—esterification of the desired products. This may be avoided by the choice of mobile phase or the use of weaker acids—acetic acid in place of trifluoroacetic acid as one example. While the use of sulfonic acid additives can give high selectivity for some basic compounds (24), these acids are more difficult to remove and usually require a solvent extraction to eliminate them from the final product. Their use in preparative chromatography remains limited.

The use of additives can give rise to memory effects on the stationary phases. These are seen mainly when a column has been used with an additive and is subsequently used without one for a different separation. In some cases the residual additive adsorbed on the CSP can influence the subsequent separation, although most of such circumstances are seen where the second separation really needs an additive to be present to control the interactions of the solute with (generally) the silica support. Techniques exist for removal of additives from the columns (25) and this aspect of their use is not a major issue.

Optimization and Scale-Up

Scale-up of preparative separations is easy, provided some simple rules are followed. Probably the most important is that the separation should be developed and optimized on the same packing material as will be used in the preparative separation. This is important, not only because there are often some differences in selectivity between otherwise nominally identical phases with different particle sizes but also because use of larger particles can decrease the resolution between components due to their lower efficiency.

In this context, it is worth considering for a moment the general resolution equation (Eq. 6).

$$R_s = \frac{1}{4}\sqrt{N}\left(\frac{k'+1}{k'}\right)\left(\frac{\alpha}{\alpha-1}\right) \qquad (6)$$

This relates the resolution between a peak pair to the column efficiency, the retention time, and the selectivity observed in the separation. Figure 7.10 shows the relation between resolution and (*i*) retention, (*ii*) column efficiency, and (*iii*) selectivity. Where retention factors are small, the resolution may be increased dramatically by relatively small increases in retention (but this is not true for SMB separations, see below). Once the retention factor is greater than around 5, there is little to be gained in increasing retention. The resolution is proportional to the square root of the column efficiency; thus, doubling the particle size will require a column approximately four times longer in order to maintain the resolution between the peaks. In contrast, increasing selectivity will increase the resolution between the peaks dramatically and is the preferred route to higher production rates; it has been estimated (26) that for low selectivity the production rate for a separation is roughly proportional to $(\alpha-1)^3$; at higher values of selectivity the power is somewhat decreased. The production rate for a preparative separation is not entirely controlled by selectivity, however. Although the load for a separation is not especially dependent on retention, the production rate (in terms of g/min) is dependent both on the load and the cycle time (the time lapse between injections). Thus, there is a trade-off between long retention to improve resolution and short retention to decrease cycle time.

Increasing the scale of the separation makes use of the fact that the chromatographic process scales linearly with column cross-sectional area. Thus, a separation with a maximum load of (say) 4 mg in an analytical column (4.6 mm id) will scale to a load of 19 mg on a column 1 cm id (= $4 \times 10^2/4.6^2$) or 470 mg on a column 5 cm id = ($4 \times 50^2/4.6^2$), provided the column length, particle size, and the linear flow velocity remain constant.

Overlapping Injections
Injection of the sample, waiting for it to elute, collecting the fractions, and then making the next injection is not an efficient use of time. It would be better if the injections could be arranged such that after the second peak from one injection emerges from the column, the first peak from the subsequent injection starts to elute. This can be simply managed. Consider an overloaded separation as shown in Figure 7.11. The start of the first peak emerges after 5 minutes and the tail of the second one elutes at 8 minutes. Thus, the total bandwidth of the peaks on elution is 3 minutes. If we want the leading edge of the first peak from one injection to elute immediately after the tail of the second peak from the previous injection, all that is necessary is to make one injection 3 minutes after the first. Thus, one makes the first injection and 3 minutes later—although nothing has yet eluted from the column—the second injection is made. After another 3 minutes the next is made, and so on. The resulting peak train emerging from the column is shown in Figure 7.12A. Figure 7.12B shows what is going on in the column. Although the elution chromatogram seems to

PREPARATIVE ENANTIOSELECTIVE CHROMATOGRAPHY

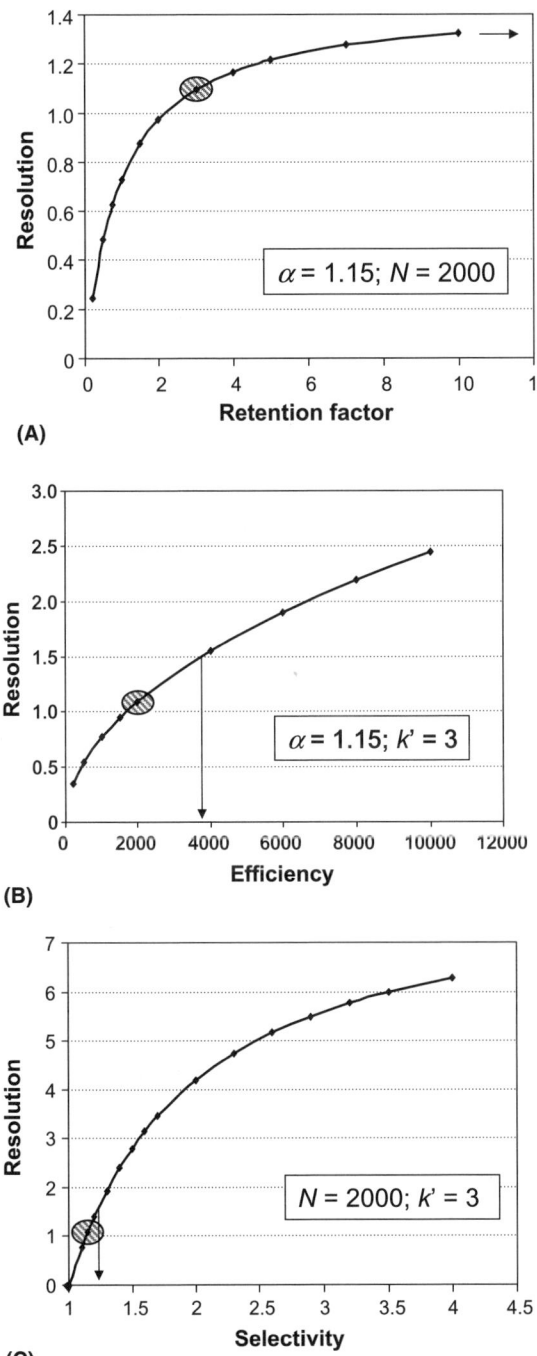

FIGURE 7.10 The figure shows the relation between resolution and (**A**) retention, (**B**) column efficiency, and (**C**) selectivity. Highlighted point: selectivity 1.15, efficiency 2000 plates, retention factor 3. Arrows correspond to resolution = 1.5 (baseline).

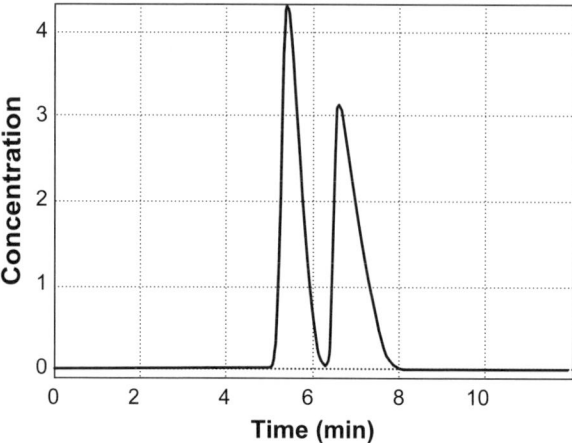

FIGURE 7.11 Generic overloaded separation.

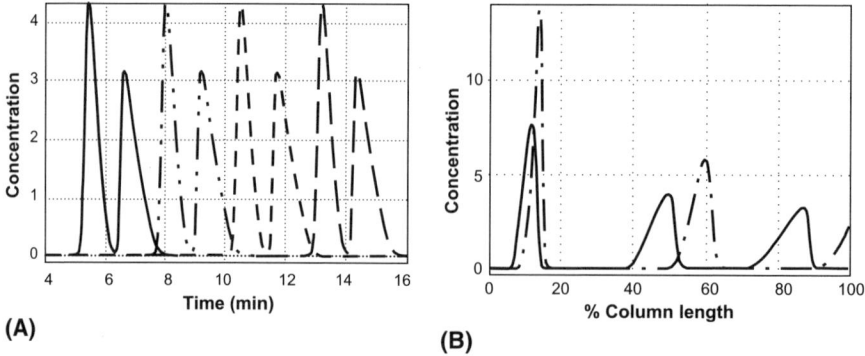

FIGURE 7.12 (A) Detector trace from overlapping injections for the separation of Figure 7.11. Solid line: first injection; dashed double dotted line: second injection made at $t = 3$ min; short dashed line: third injection made at $t = 6$ min; long dashed line: fourth injection made at $t = 9$ min. (B) Internal column concentration profiles during elution for part A. Dashed dotted line: peak 1; solid line: peak 2.

suggest that the column bed is being used efficiently, in fact there are significant spaces between successive injections, which are necessary to allow for the band spreading and to prevent the faster-eluting component from catching the tail of the slower component from the previous injection. Timing the injections is critical to preserve the peak purity and to minimize the "dead" time when nothing is eluting from the column. Figure 7.13 shows this in practice, for the separation of the enantiomers of benzyl-3-methyl-4-piperidone on a 500 × 50 mm CHIRALPAK AD column with a production rate of 10 g/hr. Conventional HPLC separations of this type have been used routinely for the

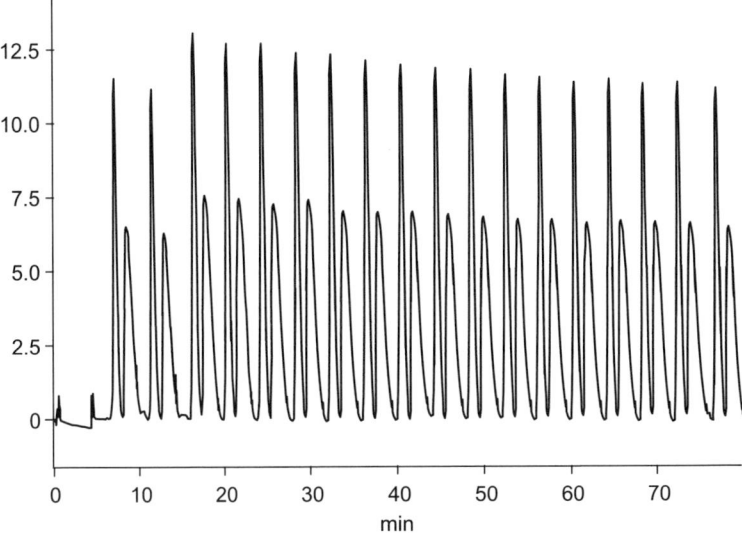

FIGURE 7.13 Separation of the enantiomers of benzyl-3-methyl-4-piperidone. Production rate: 10.1 g/hr. Column: 20 μm CHIRALPAK AD column (500 × 50 mm); mobile phase: acetonitrile; flow rate: 150 mL/min; temperature: 25°C; injected quantity: 0.675 g.

isolation of hundreds of grams to multikilogram quantities of products for toxicology and early phase 1 supplies.

Shave Recycle

In some cases, the selectivity of the separation in combination with the column efficiency is not sufficiently large to allow complete separation of the components. One solution is to use a longer column that has high enough efficiency to allow baseline resolution. Because a longer column requires more packing material and increases the operating pressure and cycle time, an alternative solution, that of recycling the peaks through the same column, was developed. The simplest implementation of this is to connect the column outlet to the inlet of the pump, sending the entire sample around the system multiple times. To allow this to happen effectively, the band spreading due to the passage of the peaks through the pumping system must be small relative to that due to the column.

A refinement of this idea is that of shave recycle. In this modification, fractions are collected from the front and rear of the solute bands each time the peaks emerge from the column. In this way, a small amount of pure product is collected with each cycle and the total load on the column is diminished. This reduction in sample load assists in the separation, allowing more of each product to be collected on subsequent cycles until all material is resolved. Figure 7.14 shows the principle of this process while Figure 7.15 shows such a recycle experiment (27). The collection points have to be set up carefully to collect the maximum quantity of product at the desired purity.

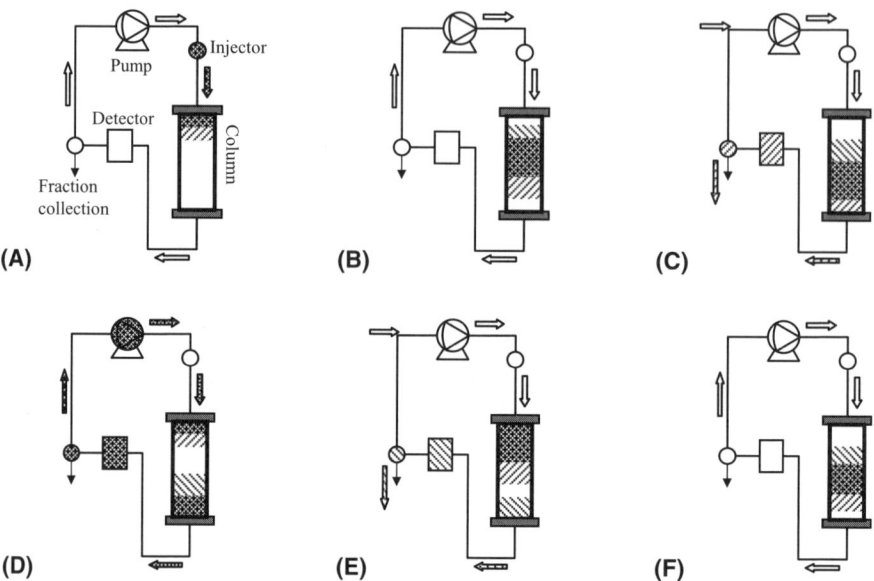

FIGURE 7.14 Principle of recycle chromatography. (**A**) Sample is injected. (**B**) Partial separation in the column. (**C**) Collection of the peak front (pure #1). (**D**) Mixed peak is recycled to the column top. (**E**) Collection of the peak tail (pure #2). (**F**) About to collect the peak front again.

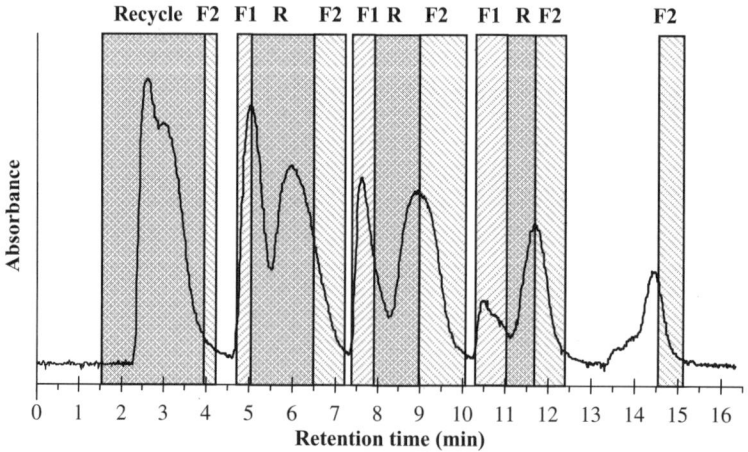

FIGURE 7.15 Shave-recycle separation of a pharmaceutical intermediate. Sample: 1.2 g on a CHIRALPAK AS-V column (200 × 50 mm) using acetonitrile as mobile phase at a flow rate of 500 mL/min. *Source*: Adapted from Ref. 27.

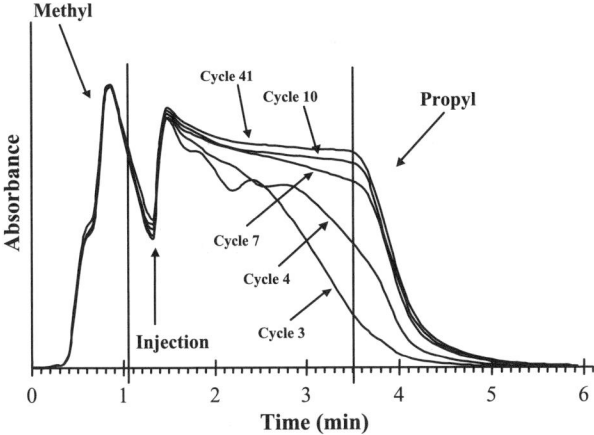

FIGURE 7.16 Development of an SSR separation. Separation of methyl and propyl p-hydroxybenzoates. *Source*: From Ref. 28.

Steady State Recycle

An improvement over shave recycling was developed by the realization that as the products were collected in a cycle the effective load in the column is decreased. If additional product is injected into the unresolved peak as it is recycled, the load per cycle can be kept constant, thus making the process semicontinuous and steady state (29). This modifies the process shown in Figure 7.14 by the addition of sample during the passage of the unseparated material through the pump by switching a loop filled with the feed into the system. The process variables include the collection points and the position at which the feed is added during the passage of the unseparated material through the pump and injection system. Figure 7.16 shows the development of an Steady-State Recycling (SSR) separation. Interested readers are referred to Ref. 28. This technique is generally employed for difficult separations of a few tens of grams to the 1 kg level. Separations of such mixtures at larger scale are more often carried out using SMB systems.

SIMULATED MOVING BED CHROMATOGRAPHY

The chromatographic processes as described above are not the most efficient use of the column. For easier separations, even where overlapping injections are employed, significant parts of the bed are not fully used. This can be seen in Figure 7.12B, which illustrates the in-column concentration profile of a train of injections. Steady state recycling has been shown to be more productive than use of overlapping injections but is still not as productive as one might wish. The solution to this is the use of countercurrent chromatography, as exemplified by SMB chromatography, which is its practical implementation. This technique is conventionally applied to the separation of multikilogram quantities up to production scale at the 100 MTA production level.

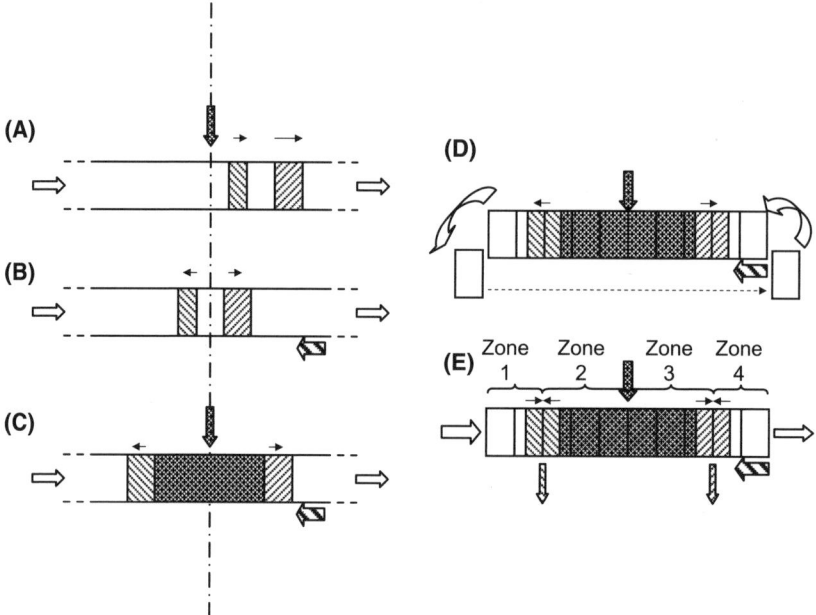

FIGURE 7.17 Principles of SMB. (**A**) Samples moving in a column with mobile phase flowing from left to right. (**B**) The column is moved at a speed intermediate between those of the solutes within the column but in the opposite direction. (**C**) Feed can be introduced to the center of the column continuously. (**D**) The introduction of the feed and the need for an infinitely long column are solved by cutting the column into sections and moving them in discrete steps rather than continuously. (**E**) Products are removed from ports between columns by increasing the flow in zone 1 to flush out the slower-moving product and decreasing the flow in zone 4 to retain the faster-moving product.

Basics of SMB

Consider a column (Fig. 7.17A) with mobile phase flowing from left to right. In a conventional chromatograph, a two-component sample introduced in the center of the column will separate into two bands, migrating at different velocities. We now suppose that the column may be moved in the opposite direction from that of the mobile phase. As it increases speed, the bands appear to slow down relative to an external observer (we must assume—for the moment—an infinitely long and transparent column) until when the column is moving in the opposite direction but at the same speed as the slowest band is moving within it; the slow band appears to our external observer to be stationary. If the column speed is increased further, the slow-moving band now appears to be moving with the column, in the direction opposite to the faster-eluting component (Fig.7.17B). In this situation, with the two bands moving in opposite directions, it is apparent that one may add feed to the center of the system continuously (Fig. 7.17C). This is the basis of a continuous separation system. There are, of course, a few flaws. The column is currently infinitely long and there is no means to introduce the feed or to remove the products. Introducing the feed may be accomplished by dividing the column into segments and instead of moving the column continuously, we can move it in discrete time segments such that the average column velocity remains the same as before. If the column segments are short enough, the two systems are essentially equivalent. We can now introduce the feed between two of the column segments through some form of valve arrangement. As the column moves, the feed position is also moved so it remains effectively stationary relative to the external observer. This division of the column removes the need for an infinite length as when a column segment reaches the end of the system it can be moved back to the starting point at the opposite end of the column train (Fig. 7.17D). The only problem now is to remove the products from the column set. The faster-moving component is moving with the mobile phase from left to right in the system. If we open a port downstream of the feed point and bleed out some of the mobile phase flow, then the flow rate downstream of this point will be reduced. At a certain point, the flow in the downstream section will be too slow to maintain the left to right movement of the solute. Thus, in section 3 (Fig. 7.17E), the solute moves from left to right, while in section 4 it now moves from right to left. It can only exit the system through the open port with that part of the mobile phase which is being removed. The mobile phase at the end of the column set is clean, as the solute is traveling away from that point. This means that the mobile phase may be recycled back to the inlet of the column set. The slower-moving component is traveling from right to left in the system. If the flow rate in section 1 is increased, there comes a point where it is carried away from the inlet of the column set, traveling from left to right. Again, we open a port to the column set between the inlet and the feed position, bleeding out the increase in flow rate, reverting the flow rate in section 2 to that originally in the system. The slower-eluting component is clearly carried toward the outlet between sections 1 and 2. This cleans the columns reaching the inlet to the system, allowing them to be switched to the outlet end (Fig. 7.17E). We now have a continuous chromatography system, which approximates closely to a true countercurrent system. The rest is engineering. In some smaller systems, the columns are moved relative to a stationary multiport valve. As it is impractical to move large-scale columns, the columns are usually held stationary and their movement is simulated by sequential operation of the valves at the various inlet and outlet positions. This is the origin of the term simulated moving bed.

The separation process can be envisaged by plotting the concentration of each of the components through the column set as indicated in Figure 7.18. Because the columns have a finite efficiency, there is a band of overlap at the feed inlet, with the faster-moving component moving clear of the overlap zone downstream of the feed point and the slower one moving clear of the overlap zone upstream. The take-off points for the products (conventionally the "Extract" port for the slow-moving component and the "Raffinate" port for the fast-moving component) are situated at the point at which the products are pure enough for the purposes of the process. The end zones are maintained as small as possible to reduce the unused part of the chromatographic bed. It is normally found that to obtain high purity and high recovery—essential for the high-cost products found in the pharmaceutical industry—the SMB system needs at least six columns. The minimum number of columns for an SMB unit is conventionally four, although with such a configuration it is not possible to reach very high product purity. As will be noted later, modifications to the

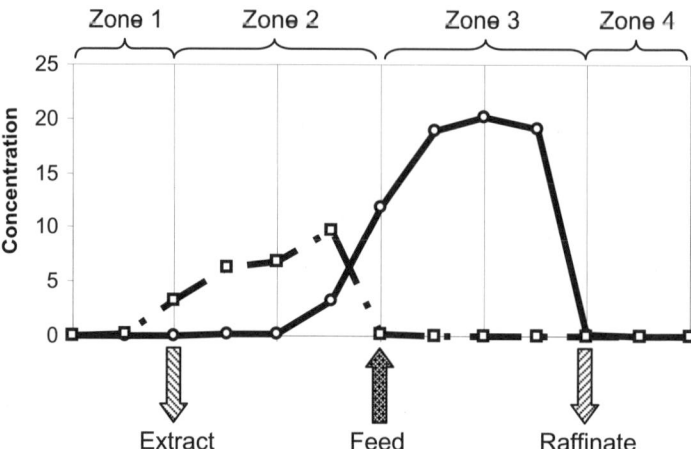

FIGURE 7.18 Internal concentration profile for the enantiomers of α-methyl-α-phenylsuccinimide. Column set: 6 × 100 × 4.6 mm CHIRALPAK IA (20μm) columns with ethyl acetate as mobile phase. Feed: 0.51 mL/min (at 200 g/L); zone 1 flow = 11.66 mL/min; extract flow = 7.09 mL/min; raffinate flow = 0.59 mL/min; switch time = 1.42 min.

conventional SMB system allow the use of five columns while attaining the high purity needed for enantiomer separations.

One of the principle advantages of SMB is that since much of the solvent is recycled and only a proportion is removed from the system to collect the two products, it uses much less solvent than single-column HPLC. Additionally, the collected fractions are isolated in much higher concentration than can be achieved by HPLC processes. At the large scale, solvent recovery of over 99% has been routinely achieved in a production process (30).

As described, it is clear that the SMB system is a binary separator. There have been several adaptations to allow isolation of more than the two components although these have not so far been adopted commercially. At smaller scale, batch chromatography is frequently more effective for these more complex separations.

SMB Method Development

In the case of low sample concentrations and linear isotherms, it is relatively easy to determine the operating parameters of the SMB system. An SMB system is controlled through four flow rates and a switch time. The flow rates are those in section 1, the extract and raffinate flows and the feed flow. In a standard SMB unit, the switch time is the time lapse between switching the valve positions from one column to the next. There are some constraints to the system, which define the relative flow rates and switch time. In section 1, both solutes must move toward the extract port. In sections 2 and 3, the slower-moving component moves toward the extract port while the faster-moving one moves toward the raffinate port. It should be noted that the flow in section 3 is higher (by the value of the feed flow rate) than in section 2. In section 4, both components move toward the raffinate port.

The velocity of a component i in section X ($u_{i,X}$) of the SMB is given by

$$u_{i,X} = \frac{u_{m,X}}{1 + k'_{i,X}} \tag{7}$$

where $u_{m,X}$ is the mobile phase velocity in that section and $k'_{i,X}$ is the retention factor for that solute.

The switch time and column length can be combined to give a stationary phase velocity (u_{col}):

$$u_{col} = \frac{L}{t_{switch}} \tag{8}$$

Thus, for a solute to move with the mobile phase toward the raffinate port, its mobile phase velocity should exceed that of the stationary phase. Equally, a solute that moves with the stationary phase must have a mobile phase velocity less than that of the stationary phase. Thus, one can write equations for each component in each section of the SMB.

In practice, there are some velocities that are critical. Section 1 flow rate is controlled by the slower-moving (extract) product. The velocity of this product has to be greater than the stationary phase velocity in the opposite direction. This ensures that the products are eluted from the columns in this section prior to their being moved into section 4. In section 2, the velocity of the faster-moving component has to be greater than that of the stationary phase. In section 3, the velocity of the slower-moving component must be less than that of the stationary phase. Finally, in section 4, the velocity of the faster-moving component has to be less than the stationary phase velocity. The critical inequalities can be written:

Section 1:

$$u_{1,1} = \frac{u_{m,1}}{1 + k'_{1,1}} > u_{col}$$
$$u_{2,1} = \frac{u_{m,1}}{1 + k'_{2,1}} > u_{col} \tag{9}$$

Section 2:

$$u_{1,2} = \frac{u_{m,2}}{1 + k'_{1,2}} > u_{col}$$
$$u_{2,2} = \frac{u_{m,2}}{1 + k'_{2,2}} < u_{col} \tag{10}$$

Section 3:

$$u_{1,3} = \frac{u_{m,3}}{1 + k'_{1,3}} > u_{col}$$
$$u_{2,3} = \frac{u_{m,3}}{1 + k'_{2,3}} < u_{col} \tag{11}$$

Section 4:

$$u_{1,4} = \frac{u_{m,4}}{1 + k'_{1,4}} < u_{col}$$
$$u_{2,4} = \frac{u_{m,4}}{1 + k'_{2,4}} < u_{col} \tag{12}$$

Finally, it has to be recognized that the internal flow rates $u_{m,2\ldots 4}$ are related to the section 1 flow rate ($u_{m,1}$) and the external feed, extract, and raffinate flows:

$$u_{m,2} = u_{m,1} - u_{extract}$$
$$u_{m,3} = u_{m,1} - u_{extract} + u_{feed} \qquad (13)$$
$$u_{m,4} = u_{m,1} - u_{extract} + u_{feed} - u_{raffinate}$$

The inequalities described by the Eqs. 9 to 12 may be solved for the operating parameters (relative to the stationary phase velocity) by inserting a parameter B, which describes the deviation from equality and by substituting the flow rate values from Eq. 13. If one assumes that the retention factors are independent of sample concentration (i.e., linear isotherms), then this gives the definitive solution for the flow rates. In practical cases, the mass overload effects reduce (usually) the retention factors at higher concentrations of the solutes in the columns that change the values thus calculated.

If the concentration profile within the columns is considered (see Fig. 7.18), it can be seen that the section 1 flow rate is still controlled by the low-concentration value of the retention factor of the more retained solute (i.e., the extract product). In the absence of displacement effects, the section 2 flow rate should be determined by the low-concentration value of the retention factor for the earlier eluting component. Usually, however, the tail of the earlier eluting component is displaced by the higher concentration of the extract product, resulting in a faster movement of the raffinate product through the section. In section 3, the critical velocity is that of the high-concentration zone of the extract product and is dependent on the retention factor of the solute at the high concentration. Equally, in section 4, the critical velocity is that of the high-concentration zone of the raffinate product. Thus, the concentrations and flow velocities are very closely interdependent and can only be calculated numerically for real systems. The usual solution is to determine the adsorption isotherm parameters and to use a computer simulation to determine the best operating conditions.

An alternative procedure was developed by Morbidelli and coworkers, which assumes the ideal model (zero dispersion) and that the separation can be modeled with the Langmuir isotherm (31). It consists of a plot of the ratios of the mobile phase to stationary phase flows in sections 2 and 3 of the SMB. They developed equations that define a roughly triangular space on the plot where both extract and raffinate streams are pure. Operating conditions corresponding to a point within this "Morbidelli Triangle" will give pure products at a production rate depending on the position in the triangle that is chosen. Conditions corresponding to points outside the triangle will give either one or the other product stream—or neither—that is pure. Although this does not give values for the flows in sections 1 or 4 (these are relatively easy to establish), this is a very useful method to reach starting conditions for subsequent optimization empirically or by computer simulation.

The computer simulation approach works very well where the adsorption isotherms can be measured accurately and where they fit manageable equations. If the adsorption isotherms do not fit convenient models or the parameters contain significant error, then the results of the computer simulation may be very misleading. In such a case, an empirical determination of the experimental parameters becomes necessary.

Empirical Determination of SMB Parameters

The discussion above gives a basis for an experimental development of SMB methods. We choose an arbitrary value of switch time (e.g., 1 minute) and this, from the column length, gives a stationary phase flow velocity. Bearing in mind that the mobile phase flow velocity in section 1 depends on the retention factor of the second eluting peak, we use the inequality of Eq. 9 to obtain the minimum mobile phase flow velocity in that section. Similarly, ignoring for the moment the probable displacement effects in section 2, we use the low-concentration retention factor for the raffinate product to obtain a minimum flow velocity in the second section. These minimum values of flow rate are usually increased by a few percent to ensure that they are adequate to prevent the carriage of the products in the wrong direction through the system. They also may need to be adjusted (along with the switch time) to ensure that they do not exceed practically attainable values. Once this is done, a low feed flow is introduced, increasing the raffinate flow to eliminate any loss of product from the outlet of section 4 and monitoring the extract and raffinate stream purity (it is convenient in this context, but not essential, to have a system with an external recycle that can be directed to waste during this process in case the raffinate flow is not set high enough). The section 2 flow rate is reduced to the point where the extract purity is just within specification. The feed flow is incremented until the raffinate purity decreases due to the extract product breakthrough in section 3. If by this point the extract purity has increased due to displacement effects, the section 2 flow rate can again be decreased until the extract stream is just within the desired specification. The feed flow is adjusted to the point where the raffinate purity is at the desired value. The final optimization is to hold the flow rates in sections 2 and 3 constant and to decrease the extract and raffinate flows to the point where the products are no longer retained in sections 1 and 4, respectively. A small increase in the flows to eliminate the bleeding completes the process. When optimized in this way, the separation is close to the maximum feed flow possible; for the sake of long-term robustness, the feed flow may be reduced a little to allow for changes in temperature, mobile phase composition, etc. It should be noted that even where computer simulations are possible to calculate the operating conditions, these may not be of high enough accuracy to be used without a similar empirical optimization approach. In this case the starting conditions used are those suggested by the simulation but with a 50% lower feed flow. The parameters are then adjusted following the procedures outlined above.

The concentration profile within the column set gives very useful information, not only during method development but also in normal operation as a means to detect changes in the system before they adversely affect the product purity. A conventional way to do this is to install a six-port injection valve between two of the columns in the set and use this periodically to remove small samples for subsequent analysis. If this is done over the entire cycle, the internal profile can be built up as the columns move through the set. Taking multiple samples through a single switch period can give a more detailed profile than one taken only at the end of each switch. An alternative is to use a detector between two of the columns to monitor the concentrations continuously. Ideally, the use of a UV detector in series with a polarimeter can be used to measure the concentrations of both components after deconvolution of the signals. In practice, this is complicated by the band spreading due to the detector cells (32) and it has been found that very useful data can be extracted simply with a polarimeter detector between two columns.

SMB Separations

The earliest example of an enantioseparation by SMB was reported by Negawa and Shoji (33). This was a rather low-productivity process; over time the productivity of SMB processes has grown significantly. A recent report (34) on the enantioseparation of glutethimide has given a productivity of well above 5 kg (enantiomer)/kg (CSP)/day. SMB is often used for the isolation of phase 1 materials at the 5- to 50-kg scale.

The development of a chromatographic manufacturing process for an enantiomerically pure drug has been described elsewhere (30,35). This very detailed description of the study of the very many parameters that impact the process development should be required reading for anyone contemplating the implementation of larger-scale chromatographic processing.

SMB-Related Processes

There have been a number of modifications to the basic SMB processes that were developed to increase the production rate or the productivity (it should be noted that these are not synonymous) of the process. The best known is the Varicol process (36), developed by Novasep (Pompey, France). This abandons the synchronous switching of the valves such that all columns "move" at the same time as in conventional SMB. It makes use of the fact that the column set is split into a finite number of individual columns and that the concentration steady state is cyclic. The Varicol technique takes advantage of the fact that in a conventional SMB unit the stationary phase is still not used as optimally as it could be (in a true moving bed—TMB—system). There are points where a column is almost empty of solute during parts of the cycle and the asynchronous switching allows these parts of the columns to be utilized. Thus, the number of columns used in the SMB process can be reduced (usually from six to five) with a corresponding increase in productivity although the effect on production rate is not necessarily so large.

Other modifications to the process have been proposed. These, too, are used to bring the SMB unit closer to TMB performance. One (Power Feed) modulates the feed flow depending on the position through the cycle (37), while the other (Modicon) relies on changing the feed concentration through the cycle (38). Both take advantage of the pseudo–steady state operation of SMB where the separation conditions vary through each switch due to the finite column length. Neither has been used to date at large scale.

SUPERCRITICAL FLUID CHROMATOGRAPHY

The first step toward the use of supercritical fluid chromatography (SFC) in preparative separations was carried out in the early 1970s (39). SFC uses a supercritical fluid as its mobile phase rather than the liquid phase used in HPLC. The supercritical phase has several advantages over the HPLC mobile phases, mainly because the supercritical phase has properties intermediate between liquid and gas phases. A supercritical phase has a lower viscosity than a liquid. In chromatographic terms, this means that the pressure drop across a column for a given flow rate is less in SFC than in HPLC. More importantly, the rate of diffusion of molecules in a supercritical phase is faster due to this lower viscosity. This directly affects the column efficiency and typically one can operate an SFC column at a flow rate between three and five times higher than in HPLC. The mobile phase usually used in SFC is a mixture of carbon dioxide with a polar modifier. Carbon dioxide is a low polarity solvent and it is necessary for the average pharmaceutical product to add an alcohol modifier to the mobile phase (usually in a 5–30% concentration) to allow elution of the solutes. Thus, the use of organic solvents in SFC is much lower than in HPLC and the technique is considered to be "green," especially where the CO_2 is recycled rather than vented to the atmosphere.

An SFC system is similar in design to an HPLC unit with the exception of a few details. A schematic is shown in Figure 7.19. Solvents reach supercriticality when heated under pressure. For CO_2, the conditions to reach the supercritical state are relatively mild: a pressure of 70 bar and a temperature of 31°C are required. These values are modified a little when an organic solvent is used as modifier, but conventional SFC separations are usually carried out at a pressure of 100 bar (at the column outlet) and a temperature of 35°C. Often, temperatures below the critical point are used, and although this can result in a higher noise level, there are few differences in the separations achieved using these "sub-critical" conditions. For preparative chromatography, the only difficulty lies in the collection of fractions. Fraction collection is easily done by depressurizing the mobile phase after passage through the detector. Once the pressure is decreased, the mobile phase splits into a mixture of a gas (CO_2) and a liquid (the organic modifier). Normally, the product is dissolved in the modifier (the solubility of organic compounds in gases such as CO_2 is extremely low) and all that is necessary is to collect the liquid phase. This is normally done in a cyclone, which uses the centrifugal force generated by tangential injection of the mobile phase to separate out the droplets of the modifier. This is less convenient than the collection from HPLC, although with the correct design of cyclone, the recovery of the products can be high.

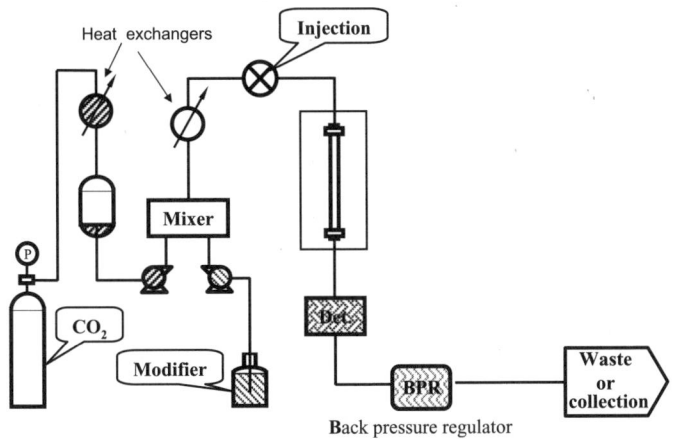

FIGURE 7.19 Schematic of a supercritical fluid chromatograph.

SFC Method Development

Method development in SFC follows the current practices in HPLC development. The low viscosity of SFC mobile phases allow the use of small particle packings at high flow rates and the use of 5 µm particles is normal for semipreparative SFC. It should be noted that there are good arguments for moving to larger particle sizes where it is possible to increase the flow rate and therefore the production rate at a given pressure. There is also a trend for the use of shorter columns in preparative SFC than in HPLC. This reflects the higher efficiency of the small particle columns and generally allows an increase in production rate. This is because the column efficiency available generally exceeds that required for the separation. Thus, the column can be shortened, reducing the number of theoretical plates but also reducing the operating pressure so that a higher flow rate can be used. This, in turn, improves the rate of production of the products. Of course, the use of the combination of short columns and very high flow rates is ultimately limited by the cycle time possible from the SFC hardware; the injection system has to be ready to make the following injection before the total cycle time for the separation has elapsed.

Screening

Screening in SFC follows the same procedures as in HPLC (19). A column switching valve and a mobile phase modifier selection valve are conventionally added to an analytical SFC system and samples are run sequentially through the columns. Optimization is carried out by adjustment of the modifier concentration and composition. Typically modifiers such as methanol and 2-propanol are used; ethanol frequently has selectivity somewhere between the two. In some cases acetonitrile is used as modifier; this has a lesser solvent strength in SFC relative to the case in HPLC and, being aprotic, has some differences in selectivity from the alcohols. When immobilized CSPs are used in SFC, it has been shown that the use of THF and MTBE-based modifiers, often with the addition of a small percentage of an alcohol to adjust solvent strength, may be a useful addition to the range of solvents employed (40).

Choice of Mobile Phase

The mobile phase modifier can have a significant effect on the production rate in SFC, just as in HPLC (see section "Selection of Solvents") (23). A limited study of the effect of mobile phase modifier has been carried out for three solutes and a range of solvents that may possibly be used in SFC. Again, marked differences are seen between the production rates attainable by the different solvent systems, although the peak shapes indicated that the adsorption behavior was more regular than that seen in the HPLC experiments.

Additives

Additives in SFC play similar roles in improving separations as they do in HPLC. Probably the major difference in the case of SFC is in the chromatography of carboxylic acids where in contrast to HPLC, an acidic additive is generally not needed. This is because the combination of the CO_2 and an alcoholic modifier is sufficiently acidic to prevent the unwanted interactions of the acids with the stationary phase. This has an advantage in that often one observes esterification of the acids during product recovery when separating them under HPLC conditions using TFA as additive. It should be noted that if aprotic modifiers such as acetonitrile are used, the acid additive becomes necessary.

Scale-Up

Scale-up in SFC is similarly close to that in HPLC. Flow rates and sample size are scaled according to the column cross-sectional area; typically no problems are found in moving from the analytical scale to the preparative.

Problems of Injection Volume

The disadvantage of SFC lies in the solubility of the solutes to be separated and the practicality of injecting samples dissolved in the supercritical fluid. Conventionally, the samples are dissolved in the organic component of the mobile phase and are injected at high pressure into the mobile phase stream. Although this apparently solves many of the difficulties arising from sample solubility in HPLC, it can result in some problems.

It is well documented for HPLC that introduction of the sample in a solvent that is not the mobile phase can give rise to peak distortion. This is because of the slow diffusion rates in liquids where the sample solvent does not instantaneously mix with the mobile phase. Thus, where the sample is dissolved in a solvent with higher eluting power than the mobile phase, part of the sample remains in the stronger solvent while it is traversing the column and therefore moves more quickly than the remainder of the sample that has already mixed with the mobile phase. This results in significant peak distortion. As the sample is of necessity dissolved in the polar modifier for SFC, this phenomenon is often observed. Figure 7.20 shows the peak profiles for devrinol observed on injecting increasing sample volumes. The distortion due to the use of the strong injection solvent on the front of the peaks is clearly evident at the larger injection volumes. This limits the possible sample load in the column.

Usually the solubility of the sample in the supercritical fluid is not known. Where there is a significant difference in solubility, it is possible for the solute to precipitate from the injection plug as it mixes with the mobile phase. In the worst case, the sample precipitates before it even reaches the column, resulting in blockage of the inlet frit. This process can be progressive, leading to increase in operating pressure with each subsequent injection, or it can be catastrophic. Although the SFC systems are equipped with pressure cutoff switches, these take a finite time to operate. At the high flow rates conventionally used in SFC systems, the instantaneous pressure can reach high values, enough to distort the inlet frits of the SFC columns. An example of this is shown in Figure 7.21. When developing SFC methods, the pressure excursions on injection should be carefully measured to avoid such failure.

At present, SFC instruments with a flow capacity to allow the effective use of columns up to 5 cm id are available. Only a few systems with columns above this size have been produced, thus practically limiting the use of SFC for smaller-scale separations. The high pressures required for the hardware, together with the necessary safety measures needed make large-scale SFC units relatively expensive, and it is probable that the technique will only go to larger scale should a "killer" application be developed.

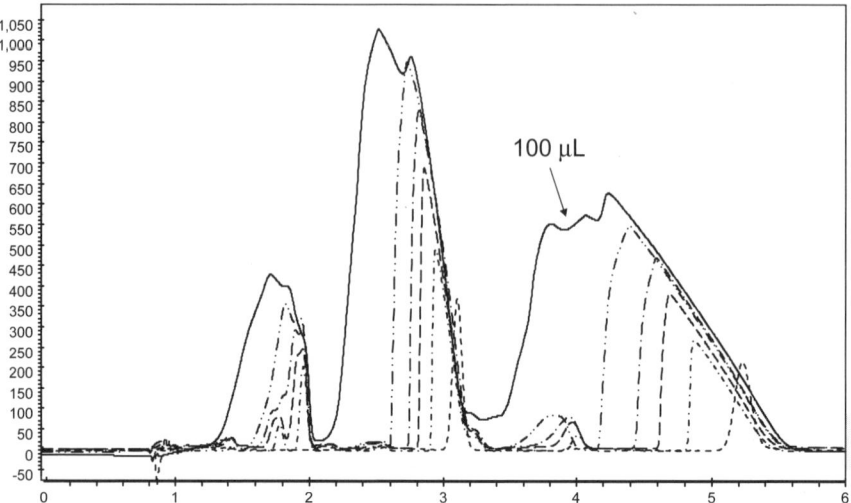

FIGURE 7.20 Peak profiles from a loading study with Devrinol. Column: CHIRALPAK IC (250 × 4.6 mm). Mobile phase: 30% of a 20% methanol in dichloromethane mixture. Sample Devrinol, 100 g/L, injection volumes 10, 20, 30, 50, and 100 mL. Short dashed line: analytical injection. Flow rate: 4 mL/min; back pressure: 100 bar; temperature: 35°C.

FIGURE 7.21 Inlet frit from a 250 × 50 mm preparative SFC column following sample precipitation on injection.

SMB-SFC

The combination of SFC and SMB has been the subject of some research (41,42). Advantages over conventional SMB have been noted since it is possible to adjust the solvent strength by changing the pressure in a zone of the SMB. The pressure modulation can be used to compensate for the higher flow rate downstream of the feed port, thus allowing a higher feed input over the isobaric system. Such equipment has so far not emerged from the R&D environment, however, mainly due to its complexity. As conventional SMB significantly reduces the solvent consumption and the energy requirements for its recovery, it is difficult to see if SFC-SMB can be cost-competitive in reducing the solvent costs further while incurring additional costs due to the phase changes and equipment complexity required for supercritical operation (43).

CONCLUSION

Preparative chromatography is an integral part of the scale-up and development of chiral pharmaceutical products. This starts at the earliest level, in the isolation of the initial few milligrams needed for testing, through the intermediate stages to the isolation of quantities sufficient for toxicology and in an increasing number of cases for phase 1 clinical trials. As the product moves forward through phases 2 and 3 toward commercialization, the final manufacturing process is designed. At this point, the pathway that leads to production of the enantiomer at the lowest cost is chosen; this may be through crystallization, biotransformation, asymmetric synthesis, or chromatography. Large-scale enantioselective chromatography has been shown on several occasions to be the more economic process and has been adopted.

REFERENCES

1. Ruthven DM. Principles of adsorption and adsorption processes. New York: Wiley-Interscience, 1984.
2. US Department of Agriculture, Economic Research Service. 2008 April. Table 52 – High fructose corn syrup: estimated number of per capita calories consumed daily, by calendar year.
3. Kroeff EP, Owens RA, Campbell EL, et al. Production scale purification of biosynthetic human insulin by reversed-phase high-performance liquid chromatography. J Chromatogr A 1989; 461:45–61.
4. Mann G, Renig D. Considerations about HPLC separation process development routes. 9th International Symposium, Exhibit and Workshops on Preparative/Process Chromatography (PREP 96), Washington, D.C., 1996.
5. Nicoud RM. The separation of optical isomers by simulated moving bed chromatography. Pharm Technol Eur 1999; 11:28–34.
6. Henderson GM, Rule HG. A new method of resolving a racemic compound. J Chem Soc 1939; 1568.
7. Dent CE. A study of the behaviour of some sixty amino-acids and other ninhydrin-reacting substances on phenol-'collidine' filter-paper chromatograms, with notes as to the occurrence of some of them in biological fluids. Biochem J 1948; 43:169–180.
8. Hesse G, Hagel R. A complete separation of a racemic mixture by elution chromatography on cellulose triacetate. Chromatographia 1973; 6:277–280.
9. Lough J. This book, chapter 6, HPLC chiral stationary phases.
10. FDA Policy Statement for the Development of New Stereoisomeric Drugs. 1 May 1992.
11. Schulte M, Wekenborg K, Strube J. Continuous chromatography In the downstream processing of products of biotechnical and natural origin. In: Subramanian G, ed. Bioseparation and Bioprocessing. Weinheim: Wiley-VCH, 2007.
12. Lembke P. Production of high purity n-3 fatty acid ethyl esters by process scale supercritical fluid chromatography. In: Anton K, Berger, C, eds. Supercirtical Fluid Chromatography with Packed Columns. Chromatographic Science Series. Vol. 75. New York: Marcel Dekker, 1998, Chapter 15.
13. Golshan-Shirazi S, Guiochon G. Analytical solution for the ideal model of chromatography in the case of a Langmuir isotherm. Anal Chem 1988; 60:2364–2374.
14. Pirkle WH, Sikkenga DL. Resolution of optical isomers by liquid chromatography. J Chromatogr 1976; 123:400–404.
15. Okamoto Y, Yashima E. Polysaccharide Derivatives for Chromatographic Separation of Enantiomers. Angew Chem Int Ed 1998; 37:1020–1043.
16. Perrin C, Vu VA, Matthijs N, et al. Screening approach for chiral separation of pharmaceuticals: Part I. Normal-phase liquid chromatography. J Chromatogr A 2002; 947:69–83.
17. Xiao LS, Han X, Murphy JB, et al. A comparative study of fast orthogonal chiral screening methods by SF C and normal phase HPLC. First International Conference on SFC, Pittsburgh, 2007.

18. Huybrechts T, Torok G, Vennekens T, et al. Multimodal HPLC Screening of Polysaccharide-based Chiral Stationary Phases. LCGC Europe 2007; 20(6):320–335.
19. Cox GB, Khattabi S, Lee JK. (Inpreparation).
20. Dingenen J. Columns and Packing Methods. Analusis 1998; 26:18.
21. Suteu C. Method development for preparative enantioselective chromatography. In: Cox GB, ed. Preparative Enantioselective Chromatography. Oxford: Blackwell, 2005.
22. Khattabi S, Cox GB. SMB Process Development and Optimization using Immobilized Chiral Stationary Phases: Chiral Separation of 1-Methyl-1-phenylsuccinimide using a THF-based mobile phase. Poster, 21st International Symposium, Exhibit and Workshops on Preparative/Process Chromatography (PREP 2008), San Jose, CA.
23. Cox GB. Solvent and stationary phase selectivity in enantioselective SFC. 1st International Conference on SFC. Pittsburgh, September 2007.
24. Stringham RW, Ye YK. Chiral separation of amines by high-performance liquid chromatography using polysaccharide stationary phases and acidic additives. J Chromatogr A 2006; 1101:86–93.
25. Ye YK, Lord B, Stringham RW. Memory effect of mobile phase additives in chiral separations on a Chiralpak AD column. J Chromatogr A 2002; 945:139–146.
26. Guiochon G, Felinger A, Golshan-Shirazi D, et al. Fundamentals of Preparative and Nonlinear Chromatography. 2nd ed. San Diego: Elsevier, 2006:889.
27. Dingenen J. Scaling-up of preparative enantiomer separations. In: Cox GB, ed. Preparative Enantioselective Chromatography. Oxford: Blackwell, 2005.
28. Grill CM, Miller LM. Steady state recycling and its use in chiral separations. In: Cox GB, ed. Preparative Enantioselective Chromatography. Oxford: Blackwell, 2005.
29. Grill CM. Closed-loop recycling with periodic intra-profile injection: a new binary preparative chromatographic technique. J Chromatogr A 1998; 796:101–103.
30. Hamende M. Case study in production scale multi-column continuous chromatography. In: Cox GB, ed. Preparative Enantioselective Chromatography. Oxford: Blackwell, 2005.
31. Mazzotti M, Storti G, Morbidelli M. Optimal operation of simulated moving bed units for nonlinear chromatographic separations. J Chromatogr A 1997; 769:3–24.
32. Cox GB, Khattabi S, Dapremont O. Realtime monitoring and control of a small scale SMB unit from a polarimeter-derived internal profile. 16th International Symposium, Exhibit and Workshops on Preparative/Process Chromatography (PREP 2003). Baltimore, MD.
33. Negawa M, Shoji F. Optical resolution by simulated moving-bed adsorption technology. J Chromatogr 1992; 590:113–117.
34. Zhang T, Franco P. Analytical and preparative potential of immobilized polysaccharide-derived chiral stationary phases. In: Subramanian G, ed. Chiral Separation Techniques. 3rd ed. Weinheim: Wiley-VCH, 2007.
35. Hamende M, Cavoy E. Chiral chromatography: from analytical to production scale. Chimie Nouvelle 2000; 18:3124–3126.
36. Ludemann-Hombourger O, Nicoud R-M, Bailly M. The Varicol process: a new multicolumn continuous chromatography process. Sep Sci Technol 2000; 35:1829–1860.
37. Ziyang Z, Mazzotti M, Morbidelli M. Power Feed operation of simulated moving bed units: changing flow-rates during the switching interval. J Chromatogr 2003; 1006:87–99.
38. Schramm H, Kaspereit M, Kienle A, et al. Simulated moving bed process with cyclic modulation of the feed concentration. J Chromatogr A 2003; 1006:77–86.
39. Jenthoft RE, Gouw TH. Apparatus for supercritical fluid chromatography with carbon dioxide as mobile phase. Anal Chem 1972; 44:681.
40. Cox GB, Krueger B, Lee JK, et al. Optimisation strategies for fast preparative enantioselective SFC separations. 21st International Symposium, Exhibit and Workshops on Preparative/Process Chromatography (PREP 2008), San Jose, CA.
41. Depta A, Giese T, Johannsen M, et al. Separation of stereoisomers in a simulated moving bed-supercritical fluid chromatography plant. J Chromatogr A 1999; 865:175.
42. Denet F, Hauck W, Nicoud R-M, et al. Enantioseparation through Supercritical Fluid Simulated Moving Bed (SF-SMB) Chromatography. Ind Eng Chem Res 2001; 40:4603.
43. Johannsen M. Preparative chromatography with supercritical fluids – operation and optimisation. 2nd International Conference on SFC. Zurich, October 2008.

8. Enantioselective separations by electromigration techniques

Michał J. Markuszewski

INTRODUCTION

Since the last decade of twentieth century, electromigration techniques have been recognized as an important analytical tool. There are a number of commercially available instruments, as well as numerous scientific publications presenting already established applications of the use of electromigration techniques. A significant feature that distinguishes these techniques from others is the fact that charged compounds can be easily and efficiently separated. This can be achieved within a short period of time and be of high resolution. Moreover, in contradiction to most of already established techniques [e.g., high-performance liquid chromatography (HPLC)], analyses of strong acids and bases can be performed without common problems such as peak tailing, poor efficiencies, and resolution. For those compounds, capillary zone electrophoresis is a most suitable method.

The instrumentation for electromigration technique modes is relatively simple and consists of the separation capillary placed between two reservoirs filled with background electrolyte (BGE) solution, platinum electrodes, high-voltage power supply, and a detector. Separations are carried out in fused silica capillaries, ranging usually from 15 to more than 100 cm in length and from 25 to 75 µm in internal diameter. For analytical performance quality, the capillary might be thermostated using either a liquid or air cooling system. High-voltage power supply used in electromigration techniques operates up to 30 kV, which results in a current level of ca. 300 µA. The capillary is externally coated with polyimide layer to ensure sustainability (Fig. 8.1).

Introduction of the sample into the capillary might be carried out either electrokinetically or hydrodynamically. Electrokinetic injection is based on application of a relatively low voltage for a short time to the sample reservoir and movement of ionized sample species into the capillary according to polarity. In hydrodynamic injection mode, a controlled pressure is applied for a certain time interval thus forcing sample volume to be introduced to the system. Typically, only minute volumes of sample are introduced, usually in the range of 0.1 to 50 nL. There are many different detection modes possible to apply with electromigration techniques, such as UV-vis spectrophotometric detection, laser-induced fluorescence (LIF), mass spectrometry (MS).

This basic instrument setup can be elaborated upon with enhanced features such as autosamplers, multiple injection devices, programmable power supply, multiple detectors, fraction collection, as well as advanced and sophisticated computer programs.

In electromigration techniques, several separation modes can be distinguished, which are characterized by different separation mechanisms:

- Capillary zone electrophoresis (CZE)
- Micellar electrokinetic chromatography (MEKC)

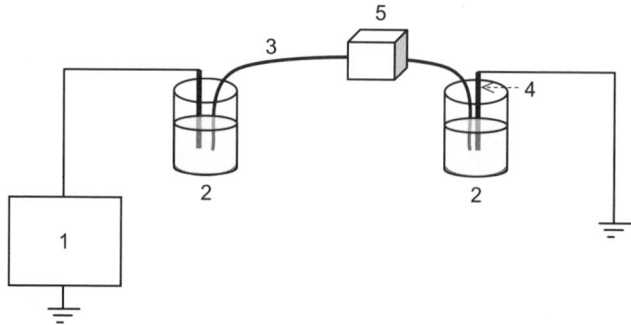

FIGURE 8.1 Schematic representation of capillary electrophoresis instrumentation: (1) high-voltage power supply, (2) electrolyte reservoirs, (3) capillary, (4) electrodes, (5) detection system.

- Capillary gel electrophoresis (CGE)
- Capillary isoelectric focusing (CIEF)
- Capillary isotachophoresis (CITP)

All of the above-mentioned separation modes can be performed using the same equipment, which makes electromigration techniques a very versatile analytical technique. Since the different modes are mainly characterized by the composition of the electrolyte system, they are easily accessed by a proper selection of the applied electrolyte solutions. In the CZE and CITP mode, the separation mechanism is mainly based on differences in effective mobility of ions. The separation mechanism in MEKC is based on differences in distribution equilibrium between an aqueous phase and a micellar phase. In CGE, differences in molecular size govern the separations of compounds and, simultaneously, in CIEF, mechanism is based on differences in isoelectric point values.

The first enantiomeric separation driven by an electrical field was performed by Yoneda et al. in 1970 (1). For this purpose the authors used paper electrophoresis. However, due to poor efficiencies and long analysis times, the method did not gain acceptance for routine analysis. Since then there has been a dramatic increase in number of studies on enantioseparations.

Generally there are two different strategies to deal with enantiomeric separations. Enantiomers can be separated either by direct or indirect separation method. Enantioseparations, which is performed without previous derivatization, is called the direct method. In this approach, the enantiomeric mixture is exposed to an optically active environment, where one of the enantiomers shows a stronger interaction with the asymmetric selector. The common procedures for further enantioseparations are the application of chiral phases in HPLC, gas chromatography (GC), and supercritical fluid chromatography (SFC). In the electromigration techniques, enantioseparations can be reached by application of chiral environment (CE, MEKC) and by modification of inner surface of capillary silica wall or use of chiral stationary phases (CEC).

The indirect separation method is based on reaction (derivatization) between the enantiomers and a chiral selector, resulting in the creation of diastereoisomers (described in detail in chap. 4). Due to different physical

TABLE 8.1 Chiral Selectors Used in Capillary Electrophoresis

Chiral selectors (as BGE additives)	Reference
Cyclodextrins	2,3,5,10–16
Macrocyclic glycopeptide antibiotics	3,17–19
Proteins	3,20–23
Crown ethers	3,8,24–27
Alkaloids	3,28
Polysaccharides (carbohydrates)	3,29,30
Imprinted polymers	31–34
Ligand exchangers	31

Source: From Ref. 9.

properties, the diastereoisomers can be separated by any analytical method (HPLC, CE). A chiral environment may be advantageous, but it is not always a requirement for diastereoisomer separations. The disadvantage of the indirect approach is that the method is time-consuming and not always exact due to the sample derivatization and the chiral derivatizing agent having to be very pure, since enantiomeric impurity might result in a few more diastereomeric products.

In capillary electrophoresis (CE) there are usually two different approaches to the use of chiral selectors for enantioseparations. One is similar to the other separation techniques such as GC or HPLC, when a chiral selector is complexed to different packing materials usually by time-consuming synthetic processes. That is also available in CE when resolution takes place on a capillary coated (from the inner side of capillary tube) with a chiral selector (2).

In contrast, the second and more common approach in CE is when chiral selectors are used as additives to a BGE. The ease of method development facilitates the search for the selector with the highest separating power.

As a chiral resolution technique, CE has been used widely for the enantiomeric resolution of drugs and pharmaceuticals (3). Several reviews have also appeared on this issue and they describe the use of many chiral compounds as chiral BGE additives (4–9).

The list of chiral selectors that are used as BGE additives in electromigration techniques is presented in Table 8.1. Commonly used chiral BGE additives are cyclodextrins, macrocyclic glycopeptide antibiotics, proteins, crown ethers, ligand exchangers, alkaloids (4–9), and so on. The most frequently used in enantioseparations in CE are α-, β-, and γ-cyclodextrins and their derivatives. This is due to their suitable water solubility and, what is equally important, properties of their various functional groups and cavities that are responsible for the formation of diastereomeric inclusion complexes with chiral compounds. A suitable chiral selector used as an additive in BGE in electromigration techniques should fulfill the following basic requirements:

- Solubility in BGE and capability of forming inclusion complexes with chiral compounds
- Sufficient chemical moieties (groups, atoms, grooves, cavities, and so on) for complexing with chiral compounds
- Specificity of chiral selector–chiral compound interactions and, accordingly, the mass transfer kinetics must be favorable in order to effectively employ the advantages offered by the electrically driven flow

- Suitable selectivity combined with enantioselectivity, as it may be used for the resolution of structurally related compounds
- Lack of UV absorbance as, most often, detection in electromigration techniques is by UV

Apart from choosing an appropriate CE chiral selector, further optimization requires consideration of many parameters that are responsible for successful enantiomeric separations.

THE OPTIMIZATION OF CE CONDITIONS FOR ENANTIOSEPARATION

In general, enantioseparation by CE can be controlled by a number of parameters. The important optimization factors may be classified as the independent and dependent ones. Those that are under the direct control of the operator are independent factors and they include type, pH, and the ionic strength of the buffer; the choice of chiral selector; the applied voltage during analysis; the temperature of the capillary compartment; the size and the diameter of the capillary; and the BGE additives. The dependent parameters, which are not under the direct control of the operator, are the field strength (V/m), the electroosmotic flow, the Joule heating, the viscosity of BGE, the sample diffusion, the sample charge, the interaction of the sample with the capillary and the BGE, the molar absorptivity. Difficulties in the optimization of the chiral resolution can be solved by varying and controlling all of above-mentioned factors.

Effect of the Composition of the BGE on Enantioseparation

The main role of the BGE is to maintain a high-voltage gradient across the capillary. This requires that the conductivity of the BGE should be higher than the conductivity of the applied sample. The use of buffers in CE is essential to control the pH of the BGE. Among the most common buffers in CE used for enantioseparations are

- Phosphate
- Acetate
- Borate
- Ammonium citrate
- Tris, 2-(N-cyclohexylamino) ethanesulfonic acid (CHES)
- Morpholinoethanesulfonic acid (MES)
- Piperazine-N,N-bis(2-ethanesulfonic acid) (PIPES)
- N-2-hydroxyethylpiperazine-N-2-ethanesulfonic acid (HEPES)

Buffers are usually used at various concentrations and at different pH values. For the optimum enantioseparation of chiral compounds, the type and specific concentration of the BGE are of utmost importance. The selection of the appropriate background electrolytes depends on the type of the chiral compounds to be resolved, on the conductivities of the buffer and sample matrix, and on the type of chiral selector used in the BGE. Another important issue is the pH of the BGE. For low-pH buffers, phosphates and citrates have commonly been used. Borate, tris, and 3-cyclohexylamino-1-propanesulphonic acid (CAPS) might be used as basic buffers. Type of detection system should be considered when

TABLE 8.2 Commonly Used Buffer Systems with Suitable pHs and Optimum Absorption Wavelength for the Enantioseparations in CE

Buffer Type	pH	UV absorption wavelength (nm)
Phosphate	1.14–3.14	195
Citrate	3.06–5.40	260
Acetate	3.76–5.76	220
MES	5.5–6.7	230
PIPES	6.1–7.5	215
Phosphate	6.20–8.20	195
HEPES	6.8–8.2	230
Tricine	7.4–8.8	230
Tris	7.2–9.0	220
Borate	8.14–10.14	180
CHES	8.6–10.0	<190

Abbreviations: CHES, 2-(*N*-cyclohexylamino)ethanesulfonic acid; MES, morpholinoethanesulfonic acid; PIPES, piperazine-*N*,*N*-bis(2-ethanesulfonic acid); HEPES, 4-(2-hydroxyethyl)-1-piperazineethanesulfonic acid.
Source: From Ref. 35.

searching for an appropriate buffer. If a UV detector is used then low-UV absorbing buffers are required. Also, volatile components are necessary in the case of MS detection methods. However, these conditions substantially limit the choice to a moderate number of electrolytes.

Effect of the pH of the BGE on Enantioseparation

The diastereomeric complexes formed between the enantiomers and the chiral selector used in BGE are affected by the pH of the buffer. By varying the buffer pH, an increase or decrease in the electroosmotic flow (EOF) might be observed that finally results in an increase or decrease of the analysis time. It is also important to note that the pH value of the buffer may be altered by an indirect manner; that is, by other parameters such as temperature or ion depletion. Suitable pH ranges for various buffers are summarized in Table 8.2 (35). In the literature reports there are indicated a wide range of pH values used for the chiral resolution of drugs. Some reports indicate chiral resolution at acidic pH values, while others indicate basic pH values, which reveals that the pH requirements depend on the chemical nature (properties) of analyzed chiral compounds, the type of the buffer used, the type of the chiral selector in BGE, and other CE conditions.

As a general rule, a low pH might be used to resolve cationic chiral compounds, while a high pH is required for the chiral resolution of anionic analytes.

Effect of the Ionic Strength of the BGE on Enantioseparation

The selection of the appropriate ionic strength depends on such factors as the capillary size (effective length and inner diameter), the applied voltage, and the efficiency of the capillary thermostating system (air-cooled system or liquid-cooled system). With an increase of ionic strength, decrease of the EOF might be observed, which in consequence means an increase in the analysis time in CE. Moreover, an increasing ionic strength also increases the current at a constant voltage, which can produce a problem with adequate thermostating of the capillary. When high buffer concentrations are used, excessive Joule heating occurs, which again affects chiral resolution. The heating problem can be solved by decreasing the applied voltage, increasing the length, and decreasing the internal diameter of the capillary. From another point of view, increasing ionic strength might decrease the formation of diastereomeric complexes as well as wall interactions. Therefore, optimization of the chiral resolution of chiral compounds may be achieved by varying the ionic strength of the BGE.

The Chiral Selectors in CE
Similarly to chromatography, a chiral environment is essential for enantiomeric resolution in the case of chiral CE. Basically, the chiral recognition mechanisms in CE are similar to those in chromatography using a chiral mobile phase additive mode. The important exception is that in the case of CE the resolution occurs through the different migration velocities of the diastereoisomeric complexes. Chiral resolution takes place due to the formation of diastereomeric complexes between the enantiomers and the chiral selector. This depends on the type and nature of the chiral selectors used and the nature of chiral compounds.

The enantioseparation of chiral compounds is mostly carried out using a chiral selector as an additive in the BGE. In general, the chiral selectors differently interact with the particular enantiomers thus leading to the enantioseparations. Therefore, the selection of an appropriate chiral selector for a specific chiral compound is a key factor. The most commonly used chiral selectors are cyclodextrins and their chemical modifications.

Cyclodextrins
Cyclodextrins (CD) are torus-shaped cyclic D-gluco-oligosaccharides produced from starch by enzymatic reaction of metabolism. Usually CD contain 6 units (α-CD), 7 units (β-CD), or 8 units (γ-CD) of D-glucopyranose and this determines the size of cavity. The interior of the cavity is relatively hydrophobic. Hydrophobic groups such as aromatic compounds or cycloalkanes can penetrate the cone and form an "inclusion" or "host-guest" complex. Alternatively, on the surface of the wider side of the CD cavity there are chiral secondary hydroxyl groups, while the opposite smaller opening is occupied by achiral primary hydroxyl groups. CD form diastereomeric complexes with several chiral compounds, due to their ability to form inclusion complexes. The formation of inclusion complexes between CD and enantiomers is controlled by a number of interactions between substituents on the asymmetric center of the analyte and the hydroxyl groups on the CD structure rim. Those possible interactions are π-π interactions, hydrogen bonding, dipole-dipole interactions, ionic bindings, and steric effects. Also, relatively good solubility of CD in aqueous buffers (exception is β-CD) and their low cost make these the chiral selectors of first choice in CE.

Chankvetadze et al. (10) have proposed an explanation of chiral recognition mechanisms in the case of CD using UV, NMR, and electrospray ionization mass spectrometric methods. Furthermore, the authors determined the structures of the diastereomeric complexes by X-ray crystallographic methods.

Besides the native CD, there are also several other so-called modified CD commercially available, which are used in CE techniques, for example, hydroxypropylated CD (HP-α-CD, HP-β-CD, HP-γ-CD) (11), in which hydroxyl groups of native CD are substituted by O-C_3H_7OH groups; these possess different stereoselectivity as well as improved solubility. Hydroxyl groups on the CD rim can also be substituted by a charged or chargeable groups (12,13). The introduction of chargeable groups will result in an increased solubility and in alteration of the separation mechanism by the introduction of electrostatic interactions. Moreover, the use of a chiral selector carrying a charge opposite to that of the analytes may significantly increase the mobility difference between the two enantiomers.

Macrocyclic Antibiotics

Macrocyclic antibiotics are similar to CD in some ways. Vancomycin, rifamycin B, and ristocetin A have been used successfully as enantioselective compounds toward the different class of analytes (17–19). Most of them contain ionizable groups and, consequently, their charge and possibly their three-dimensional conformation can vary with the pH of the BGE. Chemical structures of some macrocyclic antibiotics are presented in Figure 8.2. The complex structures of the antibiotics, which contain different chiral centers, inclusion cavities, aromatic rings, pyranose and furanose sugar moieties, rings and bridges, along with several hydrogen donor and acceptor sites and other groups, are responsible for their surprising chiral selectivities. This allows for an excellent potential to resolve a greater variety of racemates. The chiral recognition is possible mainly by π-π interactions, hydrogen bonding, inclusion complexation, dipole interactions, steric interactions, and anionic and cationic bindings. Resulting diastereomeric complexes of different physical and chemical properties become separated under applied electric field. The enantioselectivity depends on the different migration times of these complexes, which vary according to their sizes, charges, and their interactions with the capillary inner surface. As a result, such diastereomeric complexes elute at different migration times.

Since macrocyclic antibiotics have strong UV absorption, partial filling methods have been applied to overcome detection problems. Also, improvement in separation performance can be observed using vancomycin as chiral selector and dynamic coating of the capillary inner surface with poly(dimethylacrylamide) (36). Besides suppressing the EOF, adsorption of vancomycin at the capillary wall is minimized.

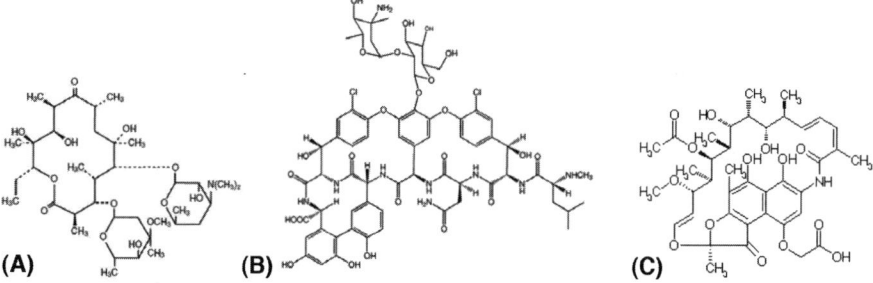

FIGURE 8.2 Chemical structures of macrocyclic antibiotics: (**A**) erythromycin, (**B**) vancomycin, (**C**) rifamycin B.

Crown Ethers

Crown ethers have been used for a long time as chiral additives in CE. They are macrocyclic polyethers capable of forming host-guest complexes with, most frequently, inorganic and organic cations (particularly primary amines). Modification of the crown ether by the introduction of four carboxylic acid groups makes it feasible to apply them in CE as chiral selectors. 18-crown-6-tetracarboxylic acid has been applied as a chiral selector in CE for the enantioseparation of amino acids, dipeptides, sympathomimetics, and other drug compounds containing primary amino groups (25). However, 18-crown-6-tetracarboxylic acid is still the only chiral crown ether used until now (8,14,25–27,37).

Proteins

The fact that binding of drugs to serum proteins takes place enantioselectively led to the idea of using proteins as chiral selectors. After investigation it appeared that proteins can also be applied successfully as chiral selectors in CE. The most important feature of proteins is the isoprotic and the isoelectric point, pI. The pI and applied pH determine the charge of the protein and play important role during the optimization of chiral selectivity. The protein will be positively charged if pH < pI, and negatively charged if pH > pI. The protein charges give them electrophoretic mobility and in effect determine their application for the separation of neutral, basic, or acidic analytes. The mechanism of chiral recognition is probably based on tertiary structure of proteins and binding by hydrophobic interactions, hydrogen bonding, and dipole-dipole interactions. A broad spectrum of proteins and enzymes has been successfully applied as chiral selectors in CE (21–30,38–40). They can be implemented in several ways for enantioselective separations by CE. The simplest way is to dissolve them in the BGE. To avoid interferences of the proteins at the detector, the partial filling technique can be used (41,42). Coating the capillary wall may avoid adsorption of proteins at the capillary wall during the CE run and additionally is responsible for EOF suppression. Proteins can also be bound to the inner surface of a capillary both covalently or in a "dynamic" way.

Polysaccharides

Apart from CD, many other linear and cyclic oligo- and polysaccharides can be applied as selectors for chiral separations in CE. Dextran sulfate was successfully applied for the chiral separation of arylglycine amides (29). Highly sulfated cyclosophoraoses possess higher enantioselectivity than the original cyclosophoraoses as chiral selectors as clearly exhibited in the example of the resolution of five basic chiral drugs (30).

Effect of the Applied Voltage and Temperature on Enantioseparation

The applied voltage is one of the most important factors in the optimization of enantioseparations by CE. In general, a voltage increase leads to an increase in the electroosmotic flow, which results in a shorter migration time. Increasing the voltage also has some disadvantages. The main problem might be related with the increasing production of Joule heat, which cannot be efficiently dissipated. This is the case when the ionic strength of the sample matrix is greater than that of EOF. Additionally, a decrease in the viscosity of the BGE is observed when an excessive Joule heating appears at high voltages, which results in reduced reproducibility. Poor reproducibility might be observed also due to an increase in the ionic mobility. The applied voltage also depends on the type of BGE used. Nelson et al. (43) reported lack of a heating effect of the capillary up to a voltage of 30 kV when borate buffer was used. With the use of CAPS and phosphate buffers, excessive Joule heating of the capillary appeared at 10 and 12 kV applied voltage.

In general, at high temperature the viscosity of the BGE decreases, which results in a short analysis time and poor resolution. Also, it is important to note that when the sample introduction is hydrostatic, the sample volume increases at a higher temperature, which sometimes results in poor resolution. At high temperature, concurrent changes in the buffer pH and peak broadening also occur. Briefly, temperature variation can be used to optimize chiral resolution by CE. However, it has not been used as a routine optimization parameter, due to the difficulty in controlling the temperature inside the capillary during experiments.

Effect of Organic Modifiers as Additives to BGE

Generally, buffers of different concentrations and pH values are used as the BGE for the chiral resolution of compounds by CE. However, for the improvement of separation conditions, the use of some organic solvents in BGE may be useful. These organic solvents in BGE are also called organic modifiers. The addition of organic modifiers may

- change the EOF;
- affect the formation of the diastereomeric complexes and the interactions of the diastereoisomeric complexes with the capillary wall;
- influence the conductivity as well as the thermal diffusion of the BGE.

Therefore, chiral resolution may be optimized by using different types of organic modifiers. The most often used organic solvents are acetonitrile, methanol, and ethanol. The organic modifiers might be used at different concentrations, but it should be remembered that at higher concentrations buffer constituents may precipitate and in effect block the capillary.

Effect of Other Parameters Used During Optimization of Enantioseparation Conditions in CE

There are few other factors that can also be utilized during optimization of enantioseparation in CE. These parameters include

- mode of polarity,
- volume of sample injected,
- EOF modifiers,
- derivatization of the chiral compounds with a suitable reagent prior to enantioseparation.

In the so-called normal CE mode, the anode (+) and cathode (−) are at the inlet and outlet ends, respectively. In this modality, the EOF always tends to travel toward the cathode where the detector is localized. In the other, the "reverse," mode, the direction of the EOF is in the opposite direction, that is, away from the detector. Therefore, only negatively charged diastereomeric complexes with an electrophoretic mobility greater than that of the EOF will approach the detector window. Such a "reverse" mode of analysis might be used only with coated capillaries, or in case when the diastereomeric complexes are all net negatively charged.

When partial resolution of chiral compounds has been observed it is very often due to sample overloading. To solve this problem and optimize the resolution, sample volume should be reduced or the concentration of the compounds in the sample decreased.

MICELLAR ELECTROKINETIC CHROMATOGRAPHY

MEKC introduced by Terabe et al. (44,45) is a mode of CE that is capable of simultaneously separating ionized and unionized compounds. The principle of the separation mechanism in MEKC is differential partitioning of the analyte between an aqueous and a micellar (pseudostationary) phase. Micelles are amphiphilic aggregates with anisotropic microenvironments, which provide both hydrophobic and electrostatic sites of interactions for analytes. Surfactants at a concentration above the critical micellar concentration (CMC) form a micellar pseudostationary phase in the solution of BGE. Finally, due to separation mechanism, it is also a suitable method for the analysis of ionic compounds according to their different hydrophobicities. The distribution of analytes between the micellar phase, regarded as a pseudostationary phase, and the surrounding aqueous phase is the principle of the separation mode (Fig. 8.3). However, to achieve appropriate enantioseparations a chiral selector is still required in the MEKC mode. There are two different approaches available for enantioseparation in MEKC. One is to use chiral surfactants, which above CMC concentrations form a micellar pseudostationary phase as a BGE. The second mode is based on the use of achiral micelles as the main component of BGE and addition of the other chiral agents to achieve enantioseparation.

Chiral surfactants include natural surfactants, monomeric synthetic and polymeric surfactants. Natural surfactants such as bile salts (e.g., sodium deoxycholate) in conjunction with CD were used in MEKC (15,46). Monomeric chiral surfactants are practically used in the form of vesicles that are bigger in size in comparison to the "normal" micelles, thus resulting in a wider migration window. Mohanty and Dey (47) first reported enantioseparation by vesicles formed from the chiral surfactant sodium N-[4-dodecyloxybenzoyl]-L-valinate in

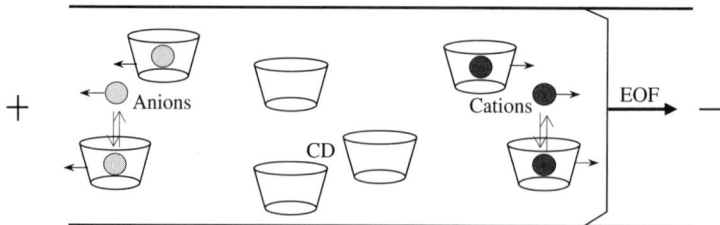

FIGURE 8.3 Schematic representation of the separation mechanism with neutral cyclodextrins.

MEKC. Demonstrating good enantioseparation potency, the vesicles are promising chiral selectors, especially for highly hydrophobic chiral compounds. Polymeric surfactants are high-molecular-mass surfactants, which are formed by covalently binding small surfactant molecules into a single-molecule micelle during the polymerization process. They are also known as molecular micelles. The advantages of use of polymeric surfactants are due to their higher stability and rigidity compared to conventional micelles. Another advantage is that molecular micelles can be used at lower concentrations. A disadvantage of the use of polymeric surfactants is poorer resolution compared to conventional chiral surfactants. This is due to a slower mass transfer resulting from the high rigidity of molecular micelles.

The first use of chiral molecular micelles was reported by Wang and Warner who developed and synthesized the polysodium N-undecenoyl-L-amino acid derivatives. In 2004 the same author (48) proposed a copolymerized surfactant from a water-soluble achiral surfactant and a chiral amino acid–based surfactant. Such a mixture allows enhancement of the water solubility. New chiral polymeric surfactants, the polymeric alkenoxy amino acid surfactants, have been proposed by Rizvi et al. (49) (Fig. 8.4).

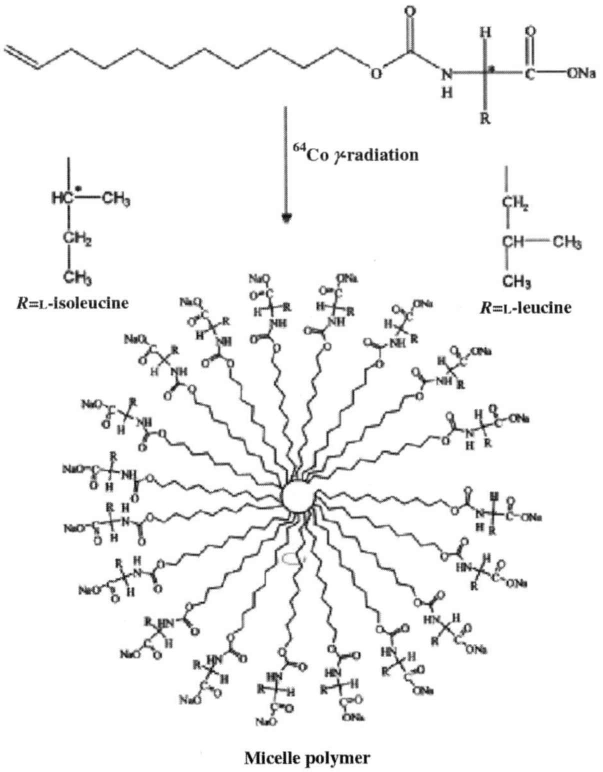

FIGURE 8.4 Structure of micelle polymer and monomer of alkenoxy surfactants. *Source*: From Ref. 49.

MICROEMULSION ELECTROKINETIC CHROMATOGRAPHY

In microemulsion electrokinetic chromatography (MEEKC), the microemulsions are solutions containing oil droplets of water-immiscible liquid (e.g., heptane), dispersed to the size of nanometers. The oil droplets are coated with a surfactant (e.g., SDS) and cosurfactant such as butanol to reduce the surface tension in the emulsion. That is why MEEKC is very often compared to MEKC.

A chiral separation in MEEKC might be achieved by

- using chiral selectors added to the microemulsion (50);
- chiral monomeric surfactant (51), also in the conjunction with CD (16);
- chiral polymeric surfactant (amino acid–based, e.g., polysodium N-undecenoyl-D-valinate) (52);
- addition of chiral alcohols (53).

All in all, MEEKC can be a very effective method for enantioseparation analysis.

CAPILLARY ELECTROCHROMATOGRAPHY

Capillary electrochromatography (CEC) is a hybrid technique between HPLC and CE that was developed in 1990 (54,55). It combines the high peak efficiency of electromigration systems with the high separation selectivity of chromatographic stationary phases. Chiral CEC can be carried out by several different approaches. Experiments can be carried out either by immobilization of the chiral selector onto the capillary wall, by packing the capillary with chiral stationary phase, in monolithic chiral stationary phases, or via imprinted chiral phases. The first approach is represented by open-tubular CEC (OT-CEC) when the chiral selector is adsorptively coated or chemically bonded to the capillary wall. In packed CEC, the capillary is packed with a stationary phase, which usually is of chiral nature. When the stationary phase is achiral, then a chiral mobile phase additive is required. Recently, since the development of monolithic stationary phase materials, the use of monolithic phases CEC has gained some interest. This approach the complicated packing of capillaries and the troublesome preparation of frits is avoided. The chromatographic and electrophoretic mechanisms work simultaneously in CEC resulting in several combinations of (enantio-) separation modes.

It is also important to note that many of chiral stationary phases, which had earlier been proved to be useful for HPLC enantioseparation, have been adapted to the CEC mode. To these belong proteins, CD and their derivatives, Pirkle-type CSP, macrocyclic antibiotics, chiral acrylamide and methacrylates, anion and cation exchangers, and polysaccharide derivatives.

Open-Tubular Capillaries

In 1992, Mayer and Schurig (56) used OT-CEC for enantioseparation. Permethylated β-CD was attached via an octamethylene spacer to dimethylpolysiloxane and coated on the capillary wall. They found also that the thickness of the coating layer affects the separation efficiency due to the mass transfer. Several other coatings and surface modifications have been reported in the literature with, for example, a three-layer coating consisting of anionic polymers and a cationic polypeptide or protein bound as the third layer (57). In the above-mentioned case of a three-layer coating phase, the chiral recognition ability is strongly dependent on the second anionic polymer layer of the coating. Another example of OT-CEC is represented by the immobilization of avidin onto the capillary wall coated with liposomes containing biotin moieties (58). The use of biotin molecules was essential to ensure the high binding affinity of liposomes for avidin. Such a prepared coated capillary gives rise to high enantioseparation efficiencies for derivatized and underivatized amino acids.

Packed Capillaries

Silica-based phases with immobilized macrocyclic antibiotics (vancomycin, teicoplanin) have been applied to a broad range of compounds of biological and pharmaceutical interest (59,60). Also, CD (e.g., permethyl-β-CD) stationary phases have been applied to the enantioseparation of some barbiturates and glutethimide (61).

Other types of packed capillaries used are polysaccharide-based phases containing cellulose or amylose derivatives adsorbed on amino-propyl modified silica gel. However, a different enantioselectivity may be expected depending on the cellulose, amylose, or other polysaccharide derivatives (62,63).

With the increasing number of racemic compounds that need to be resolved, a significant need exists to develop new chiral stationary phases. An example of such a new CSP is phase containing an L-RNA aptamer bonded to biotin and grafted to a streptavidin-modified silica gel (64). Aptamers are single-stranded nucleic acids, which due to their selectivity and affinity properties have been used in flow cytometry, sensors, ELISA type assays, CE, and affinity chromatography. Aptamer molecules can be easily synthesized with reproducibility and accuracy and at a high degree of purity. Another interesting feature of these molecules is that they may be easily changeable in their sequence in order to optimize binding selectivity. The new aptamer CSP was evaluated for the enantioseparation of a series of herbicides, demonstrating it to be stable during a long period for varied experimental electrochromatographic conditions. In addition, it was shown that the interactions of enantiomers during CEC are based solely on chromatographic mechanisms and that the electrophoresis plays only a minor role (64).

Monolithic Phases

Monolithic stationary phase capillaries have been developed to overcome the disadvantages of packed capillaries, such as difficult and time-consuming packing procedures and the need of the preparation of frits that might be easily broken. There are two approaches for preparing of the silica-based monolithic stationary phases, by the process based on in situ polycondensation of alkoxysilanes or by in situ polymerization of organic monomers (so-called continuous beds). In the first approach, chiral selectors can be incorporated by physical adsorption, encapsulation, or on-column derivatization. The second method of preparation of "continuous beds" was introduced by Hjerten in the 1980s (65). During copolymerization a chiral selector may be added to the mixture of reacting components. For example, (+)-1-(4-aminobutyl)-(5R,8S,10R)-terguride as a chiral selector was attached to the epoxy group at the surface of the monolith (66). Such a prepared monolithic phase was used for the enantioseparation of a mixture of 2-aryloxypropionic acids in the reversed-phase mode (Fig. 8.5).

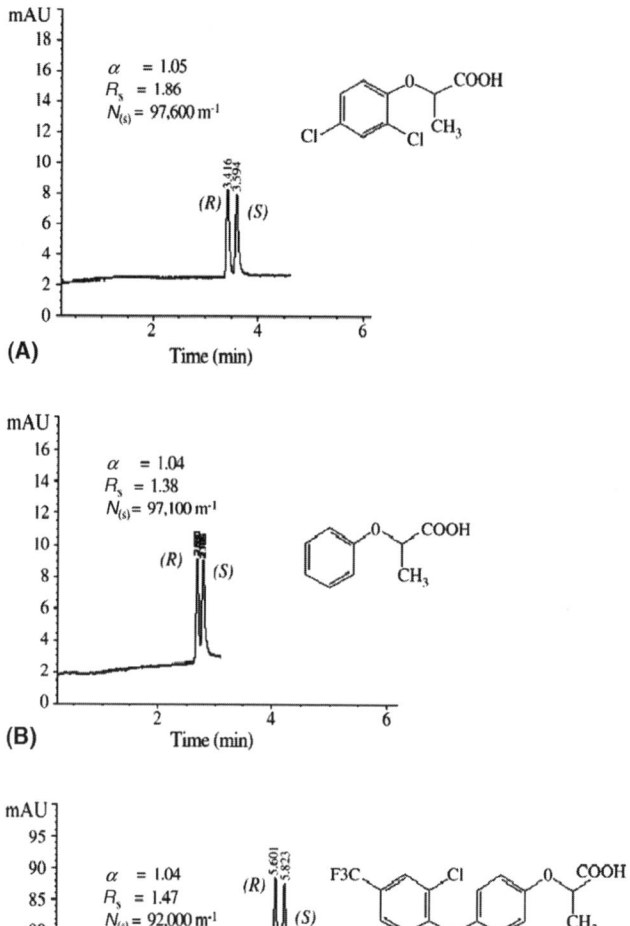

FIGURE 8.5 Enantioseparation of 2-aryloxypropionic acids on chiral porous monolithic column by capillary electrochromatography: (**A**) dichlorprop; (**B**) 2-phenoxypropionic acid; and (**C**) haloxyfop. *Source*: From Ref. 66.

Imprinted Polymers as Chiral Phases

Since more than 10 years there has been a deep research interest aimed toward the adaptation of molecularly imprinted polymers (MIP) to the capillary format and the use of these highly selective matrices for CEC (32,33). The MIP is prepared by incorporation of a template molecule into the polymerization, and with the subsequent extraction procedure a resultant polymer of a high selectivity and recognition complementary to the template molecule in size, shape, and chemical functionality is formed. MIP have been used as recognition elements in several different analytical techniques (e.g., HPLC, SPE). The merger of molecular imprinting technology and CEC has introduced several interesting polymer formats due to the adaptation of the MIP to the miniaturized capillary format. Chiral imprinted polymer phases have been developed for OT-CEC, packed capillary CEC, and monolithic phase CEC.

The cross-selectivity of the MIP also allows the recognition of structural analogues of the template molecule (34). Thus, an MIP prepared using one of the enantiomers of a chiral compound as template may be used for separations of the particular enantiomers but also of several of their structural analogues. However, the efficiency of imprinted chiral phases is relatively low.

ELECTROMIGRATION TECHNIQUES VERSUS HPLC IN THE VIEW OF ENANTIOSEPARATION FIELD

Besides the applicability and usefulness of HPLC, there are numerous disadvantages, such as

- the poor separation efficiency due to the laminar flow profile;
- possible additional interactions of enantiomers with the residual silanol groups on the stationary phase;
- the chiral selectors are fixed on the stationary phase and hence the optimization of analytical conditions by varying this parameter is not possible;
- a larger amount of solvents, in comparison to CE, is consumed (possible environmental pollutants).

On the other hand, chiral resolution in electromigration techniques is achieved using chiral selectors, beside as a CSP, also as an additive in the BGE. Electromigration techniques are rather faster and usually involve the use of simple and inexpensive buffers. The high efficiency of electromigration techniques is due to the flat flow profile. In general, the theoretical plate number in electromigration techniques is higher than in HPLC and, therefore, usually very good resolution may be achieved. There are possibilities to use more than one chiral reagent simultaneously to optimized enantioseparation.

The major problem and disadvantage of electromigration techniques is reproducibility. This makes them less appropriate for routine chiral analysis. Furthermore, detection techniques such as polarimetry and circular dichroism cannot be used as the chiral compounds elute in the form of diastereoisomeric complexes, the signals from which might not be representative of the original enantiomers. Moreover, some of the well-known chiral selectors may not be soluble in the BGE. Enantioseparation methods with use of hyphenated detection such as CE-LIF and CE-MS improve the sensitivity of analysis.

In summary, electromigration techniques are not of the first choice as compared to HPLC for enantioseparation in routine analysis due to reproducibility and sensitivity problems, but there are still new prospects for chiral electromigration techniques in the pharmaceutical and environmental field. Indeed, the scientific literature would suggest that interest in chiral CE is continuing unabated. In particular, there is evidence of more developments in the use of new chiral selectors [erythromycin lactobionate (67,68), balhimycin (69), glycogen (70,71), clindamycin phosphate (72), dipeptides (73), azithromycin (74), clarithromycin lactobionate (75)], sustained interest in single isomer CD derivatives (76,77), more on ligand-exchange CE (78–85), use of nonaqueous chiral CE, and more applications in the area of affinity separations and calculations therefrom (86,87), and biofluid cleanup (88–91). This is a testament to the versatility of CE. However, certainly in terms of chiral CE, it has failed to gain a foothold in industrial pharmaceutical R&D as it once seemed it might, not just because of reproducibility issues but because chiral LC methods will always be developed with a view to preparative work and with a view to use on manufacturing sites, where there has not been sufficient advantage in CE for it to be introduced on a widespread scale.

REFERENCES

1. Yoneda H, Miura T. Complete resolution of the racemic trisethylenediaminecobalt(III) complex into its optical antipodes by means of electrophoresis. Bull Chem Soc Jpn 1970; 43(2):574–574.
2. Jung M, Mayer S, Schurig V. Enantiomer separation by GC, SFC and CE on immobilizes siloxane cyclodextrins. LC–GC 1994; 7:340–347.
3. Chankvetadze B. Capillary Electrophoresis in Chiral Analysis. New York: Wiley, 1997.
4. Blaschke G, Chankvetadze B. Enantiomer separation of drugs by capillary electromigration techniques. J Chromatogr A 2000; 875:3–25.
5. Fanali S. Enantioselective determination by capillary electrophoresis with cyclodextrins as chiral selectors. J Chromatogr A 2000; 875:89–122.
6. Chankvetadze B. Recent trends in enatioseparation of chiral drugs, in: Molecular biology in medicinal chemistry, Eds.: Th. Dingermann, D. Steinhilber, G. Folkers, Wiley-VCH Verlag, Weinheim, 2004.
7. Ha PTT, Hoogmartens J, Van Schepdael A. Recent advances in pharmaceutical applications of chiral capillary electrophoresis. J Pharm Biomed Anal 2006; 41(1):165–175.
8. Gübitz G, Schmid MG. Chiral separation by capillary electromigration techniques. J Chromatogr A 2008; 1204:140–156.
9. Ali I, Aboul-Enein HY. Chiral Pollutants: Distribution, Toxicity and Analysis by Chromatography and Capillary Electrophoresis. Chichester: John Wiley & Sons Ltd., 2004.
10. Chankvetadze B, Burjanadze N, Pintore G, et al. Separation of brompheniramine enantiomers by capillary electrophoresis and study of chiral recognition mechanisms of cyclodextrins using NMR-spectroscopy, UV spectrometry, electrospray ionization mass spectrometry and X-ray crystallography. J Chromatogr A 2000; 875(1–2):471–484.
11. Chankvetadze B, Endresz G, Blaschke G. Enantiomeric resolution of chiral imidazole derivatives using capillary electrophoresis with cyclodextrin-type buffer modifiers. J Chromatogr A 1995; 700(1–2):43–49.
12. Dette C, Ebel S, Terabe S. Neutral and anionic cyclodextrins in capillary zone electrophrosis: Enantiomeric separation of ephedrine and related compounds. Electrophoresis 1994; 15(6):799–803.
13. Desiderio C, Fanali S. Use of negatively charged sulfobutyl ether-β-cyclodextrin for enantiomeric separation by capillary electrophoresis. J Chromatogr A 1995; 716 (1–2):183–196.
14. Elek J, Mangelings D, Ivanyi T, et al. Enantioselective capillary electrophoretic separation of tryptophane- and tyrosine methylesters in a dual system with a tetra-oxadiaza-crown-ether derivative and a cyclodextrin. J Pharm Biomed Anal 2005; 38:601–608.
15. Chen DY, Chen YY, Hu YZ. Optimized separation of cis-trans isomers and enantiomers of sertraline using cyclodextrin-modified micellar electrokinetic chromatography. Chromatographia 2004; 60(7–8):469–473.

16. Mertzman MD, Foley JP. Chiral cyclodextrin-modified microemulsion electrokinetic chromatography. Electrophoresis 2004; 25:1188–1200.
17. Armstrong DW, Tang YB, Chen SS, et al. Macrocyclic antibiotics as a new class of chiral selectors for liquid chromatography. Anal Chem 1994; 66(9):1473–1484.
18. Armstrong DW, Rundlett KL, Chen JR. Evaluation of the macrocyclic antibiotic vancomycin as a chiral selector for capillary electrophoresis. Chirality 1994; 6(6):496–509.
19. Armstrong DW, Rundlett KL, Reid GL. Use of a macrocyclic antibiotic, rifamycin B, and indirect detection for the resolution of racemic amino alcohols by CE. Anal Chem 1994; 66(10):1690–1695.
20. Haginaka J. Enantiomer separation of drugs by capillary electrophoresis using proteins as chiral selectors. J Chromatogr A 2000; 875:235–254.
21. Lloyd DK, Aubry AF, De Lorenzi E. Selectivity in capillary electrophoresis: the use of proteins. J Chromatogr A 1997; 792:349–369.
22. Hage DS. Chiral separations in capillary electrophoresis using proteins as stereoselective binding agents. Electrophoresis 1997; 18(12–13):2311–2321.
23. Ye M, Zou H, Liu Z, et al. Study of competitive binding of enantiomers to proteins by affinity capillary electrochromatography. J Pharm Biomed Anal 2002; 27:651–660.
24. Tanaka Y, Otsuka K, Terabe S. Separation of enantiomers by capillary electrophoresis-mass spectrometry employing a partial filling technique with a chiral crown ether. J Chromatogr A 2000; 875(1–2):323–330.
25. Kuhn R. Enantiomeric separation by capillary **electrophoresis** using a crown ether as chiral selector. Electrophoresis 1999; 20:2605–2613.
26. Cho SI, Shim J, Kim MS, et al. On-line sample cleanup and chiral separation of gemifloxacin in a urinary solution using chiral crown ether as a chiral selector in microchip electrophoresis. J Chromatogr A 2004; 1055(1–2):241–245.
27. Ivanyi T, Pal K, Lazar I, et al. Application of tetraoxadiaza-crown ether derivatives as chiral selector modifiers in capillary electrophoresis. J Chromatogr A 2004; 1028(2):325–332.
28. Piette V, Fillet M, Lindner W, et al. Non-aqueous capillary electrophoretic enantioseparation of N-derivatized amino acids using cinchona alkaloids and derivatives as chiral counter-ions. J Chromatogr A 2000; 875(1–2):353–360.
29. Chen Y, Lu XN, Han ZQ, et al. Chiral ion-exchange capillary electrochromatography of arylglycine amides with dextran sulfate as a pseudostationary phase. Electrophoresis 2005; 26(4–5):833–840.
30. Park H, Lee S, Kang S, et al. Enantioseparation using sulfated cyclosophoraoses as a novel chiral additive in capillary electrophoresis. Electrophoresis 2004; 25(16):2671–2674.
31. Gübitz G, Schmid MG. Chiral separation principles in capillary electrophoresis. J Chromatogr A 1997; 792:179–225.
32. Nilsson J, Spegel P, Nilsson S. Molecularly imprinted polymer formats for capillary electrochromatography. J Chromatogr B 2004; 804:3–12.
33. Nilsson K, Lindell J, Norrlöw O, et al. J Chromatogr A 1994; 680:57.
34. Hart BJ, Rush DJ, Shea KJ. J Am Chem Soc 2000; 122:460.
35. Oda RP, Landers JP. In: Landers JP, ed. Handbook of Capillary Electrophoresis. London: CRC Press, 1994.
36. Wang Z, Wang J, Hu Z, et al. Electrophoresis 2007; 28:938.
37. Nagata H, Nishi H, Kamigauchi M, et al. Guest-dependent conformation of 18-crown-6 tetracarboxylic acid: Relation to chiral separation of racemic amino acids. Chirality 2008; 20(7):820–827.
38. Valtcheva L, Mohammad J, Pettersson G, et al. Chiral separation of β-blockers by high-performance capillary electrophoresis based on non-immobilized cellulase as enantioselective protein. J Chromatogr 1993; 638(2):263–267.
39. Fanali S, Caponecchi G, Aturki Z. Enantiomeric resolution by capillary zone electrophoresis: use of pepsin for separation of chiral compounds of pharmaceutical interest. J Microcolumn September 9, 1997; 9(1):9–14.
40. Hedeland M, Isaksson R, Pettersson C. Cellobiohydrolase I as a chiral additive in capillary electrophoresis and liquid-chromatography. J Chromatogr A 1998; 807:297–305.

41. Tanaka Y, Matsubara N, Terabe S. Separation of enantiomers by affinity electrokinetic chromatography using avidin. Electrophoresis 1994; 15(6):848–853.
42. Schmid MG, Gübitz G, Kilar F. Stereoselective interaction of drug enantiomers with human serum transferrin in capillary zone electrophoresis (II). Electrophoresis 1998; 19(2):282–287.
43. Nelson RJ, Paulus A, Cohen AS, et al. Use of peltier thermoelectric devices to control column temperature in high-performance capillary electrophoresis. J Chromatogr 1989; 480(C):111–127.
44. Terabe S, Otsuka K, Ichikawa K, et al. Electrokinetic separations with micellar solutions and open-tubular capillaries. Anal Chem 1984; 56(1):111–113.
45. Terabe S, Otsuka K, Ando T. Electrokinetic chromatography with micellar solution and open-tubular capillary. Anal Chem 1985; 57:834–841.
46. Zhao S, Song Y, Liu YM. Separation of L-/D-serine enantiomers fluorescently tagged with naphthalene-2,3-dicarboxaldehyde. Talanta 2005; 67:212–216.
47. Mohanty A, Dey J. A giant vesicle forming single tailed chiral surfactant for enantioseparation by micellar electrokinetic chromatography. Chem Commun 2003; 9(12):1384–1385.
48. Kamande MW, Zhu X, Kapnissi-Christodoulou C, Warner IM. Chiral separations using a polypeptide and polymeric dipeptide surfactant polyelectrolyte multilayer coating in open-tubular capillary electrochromatography. Anal Chem 2004; 76 (22):6681–6692.
49. Rizvi SAA, Akbay C, Shamshi SA. Polymeric alkenoxy amino acid surfactants: II. Chiral separations of -blockers with multiple stereogenic centers. Electrophoresis 2004; 25:853–860.
50. Aiken JH, Huie CW. Use of a microemulsion system to incorporate a lipophilic chiral selector in electrokinetic capillary chromatography. Chromatographia 1993; 35 (7–8):448–450.
51. Pascoe R, Foley J. Rapid separation of pharmaceutical enantiomers using electrokinetic chromatography with a novel chiral microemulsion. Analyst 2002; 127: 710–714.
52. Iqbal R, Rizvi SAA, Akbay C, et al. Chiral separations in microemulsion electrokinetic chromatography: Use of micelle polymers and microemulsion polymers. J Chromatogr A 2004; 1043(2):291–302.
53. Zheng ZX, Lin JM, Chan WH, et al. Separation of enantiomers in microemulsion electrokinetic chromatography using chiral alcohols as cosurfactants. Electrophoresis 2004; 25:3263–3269.
54. Li S, Lloyd D. Direct chiral separations by capillary electrophoresis using capillaries packed with an alpha1-acid glycoprotein chiral stationary phase. Anal Chem 1993; 65:3684–3690.
55. Schurig V, Jung M, Mayer S, et al. Enantioselektive trennung von hexobarbital durch GC, SFC, LC und CEC an einer mit Chirasil-Dex belegten kapillarsäule. Angew Chem 1994; 106:2265–2267.
56. Mayer S, Schurig V. Enantiomer separation by electrochromatography on capillaries coated with chirasil-dex. Dedicated to Professor Ernst Bayer on the occasion of his 65th birthday. J High Res Chromatogr 1992; 15:129–131.
57. Kitagawa F, Kamiya M, Okamoto Y, et al. Electrophoretic analysis of proteins and enantiomers using capillaries modified by a successive multiple ionic-polymer layer (SMIL) coating technique. Anal Bioanal Chem 2006; 386:594–601.
58. Han NY, Hautala JT, Bo T, et al. Immobilization of phospholipid-avidin on fused-silica capillaries for chiral separation in open-tubular capillary electrochromatography. Electrophoresis 2006; 27:1502–1509.
59. Zheng J, Shamsi SA. Simultaneous enantioseparation and sensitive detection of eight β-blockers using capillary electrochromatography-electrospray ionization-mass spectrometry. Electrophoresis 2006; 27:2139–2151.
60. Catarcini P, Fanali S, Presutti C, et al. Evaluation of teicoplanin chiral stationary phases of 3.5 and 5 μm inside diameter silica microparticles by polar-organic mode capillary electrochromatography. Electrophoresis 2003; 24:3000–3005.

61. Wistuba D, Schurig V. Enantiomer separation by pressure-supported electrochromatography using capillaries packed with Chirasil-Dex polymer-coated silica. Electrophoresis 1999; 20:2779–2785.
62. Otsuka K, Mikami C, Terabe S. Enantiomer separations by capillary electrochromatography using chiral stationary phases. J Chromatogr A 2000; 887:457–463.
63. Chankvetadze L, Kartozia I, Yamamoto C, et al. Enantioseparations in capillary liquid chromatography and capillary electrochromatography using amylose tris(3,5-dimethylphenylcarbamate) in combination with aqueous organic mobile phase. J Sep Sci 2002; 25:653–660.
64. Andre C, Berthelot A, Thomassin M, et al. Enantioselective aptameric molecular recognition material: Design of a novel chiral stationary phase for enantioseparation of a series of chiral herbicides by capillary electrochromatography. Electrophoresis 2006; 27:3254–3262.
65. Hjerten S, Liao J-L, Zhang R. High-performance liquid chromatography on continuous polymer beds. J Chromatogr 1989; 473:273–275.
66. Messina A, Flieger M, Bachechi F, et al. Enantioseparation of 2-aryloxypropionic acids on chiral porous monolithic columns by capillary electrochromatography: Evaluation of column performance and enantioselectivity. J Chromatogr A 2006; 1120:69–74.
67. Chen B, Du Y, Wang H. Study on enantiomeric separation of basic drugs by NACE in methanol-based medium using erythromycin lactobionate as a chiral selector. Electrophoresis 2010; 31(2):371–377.
68. Xu G, Du Y, Chen B, et al. Investigation of the enantioseparation of basic drugs with erythromycin lactobionate as a chiral selector in CE. Chromatographia 2010; 72(3–4):289–295.
69. Jiang Z, Yang Z, Suessmuth RD, et al. Highlighting the possible secondary interactions in the role of balhimycin and its analogues for enantiorecognition in capillary electrophoresis. J Chromatogr A 2010; 1217(7):1149–1156.
70. Chen J, Du Y, Zhu F, et al. Glycogen: a novel branched polysaccharide chiral selector in CE. Electrophoresis 2010; 31(6):1044–1050.
71. Chen J, Du Y, Zhu F, et al. Evaluation of the enantioselectivity of glycogen-based dual chiral selector systems towards basic drugs in capillary electrophoresis. J Chromatogr A 2010; 1217(45):7158–7163.
72. Chen B, Du Y. Evaluation of the enantioseparation capability of the novel chiral selector clindamycin phosphate towards basic drugs by micellar electrokinetic chromatography. J Chromatogr A 2010; 1217(11):1806–1812.
73. Haglof J, Pettersson C. Separation of amino alcohols using divalent dipeptides as counter ions in aqueous CE. Electrophoresis 2010; 31(10):1706–1712.
74. Kumar AP, Park JH. Azithromycin as a new chiral selector in capillary electrophoresis. J Chromatogr A 2011; 1218(9):1314–1317.
75. Yu T, Du Y, Chen B. Evaluation of clarithromycin lactobionate as a novel chiral selector for enantiomeric separation of basic drugs in capillary electrophoresis. Electrophoresis 2011; 32(14):1898–1905.
76. Varga G, Tarkanyi G, Nemeth K, et al. Chiral separation by a monofunctionalized cyclodextrin derivative: From selector to permethyl-beta-cyclodextrin bonded stationary phase. J Pharm Biomed Anal 2010; 51(1):84–89.
77. Cucinotta V, Contino A, Giuffrida A, et al. Application of charged single isomer derivatives of cyclodextrins in capillary electrophoresis for chiral analysis. J Chromatogr A 2010; 1217(7):953–967.
78. Kodama S, Aizawa S, Taga A, et al. Metal(II)-ligand molar ratio dependence of enantioseparation of tartaric acid by ligand exchange CE with Cu(II) and Ni(II)-D-quinic acid systems. Electrophoresis 2010; 31(6):1051–1054.
79. Qi L, Yang G, Zhang H, et al. A chiral ligand exchange CE essay with zinc(II)-L-valine complex for determining enzyme kinetic constant of L-amino acid oxidase. Talanta 2010; 81(4–5):1554–1559.

80. Kodama S, Taga A, Yamamoto A, et al. Enantioseparation of DL-isocitric acid by a chiral ligand exchange CE with Ni(II)-D-quinic acid system. Electrophoresis 2010; 31(21):3586–3591.
81. Rizkov D, Mizrahi S, Cohen S, et al. Beta-amino alcohol selectors for enantioselective separation of amino acids by ligand-exchange capillary zone electrophoresis in a low molecular weight organogel. Electrophoresis 2010; 31(23–24):3921–3927.
82. Zhang H, Qi L, Qiao J, et al. Determination of sodium benzoate by chiral ligand exchange CE based on its inhibitory activity in D-amino acid oxidase mediated oxidation of D-serine. Anal Chim Acta 2011; 691(1–2):103–109.
83. Giuffrida A, Contino A, Maccarrone G, et al. Mass spectrometry detection as an innovative and advantageous tool in ligand exchange capillary electrophoresis. Electrophoresis 2011; 32(10):1176–1181.
84. Schmid MG, Guebitz G. Enantioseparation by chromatographic and electromigration techniques using ligand-exchange as chiral separation principle. Anal Bioanal Chem 2011; 400(8):2305–2316.
85. Kartsova LA, Alekseeva AV. Ligand-exchange capillary electrophoresis. J Anal Chem 2011; 66(7):563–571.
86. Yang F, Du Y, Chen B, et al. Enantiomeric separation of nefopamhydrochloride by affinity electrokinetic chromatography using chondroitin sulfate A as chiral selector and its chiral recognition mechanism. Chromatographia 2010; 72(5–6):489–493.
87. Asensi-Bernardi L, Martin-Biosca Y, Maria V-CR, et al. Evaluation of enantioselective binding of fluoxetine to human serum albumin by ultrafiltration and CE-experimental design and quality considerations. Electrophoresis 2010; 31(19):3268–3280.
88. Anouti S, Vandenabeele-Trambouze O, Cottet H. Heart-cutting 2D-CE with on-line preconcentration for the chiral analysis of native amino acids. Electrophoresis 2010; 31(6):1029–1035.
89. Mikus P, Marakova K. Chiral capillary electrophoresis with on-line sample preparation. Curr Pharm Anal 2010; 6(2):76–100.
90. Wang Z, Liu C, Kang J. A highly sensitive method for enantioseparation of fenoprofen and amino acid derivatives by capillary electrophoresis with on-line sample preconcentration. J Chromatogr A 2011; 1218(13):1775–1779.
91. Wang M, Cai Z, Xu L. Coupling of acetonitrile deproteinization and salting-out extraction with acetonitrile stacking in chiral capillary electrophoresis for the determination of warfarin enantiomers. J Chromatogr A 2011; 1218(26):4045–4051.

9 Alternative analytical techniques for determination or isolation of drug enantiomers

W. John Lough

As the purpose of this section of this volume is to address the analysis of chiral drugs, the focus has been on the main methods used. Therefore, in this chapter, other techniques are discussed not in depth but simply in the context of what they have to offer in terms of chiral drug analysis. That is not to say that these other techniques or approaches do not have their place or are of minor importance. They can be valuable in chiral drug analysis in certain instances and are often important in applications other than on chiral drugs. Techniques for chiral drug analysis other than liquid chromatography (LC) have previously been considered (1,2), but it is appropriate to briefly take a more current perspective.

By the incorporation of a chiral selector it is possible to adapt just about any analytical technique so that it can be used to distinguish between enantiomers. In the absence of a chiral magnetic field or a chiral selector being used in atmospheric pressure chemical ionization, mass spectrometry (MS) could have been considered as an exception to this but even here there are some studies of relevance. Schug and coworkers (3) have used electrospray ionization–mass spectrometry (ESI-MS) for evaluating enantioselective systems through the measurement of relative solution-phase binding constants via titration. However, it could be argued that this is not too much of a progression from trying to carry out MS on diastereomeric salts.

A more long-standing technique for analysing chiral compounds is polarimetry. Polarimetry is both long-standing and still in use. It is particularly prevalent in pharmacopoeial monographs (4) where it is used in identity tests for chiral drugs where zero rotation by a polarimetry test is used to confirm that a chiral drug is present as a racemate (e.g., chlorphenamine maleate). Also, it is used in identity tests for single enantiomer drugs even when a chiral LC method is used elsewhere in the monograph and for tests on natural products with multiple chiral centers (e.g., alcuronium chloride, betamethasone). In research studies, polarimetry has been used along with a range of other techniques to study the in vitro interconversion of profen enantiomers (5,6) and of L-alanine (7). In terms of the use of polarimetry for detection in LC, attempts are still being made, for bupivacaine (8), to use the ratio of UV and polarimetric signals to assess enantiomeric purity without separating the enantiomers and using an achiral column such as Luna C18. While there does not seem to be the same interest in polarimetric detection as in the early 1990s, commercial instruments continue to be available, for example, from Jasco [Tokyo, Japan (OR-2090)] and Knauer [Berlin, Germany (CHIRALYSER-MP)].

Circular dichroism (CD) detectors for LC are also available, for example, the Jasco (OR-2095) and, in use as LC detectors and stand-alone use of the technique, there is increasingly more interest in CD than in polarimetry. As with polarimetric detection, CD detection in LC can be used in conjunction with UV detection to determine enantiomeric purity of the compounds giving rise to peaks eluted from an achiral LC column. While, for UV-absorbing compounds, CD detection does not have the sensitivity problems that may be experienced with polarimetric detection, the limits of detection for trace enantiomer for this signal-ratioing strategy are ultimately limited by the precision with which the peaks can be measured. CD on its own is primarily used for the determination of absolute configuration and for studies of ligand-protein binding. Both of these are important in drug analysis. The latter is used not only for drug-protein binding studies (9) but also for the study of how analytes interact with chiral selectors that are used in chromatography (10).

The determination of absolute configuration by CD is used for natural products and increasingly for unusual chiral structures/complexes as well as for drugs. In the pharmaceutical industry, X-ray crystallography is still considered to be the definitive method for determining absolute configuration. However, in drug discovery, it is often more appropriate to use vibrational CD (11), which can be deployed much more quickly, being usable for solutions, liquids, and oils, being applicable to non-UV-absorbing compounds, and not requiring recrystallization or derivatization. Although it is used for small molecules, it is the study of biomacromolecules by vibrational CD and the related Raman optical activity that is a major analytical growth area, particularly with such molecules beginning to dominate Drug Development pipelines.

Nuclear magnetic resonance (NMR) spectroscopy is another technique that is heavily used in the study of chiral molecules, large and small. This still includes the use of chiral solvating agents in NMR for the determination of enantiomer ratios (12) as an alternative to chiral LC. The use of a chiral solvating agent or a chiral shift reagent is a very long-standing approach (13) and has always been popular among synthetic organic chemists. Interestingly, this potentially could have influenced some of the patent cases relating to single enantiomer drug patents (see chap. 15). Critical to some of these cases would be whether or not it was obvious to use chiral HPLC. Distinguished organic chemists would be among the expert witnesses used and such individuals, steeped in the use of chiral NMR, would be in the minority who would not automatically turn to chiral LC.

It was stated at the outset that just about any technique could be adapted with the use of a chiral selector to be able to be used to determine enantiomers. One of the most inventive examples of this involves fluorescence (14). Tran et al. used the room temperature ionic liquid, (S)-[(3-chloro-2-hydroxypropyl) trimethylammonium] [bis((trifluoromethyl)sulfonyl)amide], as both solvent and chiral selector in the accurate determination of the enantiomeric composition of samples of propranolol, naproxen, and warfarin.

Most of the other techniques used as an alternative to chiral LC for the determination of enantiomers are themselves also separative methods. Thin-layer chromatography (TLC) offers alternative geometry to chiral LC. For selector-selectand systems it is very similar to chiral LC in that virtually all systems used in chiral LC may be used in chiral TLC. It is less suitable than chiral LC for accurate quantitative uses but just as there is still a place for TLC in achiral drug analysis, chiral TLC still finds its uses. It may be used to monitor organic reactions, for example, and is more used by Eastern European analysts. Fairly recently (2007), a textbook has been devoted to the technique (15), and even in the very recent literature there are quite a few reports of its use. For example, the stereoselective synthesis of clopidogrel could be followed by chiral TLC using β-cyclodextrin as a chiral mobile phase additive (16). The use of chiral mobile phase additives in TLC is a simple matter and, while some products have been withdrawn, chiral TLC plates can still be had from vendors such as Regis Technologies, Inc., (Illinois, U.S.) and Sigma-Aldrich Corp. (St. Louis, USA).

Chiral gas chromatography (GC) is important in the analysis of odors (17) and is also of value in the fields of insect pheromones (18), pesticides, and fungicides. Even when it comes to less volatile compounds such as drugs, it still has a major role to play in the field of forensic analysis. Amphetamines in particular have been subject to chiral GC (19,20). Even for pharmaceutical drugs there are instances when chiral GC is used. In the early days of chiral LC it was still useful to develop a chiral GC assay for the determination of drugs in biological fluids. Davies used Chirasil-Val, a dipeptide-based GC phase, for the determination of cromakalim enantiomers (21). More recently, nonsteroidal anti-inflammatory drugs have been determined by chiral GC using diastereomeric amide formation (22). This application involved the study of drugs in the environment, a growing field. While this might seem like an unusual application, for more mainstream chiral drug analysis there is at least one instance where a pharmaceutical company has looked at developing a chiral GC screen. Clearly, this would seldom be expected to give rise to the method of choice for the active pharmaceutical ingredient (API), the same might not be the case for synthetic intermediates where, for example, chiral alicyclic compounds might be very amenable to chiral GC. GC phases used in such a screen would nowadays be mainly ones based on cyclodextrins (permethylated, etc.) but might soon include Armstrong's new cyclofructan phases (23).

Again as intimated at the outset, there is an almost endless set of possibilities for introducing a chiral selector into an analytical technique to provide a method for separating and determining enantiomers. The final possibilities to be mentioned are ones that could even be considered to be chiral LC. Countercurrent chromatography, involving centrifugal forces and two immiscible liquid phases, has its strong advocates for being a suitable means of separating enantiomers on a preparative scale. Typically a chiral selector such as a cyclodextrin would be in one of these phases. It is even possible to have countercurrent capillary electrophoresis (24). However, this innovative use of a macrocyclic antibiotic by Reilly and Risley has not found widespread use. What is becoming more common though and is likely to be one of the areas for future development in chiral analysis is the development of chiral sensors. These are usually based on membranes containing immobilized chiral selectors and, as such, borrow heavily from the technology built up over many years of development of chiral LC.

REFERENCES
1. Lough WJ, Matlin SA. Consideration of other techniques. In: Lough WJ, ed. Chiral Liquid Chromatography. Glasgow: Blackie, 1989:213–222.
2. Lough WJ. Separation of chiral compounds—from crystallisation to chromatography. In: Lough WJ, Wainer IW, eds. Chirality in the Natural and Applied Sciences. Oxford: Blackwell Publishing Ltd., 2002:179–202.
3. Schug KA, Joshi MD, Frycak P, et al. Investigation of monovalent and bivalent enantioselective molecular recognition by electrospray ionization-mass spectrometry and tandem mass spectrometry. J Am Soc Mass Spectrom 2008; 19(11):1629–1642.
4. The British Pharmacopoeia, The Stationery Office, Norwich, UK.
5. Sajewicz M, Wrzalik R, Gontarska M, et al. In vitro chiral conversion, phase separation, and wave propagation in aged profen solutions. J Liquid Chromatogr 2009; 32(9):1359–1372.

6. Sajewicz M, Gontarska M, Kronenbach D, et al. Thin-layer chromatographic and polarimetric investigation of the oscillatory in-vitro chiral inversion of S-(+)-ketoprofen. J Planar Chromatogr Modern TLC 2008; 21(5):349–353.
7. Sajewicz M, Kronenbach D, Gontarska M, et al. TLC and polarimetric investigation of the oscillatory in-vitro chiral inversion of L-alanine. J Planar Chromatogr Modern TLC 2008; 21(1):43–47.
8. Sanchez FG, Diaz AN, de Vicente ABM. Enantiomeric resolution of bupivacaine by high-performance liquid chromatography and chiroptical detection. J Chromatogr A 2008; 1188(2):314–317.
9. Hong Y, Tang Y, Zeng S. Enantioselective plasma protein binding of propafenone: mechanism, drug interaction, and species difference. Chirality 2009; 21(7):692–698.
10. Julinek O, Urbanova M, Lindner W. Enantioselective complexation of carbamoylated quinine and quinidine with N-blocked amino acids: vibrational and electronic circular dichroism study. Anal Bioanal Chem 2009; 393(1):303–312.
11. Kellenbach ER, Dukor RK, Nafie LA. Absolute configuration determination of chiral molecules without crystallisation by vibrational circular dichroism (VCD). Spectrosc Eur 2007; 19(4):15–18.
12. Merelli B, Carli M, Menguy L, et al. Enantiomeric composition of chiral beta-hydroxylamides by (1)H NMR spectroscopy using chiral solvating agent. Spectr Letts 2008; 41(7):361–368.
13. Avolio J, Rothchild R. Optical purity determination and (1)H NMR spectral simplification with lanthanide shift reagents – V. Mephenytoin, 5-ethyl-3-methyl-5-phenyl-2,4-imidazolidinedione. J Pharm Biomed Anal 2009; 2(3–4):403–408.
14. Tran CD, Oliviera D. Fluorescence determination of enantiomeric composition of pharmaceuticals via use of ionic liquid that serves as both solvent and chiral selector. Anal Biochem 2006; 356(1):51–58.
15. Kowalska T, Sherma J, eds. Thin Layer Chromatography in Chiral Separations and Analysis. Florida: CRC Press, 2007.
16. Antic D, Filipic S, Ivkovic B, et al. Direct separation of clopidogrel enantiomers by reverse-phase planar chromatography method using beta-cyclodextrin as a chiral mobile phase additive. Acta Chromatogr 2011; 23(2):235–245.
17. Konig WA. Chirality in the natural world—odours and tastes. In: Lough WJ, Wainer IW, eds. Chirality in the Natural and Applied Sciences. Oxford: Blackwell Publishing Ltd., 2002:261–284.
18. Mori K. Separation of chiral compounds: chemical communications. In: Lough WJ, Wainer IW, eds. Chirality in the Natural and Applied Sciences. Oxford: Blackwell Publishing Ltd., 2002:241–260.
19. Martins LF, Yegles M, Chung H, et al. Sensitive, rapid and validated gas chromatography/negative ion chemical ionization-mass spectrometry assay including derivatisation with a novel chiral agent for the enantioselective quantification of amphetamine-type stimulants in hair. J Chromatogr B 2006; 842:98–105.
20. Drake SJ, Morrison C, Smith F. Simultaneous chiral separation of methylamphetamine and common precursors using gas chromatography/mass spectrometry. Chirality 2011; 23(8):593–601.
21. Davies B. Development of a chiral capillary GC method for the quantitation of the enantiomers of cromakalim in biological fluids. Proc. VII International Bioanalytical Forum, Guildford, Surrey, UK, 1987.
22. Hashim NH, Khan SJ. Enantioselective analysis of ibuprofen, ketoprofen and naproxen in wastewater and environmental water samples. J Chromatogr A 2011; 1218(29):4746–4754.
23. Zhang Y, Armstrong DW. 4,6-Di-O-pentyl-3-O-trifluoroacetyl/propionylcyclofructan stationary phases for gas chromatographic enantiomeric separations. Analyst 2011; 136(14):2931–2940.
24. Reilly J, Risley, DS. The separation of enantiomers by countercurrent capillary electrophoresis using the macrocyclic antibiotic A82846B. LC-GC 1998; 16(2):170–178.

10 Stereoselective transport of drugs

Prateek Bhatia and Ruin Moaddel

The enantioselectivity of the protein-based liquid chromatography columns has been previously reported (1). The expansion to transmembrane proteins has also demonstrated that the resulting cellular membrane affinity chromatography columns can also be used to study the enantioselectivity of the immobilized proteins, c.f. transporters (2).

Transporters are responsible for the movement of small molecules across cellular membranes. This transport can be adenosine tri-phosphate (ATP)-dependent (energy-dependent) or concentration dependent (energy independent). The human genome is believed to code for up to 1200 membrane transporter proteins, which play vital roles in homeostasis by mediating the transport of essential nutrients, amino acids, bile acids, and hormones (3) and the excretion of toxins and metabolic waste products. In addition, transporters play a significant role in the disposition of therapeutic drugs (4). In fact, the low oral bioavailability of some drugs has been shown to be dependent on a combination of transporters and cytochrome P450 enzymes (5).

The major facilitator superfamily (MFS) is the largest group of transporters in existence (6) and is one of only two families of transporters that occur in every organism, namely, MFS and the ATP-binding cassette (ABC) superfamily. The MFS family includes the solute carrier family, which encompasses the majority of the amino acid transporters, the glucose transporters, organic anion transporters, and organic cation transporters, while ABC family includes frequently targeted members like P-glycoprotein (Pgp), multidrug-resistant proteins 1 and 2 (MRP1 and MRP2).

The importance of stereoselectivity in transporters is clearly seen with amino acid transporters, glucose transporters, and neurotransmitters. Examples of these include work carried out by Auclair et al., where they demonstrated that α-methyl dopa was stereoselectively transported into the aqueous humor (7). Another example is the Glut1 stereoselective transport of D-glucose (8). The stereoselectivity of this transport was studied by Lundahl using chromatographic techniques and a stationary phase containing immobilized Glut1 (8). This work introduced a novel method, namely, bioaffinity chromatography, to characterize binding sites of proteins. Although many techniques exist for studying the stereochemistry of proteins, bioaffinity chromatography or in the case of transmembrane proteins cellular membrane affinity chromatography (CMAC) has been shown to be a very important tool in addressing the stereoselectivity of drug transport (9,10). CMAC is a technique where the target protein is isolated by homogenization and solubilization of a source and immobilized onto a stationary phase, predominantly the immobilized artificial membrane-phosphatidylcholine (IAM-PC) stationary phase during a dialysis step. As will be discussed in greater detail later in the chapter, the CMAC (hOCT1) column was used for the development of a molecular model that described the stereoselective binding of competitive inhibitors with the transporter (2). Although a detailed discussion of CMAC is warranted, it is not the focus of this chapter, for interested readers, a detailed protocol has already been published (10).

The importance of stereoselective transport of drugs across the epithelia has been discussed in great detail by Ronda Ott and Kathleen M. Giacomini in the previous edition (11), which addresses the stereoselectivity of transport mechanisms, properties of epithelial cells, amino acids, organic cations and anions, and reabsorption of drugs. In this chapter, the stereoselectivity of transporters from both the MFS and ABC superfamilies will be addressed, specifically the organic cation transporter (hOCT1), Pgp, and MRP1.

SLC FAMILY

The solute carrier (SLC) transporters have 255 members in humans, of which majority are highly specific transporters. The focus of this chapter will be on polyspecific transporters, namely, the SLC22 superfamily (MFS) (12), which has 12 members in humans and rats including the organic cation transporters OCT1, OCT2, and OCT3; the carnitine transporter; and several organic anion transporters.

Transport studies with cell lines transfected with hOCT1 have established that hOCT1 mediates the bidirectional transport of small organic cations (50–350 amu) such as tetraethylammonium (TEA) and 1-methyl-4-phenylpyridinium (MPP^+) (12–14). In these studies, it was also demonstrated that hOCT1 transport can be inhibited by a diverse group of commonly administered drugs including verapamil, quinidine, quinine, and disopyramide (12,14).

The identification and characterization of interactions with the hOCT1 has been primarily accomplished using cellular uptake and *trans*-stimulation studies (13,15). In addition, CMAC has been used for the determination of hOCT1 binding interactions (15). In this technique, cellular membrane fragments obtained from a stably transfected MDCK cell line that expresses hOCT1were immobilized onto the surface of the IAM-PC stationary phase resulting in the hOCT1(+)-IAM stationary phase (16), and the binding interactions were studied using frontal affinity chromatography techniques.

The stereoselectivity of the hOCT1 transporter was initially demonstrated with the enantiomers of disopyramide (17), where the IC_{50} value of (R)-disopyramide was determined to be about twofold lower than that of (S)-disopyramide using cellular uptake studies. As disopyramide displayed enantioselectivity, the enantiomers of verapamil were studied during the initial chromatographic studies using the hOCT1(+)-IAM stationary phase and were shown to demonstrate significant enantioselectivity with (R)-verapamil having an affinity 77-fold stronger than (S)-verapamil (16). As a result, the study was expanded to a set of 22 compounds including 8 pairs of enantiomers and 3 pairs of diastereomers, resulting in a pharmacophore model, which described the observed stereoselectivity of the hOCT1 transporter. The enantioselectivity and diastereoselectivity (α) of the eight pairs of enantiomers and five pairs of diastereomers, respectively, was calculated as the K_i of the (S)-enantiomer/diastereomer divided by the K_i of the (R)-enantiomer/diastereomer, $K_i(S)/K_i(R)$ (Table 10.1). The generated pharmacophore from this data contained a positive ion interaction site, a hydrophobic site, and two hydrogen-bond acceptor sites (Fig. 10.1). Using the center of the

TABLE 10.1 The Experimentally Determined and Estimated Stereoselectivities (α) for the Chiral Compounds Used in This Study

Compound	α (Experimental)	α (Estimated)
(R,S)-Verapamil	77	150
(R,S)-Atenolol	0.5	0.8
(R,S)-Propranolol	3.0	1.4
(RR,SS)-Pseudoephedrine	1.5	1.8
(RR,SS)-Fenoterol	0.3	0.4
(RS,SR)-Fenoterol	0.5	0.7
(R,S)-Isoproterenol	1.5	1.4
(R,S)-Disopyramide	2.0	1.2
Quinine/Quinidine	1.6	1.5
(R,S)/(R,R)-Fenoterol	1.1	1.2
(S,R)/(R,R)-Fenoterol	0.5	0.9
(R,S)/(S,S)-Fenoterol	3.5	2.8
(S,R)/(S,S)-Fenoterol	1.7	2.1

For the enantiomeric pairs, the α value is defined as $K_i(S)/K_i(R)$, and for the diastereomeric pairs, the ratios are as presented in the table.
Source: From Table 2 of Ref. 2.

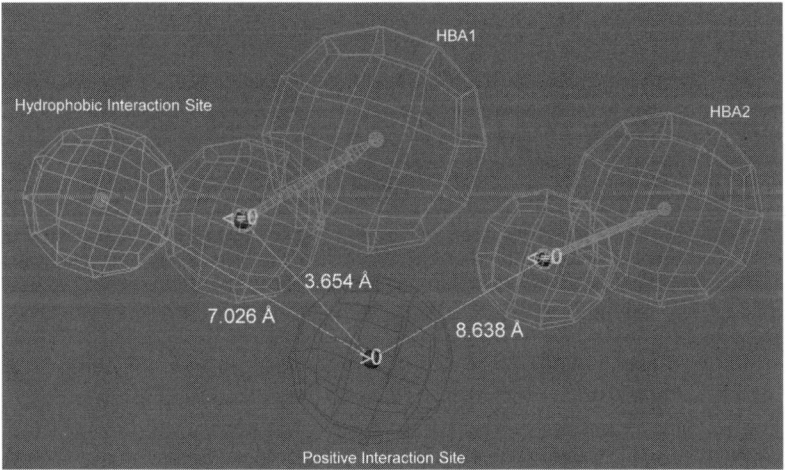

FIGURE 10.1 The pharmacophore model generated by hypothesis-1 with a positive ion interaction site, a hydrophobic interaction site and two hydrogen-bond acceptor sites, HBA1 and HBA2. *Source*: From Figure 3 of Ref. 2.

positive ion interaction site as the origin, the distances to the center of the hydrogen-bond acceptor sites were ~3.7 Å (HBA1) and ~8.6 Å (HBA2). Using the same approach, the distance to the center of the hydrophobic site was ~7 Å.

The data indicated that the mapping of the compounds to three or more of these sites correlated with the stereoselective binding to the hOCT1 and that the model reflected the relative K_i values and stereoselectivities.

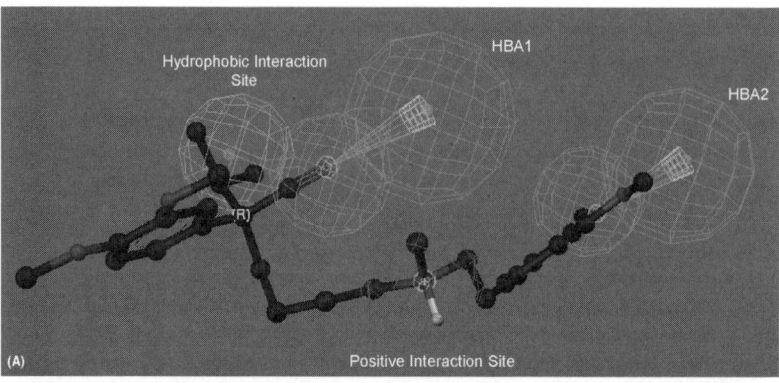
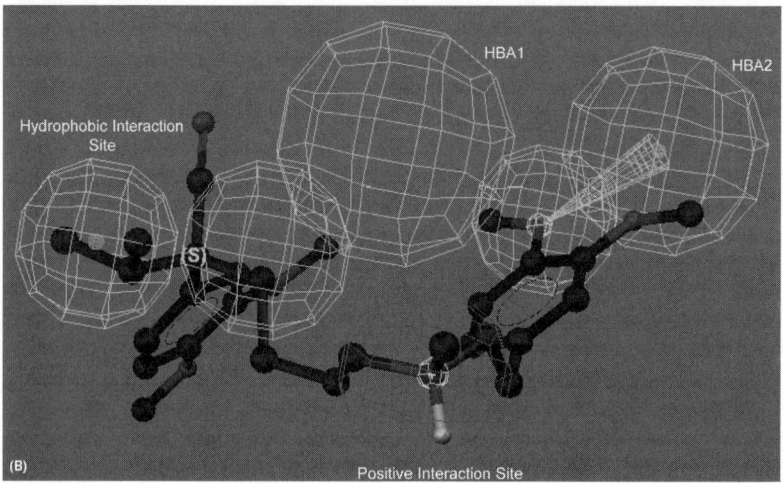

FIGURE 10.2 (**A, B**). The fit of verapamil enantiomers in the proposed pharmacophore, where: A = the mapping of (*R*)-verapamil; B = the mapping of (*S*)-verapamil. Non-polar hydrogen atoms have been omitted for clarity.

The greatest difference in enantioselectivity was observed with verapamil, where the experimentally determined enantioselectivity and estimated α's were 77 and 150, respectively (Table 10.1). When (*R*)-verapamil was fitted to the proposed pharmacophore, all the relevant functional groups of the molecule matched the hypothesis (Fig. 10.2A), while for (*S*)-verapamil, only three of the model feature sites matched (Fig. 10.2B). The model indicates that the enantioselectivity resulted from the hydrogen bonding of the nitrile moiety on verapamil with the HBA1 site. Similarly to (*S*)-verapamil, the mapping of both (*R*)- and (*S*)-propranolol to the pharmacophore model indicated that propranolol mapped to three of the four sites on the pharmacophore; however, in this case they did not interact with HBA2 site. The difference in the estimated K_i values for these enantiomers was a function of the calculated fits, which were 6.45 for (*R*)-propranolol and 6.31 for (*S*)-propranolol.

When the experimentally determined and estimated stereoselectivites of the 13 sets of enantiomers/diastereomers were compared using linear regression analysis, a significant relationship was observed, $r^2 = 0.9992$ ($p < 0.0001$). A correlation was also observed in the absence of verapamil with an $r^2 = 0.6623$ ($p = 0.0013$).

The developed pharmacophore clearly demonstrated the enantioselectivity of the hOCT1 transporter, suggesting that the three-dimensional relationship between the identified interaction sites reflects the spatial distribution of similar binding sites within hOCT1. The fact that these differences in relative fit produced the experimentally observed stereoselectivities is also consistent with previously identified chiral recognition mechanisms (18). In these mechanisms, each enantiomer of a chiral compound interacts with the same sites on the chiral selector and the differential stabilities of the resulting diastereomeric complexes is a function of the relative conformational energies required to create the complexes.

The data from this study clearly demonstrated that the pharmacophore hOCT1 transporter displays stereoselective binding of compounds. Thus, the results suggest that the enantioselectivity of new drug candidates that contain chiral centers should be determined. Even for drugs targeting polyspecific transporter, as the updated pharmacophore model generated introduced an additional hydrogen binding site, which reflected the enantioselectivity of the transport protein.

ABC TRANSPORTERS

ABC transporters hydrolyze ATP to drive drug transport. In humans 48 ABC genes have been identified, which are classified into 7 subfamilies. Most of the ABC transporters are located in the plasma membrane where they can efflux a variety of structurally diverse drugs, drug conjugates, and metabolites from the cell including phospholipids, peptides, steroids, amino acids, polysaccharides, and other xenobiotics (19).

ABC transporters consist of four domains, two nucleotide binding domains (NBD) located in the cytoplasm and two transmembrane domains (TMDs) responsible for binding and transport of substrates and/or drugs (20). Two sequence motifs located at 100 to 200 amino acids apart in each NBD, designated Walker A and Walker B, are conserved in all ABC transporters.

The mechanism of unidirectional transport seems to be the ATP-driven conversion of the inward-facing state to the outward-facing state. Binding of two ATP molecules at the interface of the NBDs closes the gap between the conserved ATP-binding motifs. As this gap closes, the TMDs flip from the inward-facing to the outward-facing conformation. ABC exporters such as Pgp efflux the bound drugs by presenting a cavity of low affinity to the extracellular medium (21).

In addition to the roles ABC transporters play in maintaining homeostasis, they have become an increasingly important therapeutic target for the treatment of cancer. In fact, breast cancer resistance was initially shown to be caused by an increased expression of Pgp transporters in patients receiving chemotherapy. While a vast number of the ABC transporters have been discovered, the predominant members targeted therapeutically are Pgp, MRP1, MRP2, and breast cancer resistance protein (BCRP) (18–21).

Pgp

Cancer cells are frequently resistant or develop resistance to anticancer agents during treatment. One form of drug resistance is observed against a variety of chemically unrelated agents and is known as multidrug resistance (MDR). The MRPs actively efflux anticancer agents, decreasing drug accumulation and allowing the malignant cells to survive in the presence of toxic levels of chemotherapeutic agents. One of the most frequently targeted ABC proteins in MDR in solid tumors and hematological malignancies is the Pgp transporter (22). In order to formulate successful pharmaceutical entities specific to the inhibition of this phenomenon, several pharmacophore models of inhibitors have been constructed and validated with molecules excluded from model building, suggesting these models are certainly adequate for ranking new molecules and interactions with Pgp in early drug discovery (23).

The Pgp transporter is the best characterized member of the ABC family. It is encoded by the MDR1 gene and is expressed as a single polypeptide that is organized in two domains, joined by a short 60 amino acid segment termed the linker region. The two halves share a similar structure and the arrangement of the domains in the primary sequence is TMD1-NBD1-TMD2-NBD2. The N- and C-terminal halves of Pgp show strong sequence homology to each other and in bacterial ABC transporters the functional domains are often composed of pairs of identical separate polypeptides (24).

Transmembrane helices 4–6 and 10–12 have been shown to be involved in substrate binding. Each hydrophobic domain is followed by a hydrophilic domain (NBD) containing a nucleotide binding site that is located at the cytoplasmic face of the membrane and couples ATP hydrolysis to the transport process (25).

Recently, several Pgp pharmacophores have been used for database screening and identification of potential Pgp substrates and inhibitors. The first such report appeared in 2003, where Rebitzer et al. used a propafenone derivative (26) MDR modulator-based pharmacophore model to screen Derwent World Drug Index. Among the returned 28 hits, 9 were previously described MDR modulators.

In another study, Elkins et al. used an in silico approach to model in vitro data derived using structurally diverse inhibitors to create a three-dimensional quantitative structural activity relationship of Pgp inhibitors. The study generated four computational models that were able to predict how other molecules might inhibit Pgp solely by providing the structure to the software (23). An extension of this study suggested a fifth pharmacophore model based on verapamil binding (27). Each of the models was used to predict the IC_{50} values of verapamil and its metabolites; all of them were able to rank the compounds as potent Pgp inhibitors. From this study it was possible to create five different computer-generated pharmacophore models for predicting Pgp substrates and inhibitors (c.f. Figure 4 from Ref. 26); although some of the compounds used in the study seem to be chiral, only racemic mixtures were used to generate the models.

Although the generated pharmacophores have been limited in the extent of stereochemistry studied, there has been some evidence of stereoselectivity at the Pgp transporter. For example, a study was performed to evaluate and compare the ability of P-glycoproteins to transport hydrophilic monoquaternary drugs. The N-methylated derivatives of the diastereomers quinidine and quinine, the monoquaternary compounds N-methylquinidine and N-methylquinine, were used as model substrates. Vincristine, an established MDR1 substrate, was used as a reference. It was observed that there was an ATP-dependent uptake of all drugs studied. The clearance of N-methylquinidine was greater than those determined for N-methylquinine. This stereoselective difference was also evident from differential inhibitory studies with the two isomers. It was apparent from the above study that Pgp displays stereospecificity (28). In addition to these studies, Neuhoff et al. demonstrated that omeprazole, a β2-adrenoreceptor antagonist, had stereoselective affinity for the verapamil binding site on the Pgp transporter (29). The stereoisomers of verapamil have also been shown to selectively bind to MRP1 in a separate study (30). Finally, a study investigating the Pgp antagonistic effects of enantioselective mefloquine (MQ) using CMAC. Membranes isolated from MDA435/LCC6MDR1 were immobilized on an IAM stationary phase. Frontal affinity chromatographic studies were carried out using vinblastine, a known Pgp substrate as the marker ligand. Interactions of the enantiomers of MQ did not produce stereoselective inhibition, but the binding of MQ did result in an anti-cooperative allosteric interaction that effected vinblastine binding. In this process, MQ binds to a contiguous or separate site on Pgp producing secondary effects at the vinblastine binding site that reduces its ability to bind to that site. However, when the marker ligand was replaced by [^{3}H]-cyclosporin A (CsA), a known Pgp inhibitor, significant enantioselective binding was observed. In the presence of 3-mM ATP in the mobile phase (+)-MQ showed dose-dependent displacement of [^{3}H]-CsA frontal chromatogram, while increasing concentrations of (−)-MQ did not competitively displace [^{3}H]-CsA. This indicated the existence of a competitive binding site that is stereoselective (31). All these studies indicate that the stereoselectivity of a drug plays an important role in its binding properties to the Pgp transporter. As previously shown, with the hOCT1 transporter, where the consideration of stereoselective binding (2) modified the existing pharmacophore that did not consider stereoselectivity (15). In the same manner, the existing pharmacophore model for the Pgp without a stereoselective perspective might not be accurate and might not be a true predictor of spatial requirements for eventual development of Pgp substrates/inhibitors.

MRP1

The human MRP1 is one of the most extensively researched members of the ABC protein family. MRP1 is a 190-kDa protein encoded by the mrp1 gene and is constituted by 1531 amino acids distributed in 3 domains and 17 transmembrane helices; among them helices 10, 11 and 16, 17 are involved in drug recognition (19,32). It is an efflux pump able to transport anionic conjugate compounds out of cells in an ATP-dependent manner (23).

It confers resistance to diverse chemotherapeutic agents when overexpressed in the plasma membrane of cells. The proposed role of MRP1 in the therapeutic resistance phenomena in anticancer treatment has triggered many studies describing its function and mechanism, substrate specificity, and search for inhibitors (33). A look at the amino acid sequence of MRP1 shows that its core structure is similar to all ABC transporters. Like MDR1, MRP1 contains two membranes spanning domains followed by one nucleotide binding domain (34). Each membrane-spanning domain consists of six transmembrane α-helices, and two membrane-spanning domains are located on the cytoplasmic side of the plasma membrane (35).

Unlike Pgp, which transports hydrophobic compounds, MRP1 has been shown to transport a wide variety of hydrophilic compounds (36). These proteins have also been shown to transport neutral compounds conjugated with glutathione, glucuronide, and sulfate (37). It has been reported that if the structural activity relationship of this protein is deduced from its inhibitors and substrates, the presence of two aromatic nuclei and two nitrogen groups are a must (36).

In another study to determine the quantitative structure-activity relationship (QSAR) of MRP1, the effect of a number of flavonoids was studied. The data generated were used to elicit structural requirement of flavonoids to effectively inhibit this protein. Based on the findings of this study, the QSAR equation created in this study showed that there are three structural characteristics that are important for inhibition of MRP1: the total number of methoxylated moieties, the total number of hydroxyl groups, and the dihedral angle between the B- and C-ring (38).

Five different MRP1 inhibitors (LY329146, LY402913, dehydrosilybin, indolopyrimidine, and phenoxymethylquinoxalinone II) were used to develop a Catalyst HIPHOP model. LY329146 and LY402913 are raloxifen derivates developed by Eli Lily (36) (c.f. Figure 3 from Ref. 38). The model contains three-ring aromatic and three HBA features. The model was applied to screen a literature database of over 500 clinically used drugs and retrieved 8 hits (candesartan, eprosartan, fexofenadine, losartan, sulfasalazine, telmisartan, vancomycin, and zafirlukast); as a proof of the accuracy of the model, losartan, sulfasalazine, and zafirlukast have been shown to be high-affinity substrates of MRP1 (23).

Although there are various reports where expression and subsequent functionality of the MRP1 transporter have been shown by different models, very little stereoselectivity has been observed. One example where MRP1 has shown to exhibit different patterns is its binding to (R)- and (S)-verapamil. Verapamil-induced apoptosis in cells overexpressing MRP1 was stereoselective with the (S)-enantiomer being much more potent than the (R)-enantiomer. Furthermore, it is interesting to note that the (S)-verapamil was more potent than the racemic mixture, which might indicate that the (R)-enantiomer works as an antagonist of the (S)-enantiomer and decreases the intracellular concentration (23).

Most inhibitors of MRP1 work via GSH (intracellular glutathione) depletion and GST (glutathione (S)-transferase) activity. In a recent report it was indicated that not only a newly synthesized chiral compound tetraisohydroquinoline (HZ08) bound stereoselectively to Pgp and MRP1 but also that it brought about reversal to MDR by interacting with MRP1 directly, independent of GSH or GST activity. (R,S)-HZ08 was shown to reverse resistance to adriamycin and vincristine in human leukemia cells overexpressing both Pgp and MRP1 (40).

All these studies indicate that the stereoselectivity of transporters is of key importance in drug development and drug discovery programs. Overlooking the stereoselective binding in transporters can result in wasted resources and time, by generating models that do not reflect the actual binding site, and could result in erroneous leads.

ACKNOWLEDGMENTS

The research described in this chapter includes discussion of studies that were supported entirely or in part by the Intramural Research Program of the NIH, National Institute on Aging.

REFERENCES

1. Noctor TAG. Enantioselective binding of drugs to plasma proteins. In: Wainer IR, ed. Drug stereochemistry: analytical methods and pharmacology. 2nd ed. New York: Marcel Dekker; 1993:337–364.
2. Moaddel R, Ravichandran S, Bighi F, et al. Pharmacophore modelling of stereoselective binding to the human organic cation transporter (hOCT1). Br J Pharmacol 2007; 151(8):1305–1314.
3. Jonker JW, Schinkel AH. Pharmacological and physiological functions of the polyspecific organic cation transporters: OCT1, 2 and 3 (SLC22A1-3). J Pharmacol Exp Ther 2004; 308:2–9.
4. Ho RH, Kim RB. Transporters and drug therapy: implications for drug disposition and disease. Clin Pharmacol Ther 2005; 78(3):260–277.
5. Oostendorp RL, Beijnen JH, Schellens JH. The biological and clinical role of drug transporters at the intestinal barrier. Cancer Treat Rev 2009; 35(2):137–417.
6. Saier MH Jr, Beatty JT, Goffeau A, et al. The major facilitator superfamily. J Mol Microbiol Biotechnol 1999; 1(2):257–279.
7. Auclair E, Laude D, Wainer IW, et al. Comparative pharmacokinetics of D- and L-alpha methyldopa in plasma, aqueous humor, and cerebrospinal fluid in rabbits. Fundam Clin Pharm 1988; 2(4):283–293.
8. Lu L, Brekkan E, Haneskog L, et al. Effects of pH on the activity of the human red cell glucose transporter Glut 1: transport retention chromatography of D-glucose and L-glucose on immobilized Glut 1 liposomes. Biochim Biophys Acta 1993; 1150(2):135–146.
9. Moaddel R, Wainer IW. Development of immobilized membrane-based affinity columns for use in the online characterization of membrane bound proteins and for targeted affinity isolations. Anal Chim Acta 2006; 564(1):97–105.
10. Moaddel R, Wainer IW, The preparation and development of cellular membrane affinity chromatography columns. Nat Protoc 2009; 4(2):197–205.
11. Wainer IW. Drug Stereochemistry: Analytical Methods and Pharmacology. New York: Marcel Dekker Inc., 1993.
12. Koepsell, H, Schmitt, BM, Gorboulev V. Organic cation transporters. Rev Physiol Biochem Pharmacol 2003; 150(1):36–90.
13. Zhang L, Gorset W, Dresser MJ, et al. The interaction of n-tetraalkylammonium compounds with a human organic cation transporter, hOCT1. J Pharmacol Exp Ther 1999; 288(3):1192–1198.
14. Dresser MJ, Leabman MK, Giacomini KM. Transporters involved in the elimination of drugs in the kidney: organic anion transporters and organic cation transporters. J Pharm Sci 2001; 90(4).
15. Bednarczyk D, Ekins S, Wikel JH, et al. Influence of molecular structure on substrate binding to the human organic cation transporter, hOCT1. Mol Pharmacol 2003; 63(3):489–498.
16. Moaddel R, Yamaguchi R, Ho P, et al. Development and characterization of an immobilized human organic cation transporter based liquid chromatographic stationary phase. J Chromatogr B 2005; 818(2):263–268.
17. Zhang L, Schaner ME, Giacomini KM. Functional characterization of an organic cation transporter (hOCT1) in a transiently transfected human cell line (HeLa). J Pharmacol Exp Ther 1998; 286(1):354–361.
18. Booth TD, Wainer IW. Mechanistic investigation into the enantioselective separation of mexiletine and related compounds, chromatographed on an amylose tris

(3, 5-dimethylphenylcarbamate) chiral stationary phase. J Chromatogr A 1996; 741(2):205–211.
19. Deeley RG, Westlake C, Cole SP. Transmembrane transport of endo- and xenobiotics by mammalian ATP-binding cassette multidrug resistance proteins. Physiol Rev 2006; 86(3):849–899.
20. Klopman G, Shi LM, Ramu A. Quantitative structure-activity relationship of multidrug resistance reversal agents. Mol Pharmacol 1997; 52(2):323–334.
21. Hollenstein K, Frei DC, Locher KP. Structure of an ABC transporter in complex with its binding protein. Nature 2007; 446:213–216.
22. Ishikawa T, Hirano H, Onishi Y, et al. Functional evaluation of ABCB1 (P-glycoprotein) polymorphisms: high-speed screening and structure-activity relationship analyses. Drug Metab Pharmacokinet 2004; 19(1):1–14.
23. Ekins S, Kim RB, Leake BF, et al. Three-dimensional quantitative structure activity relationships of inhibitors of P-glycoprotein. Mol Pharmacol 2002; 61:964–973.
24. Lee JY, Urbatsch IL, Senior AE, et al. Projection structure of P-glycoprotein by electron microscopy. Evidence for a closed conformation of the nucleotide binding domains. J Biol Chem 2002; 277(42):40125–40131.
25. Lai Y, Xing L, Poda GI, et al. Structure-activity relationships for interaction with multidrug resistance protein 2 (ABCC2/MRP2): the role of torsion angle for a series of biphenyl-substituted heterocycles. Drug Metas Dispos 2007; 35(6):937–945.
26. Rebitzer S, Annibali D, Kopp S, et al. In silico screening with benzofurane-and benzopyrane-type MDR-modulators. Farmaco 2003; 58(3):185–191.
27. Ekins S, Kim RB, Leake BF, et al. Application of three-dimensional quantitative structure-activity relationships of P-glycoprotein inhibitors and substrates. Mol Pharmacol 2002; 61(5):974–981.
28. Hooiveld GJ, Heegsma J, Van Montfoort JE, et al. Stereoselective transport of hydrophilic quaternary drugs by human MDR1 and rat Mdr1b P-glycoproteins. Br J Pharmacol 2002; 135(7):1685–1694.
29. Neuhoff S, Langguth P, Dressler C, et al. Affinities at the verapamil binding site of MDR1-encoded P-glycoprotein: drugs and analogs, stereoisomers and metabolites. Int J Clin Pharmacol Therap 2000; 38(4):168–179.
30. Perrotton T, Trompier D, Chang XB, et al. (R)-and (S)-Verapamil differentially modulate the multidrug-resistant protein MRP1. J Biol Chem 2007; 282(43): 315–342.
31. Lu L, Leonessa F, Baynham MT, et al. The enantioselective binding of mefloquine enantiomers to P-glycoprotein determined using an immobilized P-glycoprotein liquid chromatographic stationary phase. Pharm Res 2001; 18(9):1327–1330.
32. Lohoff M, Prechtl S, Sommer F, et al. A multidrug-resistance protein (MRP)-like transmembrane pump is highly expressed by resting murine T helper (Th) 2, but not Th1 cells, and is induced to equal expression levels in Th1 and Th2 cells after antigenic stimulation in vivo. J Clin Invest 1998; 101(3):703–710.
33. Ramaen O, Leulliot N, Sizun C, et al. Structure of the human multidrug resistance protein 1 nucleotide binding domain 1 bound to Mg2+/ATP reveals a non-productive catalytic site. J Mol Biol 2006; 359(4):940–949.
34. Rosenberg MF, Mao Q, Holzenburg A, et al. The structure of the multidrug resistance protein 1 (MRP1/ABCC1). Crystallization and single-particle analysis. J Biol Chem 2001; 276(19):16076–16082.
35. Colmenarejo G. In silico ADME prediction: data sets and models. Curr Comput Aided Drug Design 2005; 1(4):365.
36. Boumendjel A, Baubichon-Cortay H, Trompier D, et al. Anticancer multidrug resistance mediated by MRP1: recent advances in the discovery of reversal agents. Med Res Rev 2005; 25(4):453–472.
37. Van Tellingen O, Buckle T, Jonker JW, et al. P-glycoprotein and Mrp1 collectively protect the bone marrow from vincristine-induced toxicity in vivo. Br J Cancer 2003; 89(9):1776–1782.

38. Van Zanden JJ, Wortelboer HM, Bijlsma S, et al. Quantitative structure activity relationship studies on the flavonoid mediated inhibition of multidrug resistance proteins 1 and 2. Biochem Pharmacol 2005; 69(4):699–708.
39. Chang C, Ekins S, Bahadduri P, et al. Pharmacophore-based discovery of ligands for drug transporters star, open. Adv Drug Deliv Rev 2006; 58(12–13):1431–1450.
40. Yan F, Jiang Y, Li YM, et al. Reversal of P-glycoprotein and multidrug resistance-associated protein 1 mediated multidrug resistance in cancer cells by HZ08 Isomers, tetrataisohydroquinolin derivatives. Biol Pharm Bull 2008; 31(6):1258–1264.

11 Enantioselective binding of drugs to plasma proteins

Thomas H. Kim

INTRODUCTION

Early evaluation of absorption, distribution, metabolism, excretion, and toxicity (ADME/Tox) parameters is essential in the drug discovery and development process since information on these parameters possibly eliminates wasted developmental effort and cost on unsuitable compounds, and directs the focus of drug optimization toward more "drug-like" compounds, which eventually become successful drug candidates.

Plasma protein binding is one of those parameters that predicts the distribution of drugs in the body, and quantitative determination of this property is essential in clinical drug development process because the therapeutic effect of drugs in the systemic circulation generally depends on the degree of its interaction with various plasma proteins in blood. In some cases, the binding of these solutes will occur in a specific manner, as is the case with interactions of the hormone L-thyroxine with thyroxine-binding globulin or the binding of corticosteroids and sex hormones to various steroid-binding globulins (1). In other situations these interactions will occur with a general ligand, such as the blood transport proteins human serum albumin (HSA) and α_1-acid glycoprotein (AGP) (2–5).

HSA and AGP are two most abundant plasma proteins responsible for drug bindings. These proteins act as a reservoir and the degree of protein-drug binding in plasma can have a significant effect on the pharmacokinetic and pharmacodynamic properties of a drug. This is due to the fact that the unbound drug can penetrate the wall of the blood vessel and become therapeutically effective once it finds a target receptor. It is also the fraction that may be metabolized and excreted, and the metabolism and excretion of unbound drug can affect the drug's biological half-life in the body since the bound portion is slowly released in order to maintain equilibrium.

The direct or indirect competition of two solutes for the same binding proteins can be an important source of displacement, which can alter the availability of one solute relative to the other. An example of this is the displacement of phenytoin from HSA by valproic acid or the displacement of disopyramide from AGP by mono-N-dealkyldisopyramide (6,7). Competition between drugs and endogenous compounds, such as the displacement of various drugs from HSA by fatty acids or bilirubin, is another source that can alter the availability of a drug (8).

The plasma protein-drug binding is often stereoselective due to the inherent chirality of plasma proteins like HSA and AGP. This chiral discrepancy often leads to the difference in affinities to these proteins for the individual stereoisomers, resulting in differing free fractions. This topic has been of particular interest for some years now with agencies such as the U.S. Food and Drug Administration increasing rules and regulations involving the marketing and use of chiral drugs (see chap. 13). This has also resulted in increased research aimed at resolving such agents, some work of which has involved the use of proteins as chiral ligands in chromatographic and electrophoretic methods.

INTERACTION OF DRUGS WITH PLASMA PROTEINS

The binding of a plasma protein to a drug, regardless of whether the drug is chiral, is often described by the reaction model shown in Eqs. 1 to 4, where P_1 through P_n are the binding regions on the protein and D is the drug, and P_1-D and P_n-D, etc., are the resulting protein-drug complex.

$$P_1 + D \leftrightarrow P_1 - D \qquad (1)$$

$$K_{a1} = \frac{k_{a1}}{k_{d1}} = \frac{[P_1 - D]}{[P_1][D]} \qquad (2)$$

$$P_n + D \leftrightarrow P_n - D \qquad (3)$$

$$K_{an} = \frac{k_{an}}{k_{dn}} = \frac{[P_n - D]}{[P_n][D]} \qquad (4)$$

In these reactions, [] represents the molar concentration of P and D in solution, K_{a1} through K_{an} are the association equilibrium constant for the drug at each of its binding regions on the protein, k_{a1} through k_{an} are the second-order association rate constants for protein-drug binding, and k_{d1} through k_{dn} are the first-order dissociation constants.

The above equations assume that each binding region on the protein is independent. This means there are no allosteric interactions present in drug-protein interactions, and association constants for individual binding regions are constant and independent of each other. Furthermore, a reversible, single-step process describes the binding of drug to individual binding region on a protein in above equations. This is hardly true in reality, and multiple steps are often involved in such reaction processes (see chap. 3). The diffusion of drug to the protein and the conformational change in protein as a result of its binding to the drug are typical multistep process examples in such reaction processes (9–11). There are many different functional groups present on protein and these functional groups, as an individual moiety or group, are participated in a drug-protein complex stabilizing processes through van der Waal's force and electrostatic interaction as well as hydrogen bonds. These interactions can lead to either specific or nonspecific interactions between the protein and drug. The well-defined three-dimensional binding regions on a protein in combination with protein-drug stabilizing forces mentioned above make it possible for the chiral discrepancy of protein to an individual stereoisomer.

ENANTIOSELECTIVE DRUG BINDING TO PLASMA PROTEINS

There are a number of proteins responsible for binding of drug in blood plasma. Some proteins such as serum albumin or globulins have 8-order or more abundance than scarce proteins. Virtually all proteins in plasma participate in binding of endo- or exogenous molecules to some degree. However, these interactions are often nonspecific and have no significant clinical implications associated with it.

Some transport proteins and their interactions with other molecules (small or large) in plasma are clinically important since their binding can affect overall bioavailability and thus pharmacokinetics of drugs (see chap. 10).

In this section, the enantioselective bindings of serum album and acid glycoprotein, two proteins in plasma most responsible in drug binding actions, are mainly discussed. Other proteins with little clinical implication will be briefly discussed at the end of the section.

Human Serum Alubmin

HSA is the most abundant protein in serum, with typical blood concentration of 50 g/L. It is probably the most extensively studied of all proteins (12). The crystal structure of HSA indicates that this protein is composed of three homologous domains (referred to as I, II, and III), with each domain being divided into two subdomains (referred to as A and B). Many low-mass solutes show reversible binding to HSA, including both endogenous agents (e.g., long chain fatty acids, steroids, and bilirubin) and exogenous compounds (e.g., salicylate, diazepam, and warfarin). The binding of HSA with these substances occurs at relatively well-defined regions on this protein. For drugs, the two most important binding regions on HSA are Sudlow sites I and II. These are also referred to as the warfarin-azapropazone and indole-benzodiazepine sites, respectively, due to their affinities for solutes belonging to these groups of compounds. Sudlow site I is located in the IIA subdomain of HSA and specifically binds to warfarin and coumarins, along with salicylate and many other drugs. Sudlow site II is located in the IIIA subdomain of HSA and interacts with such agents as L-tryptophan, diazepam, and naproxen (1,6,7,12). The presence of these two sites has been confirmed in many studies, including crystallographic work that has identified the location of these two regions (13–16). The enantioselective binding of drug to isolated HSA is presented in Table 11.1.

TABLE 11.1 Drugs Exhibiting Enantioselective Binding to Isolated HSA

Drug	Association constant $n_1 k_1$ (M^{-1})	References
(R)-warfarin	3.40×10^5	17
(S)-warfarin	2.60×10^5	17
(R)-ibuprofen	5.3×10^5	18
(S)-ibuprofen	1.1×10^5	18
(R)-carprofen	6.11×10^6	19,20
(S)-carprofen	3.85×10^6	19,20
(R)-amlodipine	1.0×10^4	21
(S)-amlodipine	1.0×10^5	21
(R)-flurbiprofen	3.39×10^6	22
(S)-flurbiprofen	7.31×10^6	22
(R)-bimoclomol	1.3×10^4	23
(S)-bimoclomol	1.0×10^5	23
(R)-profafenone	2.05×10^3	24
(S)-profafenone	2.08×10^3	24
(R)-sulbenicillin	4.73×10^3	25
(S)-sulbenicillin	2.06×10^3	25
(+)-pirprofen	3.9×10^6	26
(−)-pirprofen	3.7×10^6	26
(R)-alprenolol	6.3×10^5	27
(S)-alprenolol	9.7×10^5	27

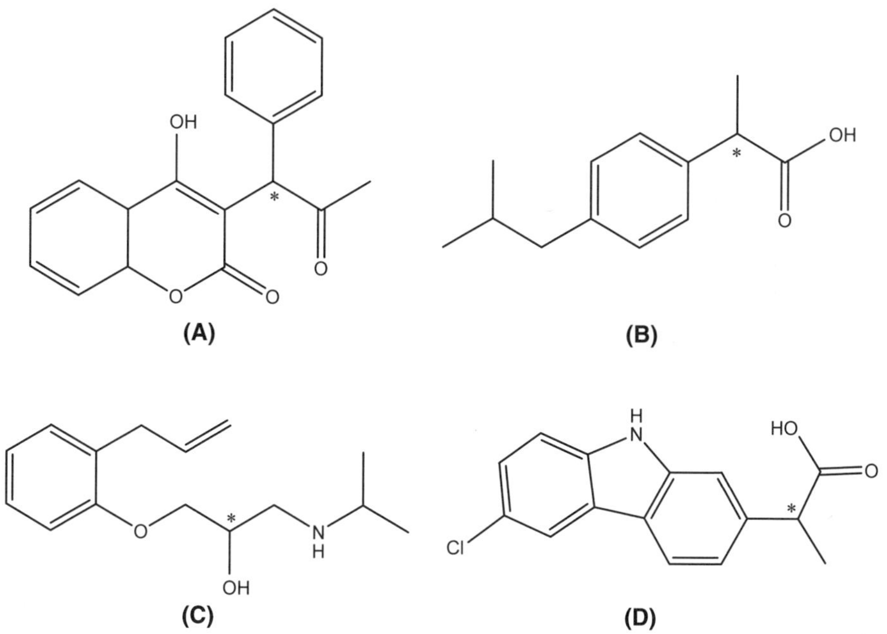

FIGURE 11.1 Compounds having enantioseletivity to HSA: **(A)** warfarin, **(B)** ibuprofen, **(C)** alprenolol, and **(D)** carprofen. The asterisk shows the chiral center.

Warfarin is a coumarin derivative anticoagulant agent and the structure of warfarin is shown in Figure 11.1. Warfarin exists in two chiral forms, (R)- and (S)-warfarin, and has been widely used as a site I probe with (S)-warfarin having several times the pharmacological activity of the (R)-warfarin in humans. Although they are believed to bind at the same site on HSA in solution, (S)-warfarin has slightly higher association constants than (R)-warfarin on the Sudlow site I while (R)-warfarin appears to have a larger number of binding sites (17,28–31). The thermodynamic study (ΔS) of chiral warfarin at temperature ranges from 4°C to 45°C showed (R)-warfarin mainly interacting with the binding site interior while (S)-warfarin interacted more with the site's outer surface (17).

It is widely accepted that the source of enantioselectivity in binding of coumarin to HSA is due to the differential binding behavior at different sites. Noctor et al. determined that the enantioselective binding for warfarin to HSA accounts for over 70% of the total binding of the drug to albumin at site I (32). The differential binding of acenocoumarol on serum albumin was performed by affinity chromatography on HSA-sepharose column and the result showed that the (R)-enantiomer has more than 2.5 times greater the binding constant than (S)-enantiomer while (S)-phenprocoumon exhibits two times higher enantioselective binding over its antipode at the Sudlow site 1 (31,33).

Verapamil also enantioselectively binds to Sudlow site I. Mallik et al. suggested that (R)- and (S)-verapamil have direct competition with (S)-warfarin at Sudlow site I with the average association constant of 1.4×10^4 M^{-1} (107). Several warfarin alternatives for probing Sudlow site I have been suggested due to a slow change in structure of warfarin in common buffers used for binding studies (34,35). Joseph et al. investigated four compounds (coumarin, 7-hydroxycoumarin, 7-hydroxy-4-methylcoumarin, and 4-hydroxycoumarin) structurally related to warfarin for possible alternative to characterize site I using competitive zonal elution chromatography experiments and suggested that 4-hydroxycoumarin was the best alternative for replacing warfarin for probing Sudlow site I.

Indole-benzodiazepine binding site or Sudlow site II on HSA as the name suggests has distinct binding affinities to compounds containing indole groups such as L-tryptophan and benzodiazepine derivatives such as diazepam. In 1963, Sanger identified the sequence of Arg-Tyr-Thr-Arg as the site of specific labeling of bovine serum albumin and the tyrosine residue was recognized to Tyr 411 of HSA in the IIIA subdomain of HSA near the tip of long loop 7 (12). Moravek et al. found Tyr-411 to be the tyrosine most susceptible to nitration and the nitration of this particular amino acid inhibited tryptophan and diazepam binding to site II (36).

The binding strength of albumin for L-tryptophan rises with decreasing temperature as for many hydrophobic compounds, and modification of the tryptophan molecule alters the affinity as substitution of a methyl group for the α-hydrogen blocks binding while decarboxylation to tryptamine reduces it over 20-fold (37,38). Seedher and Bhatia investigated the binding of the nonsteroidal anti-inflammatory drugs (NSAIDs) etoricoxib and parecoxib with HSA using fluorescence spectroscopy and found there was only one class of binding site and the displacement studies using the site-specific probe, dansylsarcosine, showed that these drugs are bound at site II on the HSA molecule (39). The same group also investigated the mechanism of interaction between celecoxib and valdecoxib, other NSAIDs, with HSA using fluorescence spectroscopy and found, like etoricoxib and parecoxib, that they are bound at site II on HSA (40). They also suggested Stern-Volmer analysis of the fluorescence quenching data indicated that predominantly a static quenching mechanism is operative and the tryptophan residues of albumin are fully accessible to celecoxib and only partially accessible to valecoxib. Kim and Hage utilized high-performance affinity chromatography (HPAC) to characterize the binding of carbamazepine to an immobilized HSA column, and the frontal chromatography study indicated that carbamazepine has a single binding site on HSA with an association constant of 5.3×10^3 M^{-1} at pH 7.4 and 37°C (41,42) with selective binding at site II. Furthermore, they investigated thermodynamic parameters of binding and found that the value of free energy (ΔG) for this interaction was −5.35 kcal/mol at 37°C with an associated change in enthalpy (ΔH) of −6.45 kcal/mol and a change in entropy (ΔS) of −3.56 cal/molK.

α_1-Acid Glycoprotein

AGP or orosomucoid (ORM) has 183 amino acid residues (M_w = 36,000–44,000 depending on the number of glycans) with a pI of 2.8 to 3.8 and is normally present in plasma at concentrations about 100 times less (0.5–1.4 mg/mL) than HSA. AGP is an acute phase plasma protein in humans as well as other species and its serum concentration is subject to a large variation in health and disease states such as systemic tissue injury, inflammation, or infection causing increases in hepatic synthesis of AGP. The ratio of unbound to bound fraction of drugs can be changed because various health states can significantly alter the

TABLE 11.2 Drugs Exhibiting Enantioselective Binding to Isolated AGP

Drug	Association constant $n_1 k_1$ (M^{-1})	References
(R)-profafenone	1.30×10^6	24
(S)-profafenone	3.83×10^6	24
(R)-propranolol	6.0×10^5	45
(S)-propranolol	8.1×10^5	45
(+)-zopiclone	2.23×10^4	46
(−)-zopiclone	8.5×10^3	46
(−)-chlorpheniramine	3.04×10^3	47
(+)-chlorpheniramine	1.53×10^4	47
(R)-disopyramide	6.6×10^5	45
(S)-disopyramide	1.8×10^6	45
(R)-verapamil	2.86×10^5	45
(S)-verapamil	1.51×10^5	45
(R)-oxybutynin[a]	2.14×10^7	48
(S)-oxybutynin[a]	8.3×10^6	48

[a]Association constant to F1-S fraction of AGP.

concentration of AGP. It has been widely accepted that basic drugs tend to bind preferably to the heavily glycosylated AGP while acidic and neutral compounds tend to bind to HSA (43,44). The enantioselective binding of drug to isolated AGP is presented in Table 11.2 (Fig. 11.2).

Enantioselective binding of α-adrenergic blocking agent prazosin was studied by means of circular dichroism and equilibrium dialysis, and the interaction results in pronounced negative extrinsic Cotton effects at 255 nm and a smaller negative band at 285 nm, which are associated with the binding of prazosin to only one site of AGP (4). The association constant of prazosin was 4.4×10^4 M^{-1} and 0.85 binding sites per AGP (4). Fernandez et al. studied the binding of racemic zopiclone and of its two enantiomers to plasma proteins HSA and AGP using equilibrium dialysis and found that the binding of (+)-zopiclone to AGP was 2.6-fold greater compared to (−)-zopiclone, while 1.8-fold greater affinity was observed for binding of (+)-zopiclone to HSA (46).

The binding of anti-arrhythmic drug propafenone enantiomers with HSA and AGP, as well as plasma from different species (rat, rabbit, and cow), were studied using indirect chiral HPLC and ultrafiltration techniques. The binding of (S)-propafenone to AGP accounts for most of the drug binding to plasma proteins and two classes of binding sites in AGP were identified as a stereoselective high-affinity, small binding capacity site ($n_{1(S)} K_{1(S)} = 3.8 \times 10^6$ M^{-1}) and the other with low affinity, large binding capacity site ($n_{2(S)} K_{2(S)} = 9.95 \times 10^3$ M^{-1}). The binding mode of both enantiomers with AGP mainly involved hydrophobic interactions, as revealed by thermodynamic parameters of propafenone binding with AGP.

Hiep et al. investigated in vitro binding of chlorpheniramine, a first-generation alkylamine antihistamine drug, enantiomers to AGP by means of equilibrium dialysis, and results showed that the binding of chlorpheniramine to AGP is stereoselective with (S)-chlorpheniramine bound to AGP tighter (1.5×10^4 M^{-1}) than its antipode (3.0×10^3 M^{-1}) (47). The protein binding of chlorpheniramine as well as its enantioselectivity to AGP found in humans was opposite to those of rats and the authors suggested that these differences

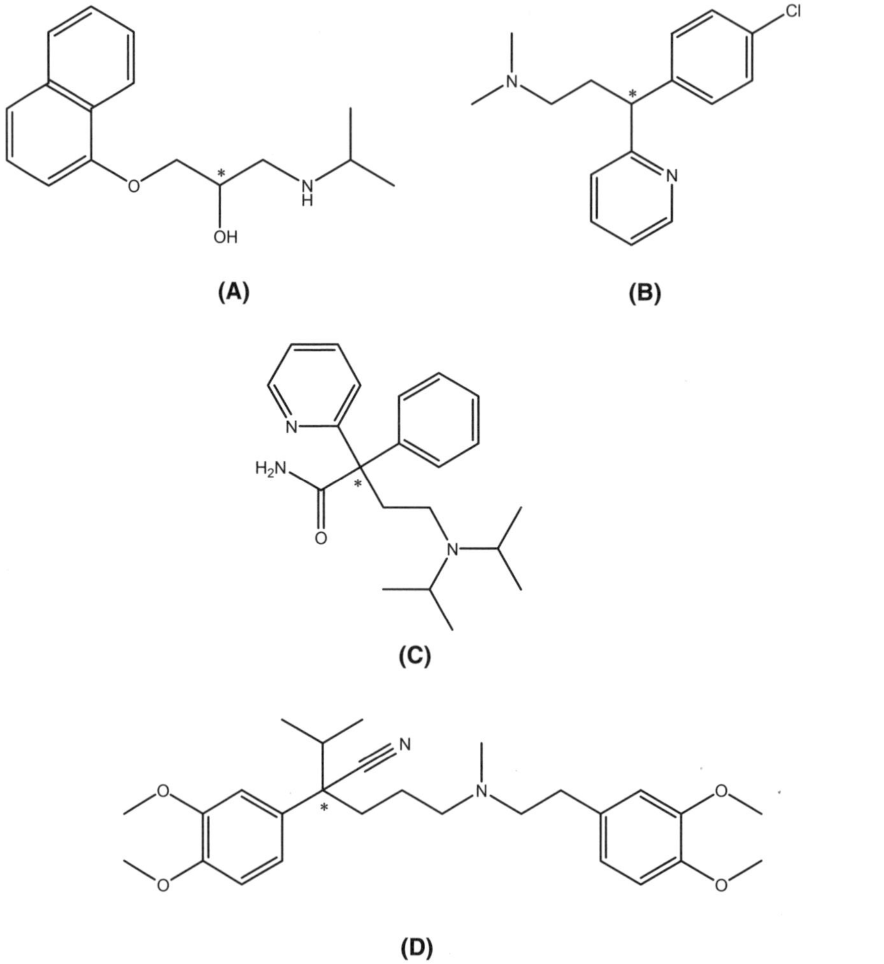

FIGURE 11.2 Compounds having enantioseletivity to AGP: (**A**) propranolol, (**B**) chlorpheniramine, (**C**) disopyramide, and (**D**) verapamil. The asterisk shows the chiral center.

could be because of differences in drug affinity and/or in the total number of binding sites per protein molecules.

A heat shock protein coinducer bimoclomol is known to enantioselectively bind to plasma protein (23). Visy et al. used ultrafiltration and equilibrium dialysis to investigate binding of biomoclomol to AGP as well as HSA. They reported that the considerably stronger binding of the (S)-biomoclomol to plasma protein is due to its binding to AGP. The association equilibrium constants for (R)- and (S)-biomoclomol to AGP are $n_{(R)}K_{(R)} = 1.3 \times 10^4$ M^{-1} and $n_{(S)}K_{(S)} = 1.0 \times 10^5$ M^{-1}, respectively. The binding of (R)- and (S)-biomoclomol to HSA was found to be weak ($nK = 5.0 \times 10^3$ M^{-1}) and not stereoselective. The number of binding site of bimoclomol was found to be about 0.5, indicating that biomoclomol binds only to a subfraction of the protein.

Eap et al. isolated the S- and F-forms of AGP variants by isoelectric focusing with immobilines (49). They performed equilibrium dialysis experiments using a multicompartmental system; a higher affinity for various basic drugs was with S- in comparison with F-AGP: amitriptyline, nortriptyline, imipramine, desipramine, trimipramine, methadone, thioridazine, clomipramine, desmethylclomipramine, and maprotiline. The selectivity (binding to S- vs. F-AGP) is the most pronounced for methadone and the lowest for thioridazine, while it is absent for the acidic drug mephenytoin (49).

The plasma concentration of AGP as well as the plasma binding of racemic, D- and L-methadone was measured in healthy volunteers to correlate the binding parameters and the concentration of AGP in plasma. The AGP phenotypes and the concentration of AGP variants were determined and a significant correlation was obtained between the binding ratio (B/F) for D- and L-methadone and the total AGP concentration with $r = 0.724$, $p \leq 0.001$ (50). Two variants of AGP, ORMs 2A and F1, were also found to play important determining roles in the binding of methadone enantiomers and the authors pointed out that the levels of AGP variants should be considered in protein binding studies (50).

Other Plasma Proteins

The relative occurrences of enantioselective drug binding by other plasma proteins are much less than those of HSA and AGP despite the amount of protein. The normal level of immunoglobulins, for example, is in the range of 1.0 to 1.5 g/dL of blood and accounts for about 35% of blood proteins. However, there have been only limited number of studies regarding enantioselective binding of drugs to antibodies (51,52). A monoclonal anti-d-hydroxy acid antibody was immobilized onto chromatographic support material to study for enantiomeric separation of free α-hydroxy acids as well as several aliphatic and aromatic members of this class of compound (53). The same group also reported the production of enantioselective antibodies using classical immunological and modern molecular biological techniques in rabbits and mice to study the stereoselectivity of antibodies as well as to apply them in immunosensors and chiral chromatography. A membrane-based optical sensor allowed detection of enantiomeric impurities at the 1/2000 level (54).

The antibodies used in these studies were artificially raised and the clinical implications of enantioselective binding of antibodies to drugs are neither known nor has thorough investigation on this issue ever been explored.

Lipoproteins are another class of blood plasma proteins with low abundance (10 mg/dL). The clinical implication of lipoprotein bindings is not known but it may be important since these proteins are able to bind a variety of classes of drug such as basic, neutral, and lipophilic compounds. Brocks and coworkers used a stereoselective liquid chromatographic assay to determine relative binding of (+) and (–) enantiomers of antimalarial agent halofantrine (55,56). The (–) enantiomer of halofantrine showed higher affinity for the lipoprotein-deficient fraction than the (+) enantiomer in human, monkey and rat plasma. A significant fraction of the (+) enantiomer of halofantrine was located in lipoprotein-rich fraction in dog and human plasma while the ratio of (+):(–) was consistent in lipoprotein-deficient fraction, showing substantial interspecies differences in the pattern of halofantrine distribution within different fractions of lipoproteins (56). The methods based on the high-performance frontal analysis (HPFA) capillary electrophoresis were used to analyze the enantioselective binding of verapamil and nilvadipine to high-density lipoprotein (HDL), low-density lipoprotein, as well as oxidized LDL (57–59). It was found that the binding of verapamil and nilvadipine to these lipoproteins is neither specific nor enantioselective, while the total binding affinity between LDL and verapamil was increased by 3.3-, 4.6-, 7.0-, and 19-fold after 0.5, 1, 2, and 12 hours oxidation, respectively, whereas the total binding affinity between nilvadipine and LDL was increased by 1.3-, 1.4-, 1.4-, and 1.7-fold in the same reaction times, respectively, indicating the LDL oxidation enhances the drug binding affinity.

NEW ANALYTICAL DEVELOPMENTS IN ENANTIOSELECTIVE PLASMA PROTEIN BINDING STUDIES

Many methods are available for the study of drug-protein binding. Equilibrium dialysis and ultrafiltration have been the most widely used for this purpose (60). Equilibrium dialysis has been the reference method for such measurements. One reason for this is its simplicity as well as its relatively low cost. It, however, suffers from several disadvantages. One disadvantage is the long period of time typically required for establishing equilibrium during the dialysis process (hours to even days). This not only makes this method inconvenient for routine analysis, but it creates problems if the analyte is unstable or if its binding is susceptible to changes with temperature or pH.

Ultrafiltration is another technique used in evaluating protein-drug binding. Its operation is similar to that of equilibrium dialysis but requires less time to establish equilibrium. This method, as well as dialysis, requires a labeled solute and is often performed using a second method for analysis, such as an immunoassay, gas chromatography (GC), or high-performance liquid chromatography (HPLC). In addition, the effects of analyte adsorption to the membrane of filter must be considered.

The problems associated with these traditional methods have accelerated the efforts to find better, faster, more efficient, and convenient approaches for the analysis of protein-drug binding. Such efforts involve the use of HPAC, affinity capillary electrophoresis (ACE), and surface plasmon resonance (SPR).

High-Performance Affinity Chromatography

HPAC is a type of HPLC in which the separation or purification is based on specific and reversible interactions between the analyte and an immobilized ligand. The retention is based on the interactions commonly seen in biological systems. Examples include the binding of a hormone with its receptor, an enzyme with its substrate, or an antibody with an antigen.

In affinity chromatography, a sample containing the analyte is applied to a column containing an immobilized affinity ligand. The analyte is then retained by this ligand due to their specific interactions while other components are quickly washed from the column. The analyte is later eluted off the column by altering the mobile phase, pH, or ionic strength, or by adding competing agents. The column is then restored to the original mobile phase conditions, allowed to regenerate, and the next analysis is performed.

Traditional affinity chromatography exploits these interactions by immobilizing one of a pair of interacting molecules onto a solid support and packing this into a column. The support material typically used in low-performance affinity chromatography is a large diameter, nonrigid material such as agarose or dextran. These materials not only show poor mass transfer properties, but they also suffer from limited stabilities at high flow rates and pressures. The low back pressure produced from these materials means that they can be operated under gravity flow, making them relatively simple and inexpensive to use for the purification or pretreatment of samples.

HPAC combines the speed and efficiency of HPLC with the specificity of affinity chromatography. These properties are produced by using support materials that consist of small, rigid particles capable of withstanding high flow rates or pressures and that possess good mass transfer properties. Examples include silica, glass, and hydroxylated polystyrene media. Compared to low-performance affinity chromatography, the separation in HPAC requires standard HPLC equipment such as high-precision pumps and valves. Although these make HPAC more expensive than low-performance affinity chromatography, the better speed and precision of HPAC make it preferable for routine work and analytical applications.

HPAC is often used to evaluate equilibrium or kinetic parameters of biological systems. The use of HPAC for this purpose is referred to as analytical affinity chromatography, quantitative affinity chromatography, or biochromatography (61,62).

The ligand of interest is immobilized onto a suitable support and an injection of analyte is made onto the column. Information regarding equilibrium or kinetic parameters can later be obtained by carefully observing the elution time or volume of analyte as it passes through the column. This approach was often used to evaluate the binding of HSA or AGP with the small drugs (63–70).

Two different techniques in affinity chromatography have been widely used to study enantioselective binding of drugs to plasma proteins. The first technique is frontal analysis in which a known concentration of pure solute is continuously applied to a column containing a fixed amount of an immobilized protein. As the amount of bound solute in the column increases, it begins to saturate the ligand and form a breakthrough curve. This is shown in Figure 11.3A. The mean point of this curve is directly related to the concentration of applied solute and the amount of immobilized ligand. The second technique is zonal elution. In this method, a known concentration of a competing agent is continuously applied in the mobile phase to a column containing an immobilized ligand while small amounts of analyte are injected. If a competing agent and an analyte have direct competition at binding sites on immobilized ligands and these interactions have fast association/dissociation kinetics, a shift in retention factor of the analyte provides information on the association equilibrium constant (Fig. 11.3B).

Kimura and coworkers used a method based on HPFA and Scatchard analysis to evaluate the binding properties of enantiomers of thyroxine to HSA and found that the binding constant (K) as well as the number of binding sites on an HSA molecule (n) were $K = 1.01 \times 10^6$ M^{-1} and $n = 1.90$ for L-thyroxine, and $K = 9.71 \times 10^5$ M^{-1} and $n = 1.97$ for D-thyroxine. Furthermore, the binding sites were identified using phenylbutazone and diazepam as site-specific probes for sites I and II, respectively, and each enantiomer was found to bind to both sites (71). The enantioselective binding characteristics of ketoprofen to HSA had been performed by using competitive zonal elution chromatography techniques

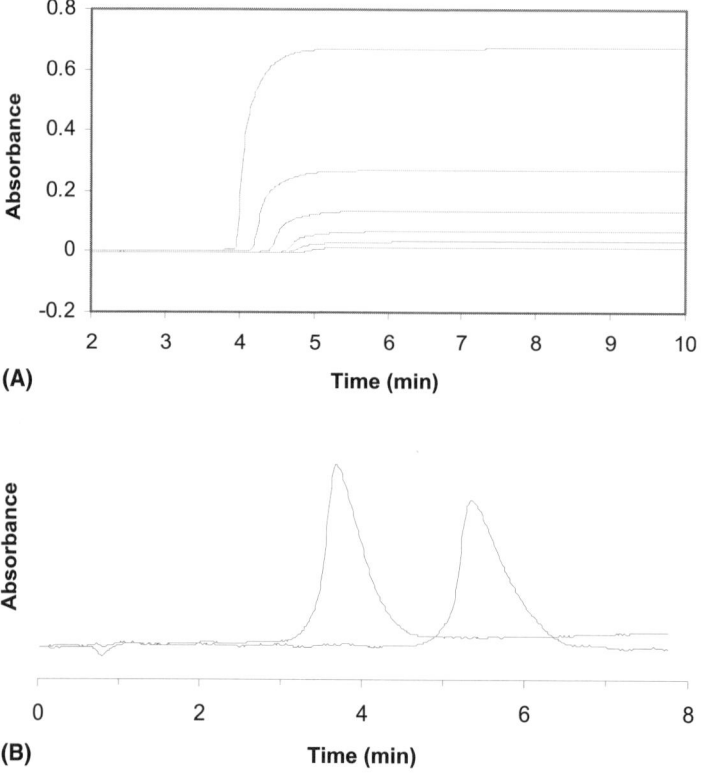

FIGURE 11.3 Typical frontal chromatogram (**A**) obtained from the immobilized protein column. The concentration of applied drug decreased from left to right. Typical zonal elution chromatogram is shown in (**B**), where the amount of competing agents in mobile phase shift the retention times of analyte from right to left.

using phenylbutazone and diazepam for site I and II and showed that both (*R*)- and (*S*)-ketoprofen display high affinity to the primary phenylbutazone and diazepam binding sites as well as low affinity to the secondary diazepam sites. The authors further characterized the binding mechanism as a stepwise binding process to the affinity binding site initiated by the binding to the primary diazepam site and followed by the attachment to the primary phenylbutazone site (72). The enantioselective binding of a drug to HSA may induce the conformational changes of protein to cause stronger or weaker binding to the secondary drug (i.e., allosteric interaction). Fitos and coworkers reported the effect of ibuprofen enantiomers on the stereoselective binding of (*S*)-lorazepam and found that the binding of (*S*)-lorazepam exhibited enhanced binding in the presence of (*S*)-ibuprofen (73). Chen and Hage quantitatively characterized allosteric interaction between drugs to HSA using equations derived from careful observation of the interactions of ibuprofen/(*S*)-lorazepam acetate, (*S*)-oxazepam hemisuccinate/(*R*)-oxazepam hemisuccinate, and L-tryptophan/phenytoin during their binding to HSA (74). Another item that can be studied using HPAC is a quantitative structure-retention relationship (QSRR) between a drug and protein in which an essential functional group of a drug responsible for binding is identified. Markuszewski and Kaliszan prepared a series of different biomacromolecules (HSA, AGP, keratin, collagen, melanin, amylose tris(3,5-dimethylphenylcarbamate), and basic fatty acid binding proteins) columns to collect the data for interpretation of structural requirements of specific binding sites on biomacromolecules (75). Mallik and coworker developed affinity silica monolith columns containing HSA and AGP and evaluated in terms of its binding, efficiency, and selectivity in chiral separations. The silica monolith column gave higher retention and higher or comparable resolution and efficiency when compared with HSA columns that contained silica particles or GMA/EDMA copolymer in analyzing (D/L)-tryptophan, (R/S)-warfarin, and (R/S) propranolol (76,77).

Affinity Capillary Electrophoresis

ACE is a more recent technique that can be used for the study of molecular interactions. This method uses shifts in electrophoretic mobilities of a solute as it binds with an affinity ligand. ACE can provide both qualitative and quantitative information on solute-protein binding. This information includes the detection of complex formation, the identification of an active component for binding in a multicomponent mixture, the identification of structural requirements for recognition, the determination of the equilibrium constant and stoichiometry for a binding reaction, and concentration measurements based on immunochemical recognition (65,78).

ACE is especially useful for evaluating solute-protein binding because it requires only small quantities of proteins (i.e., usually in less than several nanoliters). Typical analysis times are short and ACE can be easily automated. Protein adsorption inside the capillary surface, however, can be a drawback and cause nonreproducible results. This can be corrected by employing a surfaced modified capillary to prevent protein adsorption (10,79).

There are a number of ways to evaluate drug-protein interactions in ACE. One particular method used in this work is mobility shift assay. This is shown in Figure 11.4. In a mobility shift assay, the evaluation of drug-protein binding is performed by injecting a small amount of drug (D) into a capillary that contains a soluble protein (P) in the running buffer. This is represented by the reaction shown in Eq. 5.

$$D(\mu_D) + P(\mu_P) \leftrightarrow d - P(\mu_{D-P}) \tag{5}$$

In the above expression, μ_D, μ_P, and μ_{D-P} are the electrophoretic mobilities of the drug, protein, and drug-protein complex, respectively. Since the amount of drug is small compared to that of protein, the binding of drug to the protein shifts the apparent mobility of the drug. This shift can be described by using the change in migration time of the drug that is observed at different protein concentrations.

An analysis based on the above equation makes several assumptions: (*i*) equilibrium is established between the bound and unbound species and (*ii*) the association and dissociation kinetics of drug-protein binding is fast relative to the time for the analysis (80). If the electrophoretic mobilities of the drug and protein are similar to each other, the observed mobility shift of the drug will be small or difficult to analyze. In this case, an additional running buffer additive

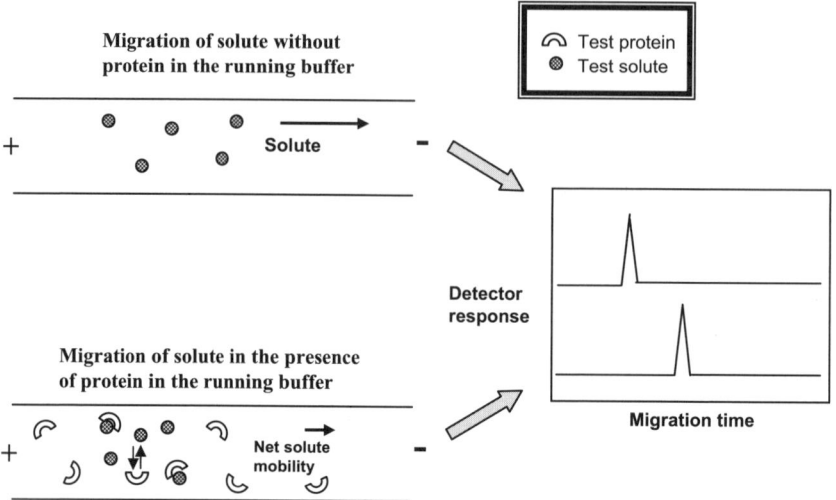

FIGURE 11.4 The principle of mobility shift assay in affinity capillary electrophoresis. The migration time of analyte is retarded in the presence of protein in the running buffer.

such as dextran can be used to alter the electrophoretic mobility of the protein while not affecting the small drug.

Serum albumin and AGP are two most widely used chiral selectors in ACE. The stereoselective binding of antihistamines (brompheniramine, chlorpheniramine, hydroxyzine, orphenadrine, and phenindamine), phenothiazines (promethazine and trimeprazine), and a local anesthetic (bupivacaine) to human plasma proteins were evaluated using HSA as chiral selector and the partial filling technique. The binding of phenindamine to HSA showed the highest stereoselectivity followed by trimeprazine and promethazine (81). Zhu and coworkers reported the enantioseparation of three basic drugs ofloxacin, propranolol, and verapamil in ACE using HSA and BSA as chiral selectors in phosphate buffer at pH 7.4. Interestingly, ofloxacin was only separated in the presence of BSA, and verapamil only with HSA, while propranolol was separated with either HSA or BSA showing high species dependence on chiral separation. They also used two displacers (ketoprofen and warfarin) known to bind serum albumin at a specific location (site II and I) to probe binding sites of ofloxacin, propranolol, and verapamil on serum albumin and found that verapamil binds to both sites in HSA while the binding of ofloxacin to BSA occurs in site I (82).

Oravcova and coworkers compared the Hummel-Dreyer method in capillary zone electrophoresis for studying the interaction of racemic carvedilol and its individual enantiomers with isolated AGP and HSA to the corresponding HPLC method. The binding parameters characterizing the high-affinity binding site of AGP evaluated by using capillary electrophoresis were in good accordance with those obtained by HPLC, and it was concluded that the Hummel-Dreyer ACE method is an efficient and fast technique for quantitative description of binding parameters of hydrophobic drugs (83). The evaluation of association constants between drugs and AGP using the ACE partial filling technique as well as conditions for maximizing chiral separation of drugs such as effective plug length (i.e., the amount of chiral selector) and optimum temperature were well described using disopyramide and remoxipride as model drugs. Authors found that the affinity between the enantiomers of disopyramide or remoxipride and AGP varied with increasing temperature, as shown by determined association constants. It was found that the association between the enantiomers and AGP was strongest at 25°C and decreased at both lower and higher temperatures. This unexpected finding may indicate conformational changes of the protein with temperature variations (84). The effect of glycosylation on chiral separation of basic drugs was also studied using AGP as a model protein in HPFA combined with capillary electrophoresis (85). AGP was fractionated into two portions, bound and unbound fraction, by concanavalin A lectin affinity chromatography and the results suggested that the (S)-propranolol is bound to unbound fraction and bound fraction of AGP more strongly than (R)-propranolol while the reverse applies to verapamil enantiomers. This suggested that the glycosylation of AGP does not play a significant role in the chiral recognition in binding of these drugs (85). Human serum transferrin and lipoproteins have been used in ACE for chiral separation of small molecules. Studies on chiral separation of 15 compounds (β-blockers and H1-antihistamine agents) in ACE partial filling technique using human serum transferrin as chiral selector suggested that the enantioselectivity is compound-specific (6 out of 25 analogues) and the very small differences in the chemical structure of the compounds resulted in significant changes in the chiral recognition. For example, one methyl group difference in mepindolol and pindolol or one oxygen difference in alprenolol and exprenolol suggested the essential role of these extra-functional groups in chiral selectivity by transferrin (86). Chiral recognition of drugs can be used to characterize surface interaction sites on iron-free transferrin in ACE. Separation of enantiomers of drugs including tryptophan derivatives was investigated in ACE using the partial filling technique, and the study suggested that the variations in migration times of these drugs reflected the possible interactions with the protein, suggesting (R)-enantiomers were always retained more strongly by transferrin. This was further confirmed by molecular modeling on the docking of tryptophan derivatives on transferrin showing that the (R)-enantiomers possess a stronger complexation with transferrin, whereas the (S)-enantiomers are bound by weaker interactions (87). HPFA with CE was applied to the study of enantioselective binding of verapamil to plasma lipoproteins. The drug-lipoprotein mixed solution, which had been in the binding equilibrium, was hydrodynamically introduced into a noncoated fused-silica capillary and the bindings of verapamil to HDL, LDL, and oxidized LDL were neither site-specific nor enantioselective (59). The same authors also pointed out that the partition-like binding to the lipid part of these lipoproteins seemed to be dominant and the total binding affinities of LDL to verapamil were about seven-times stronger than those of HDL; also the oxidation of LDL by copper ion enhanced the binding affinities significantly (59).

Surface Plasmon Resonance

The surface plasmon resonance (SPR) occurs when polarized light coming from the side of higher refractive index is partly reflected and partly refracted at an interface between media of different refractive index (glass and buffer). The total internal reflection is observed when no light is refracted across the interface at above a certain angle of incidence. An electromagnetic field component, known as an evanescent wave, penetrates a short distance into a medium and interacts with free electrons in the gold layer coating between the media generating electron charge density wave called plasmons and causing a reduction in the intensity of the reflected light. The resonance conditions are influenced by the material adsorbed onto the thin metal film. Biological macromolecules such as protein and DNA are often immobilized onto the surface and a linear relationship is often found between resonance energy and mass concentration of these molecules on the surface. The SPR signal (expressed in response unit, RU) is therefore a measure of mass concentration at the sensor surface and the drug and protein binding characteristics (rate constants and equilibrium constant) can be determined.

There are only limited studies of SPR on characterization of enantioselectivity plasma proteins (88–91). Sandblad et al. discussed guidelines for reliable experimental and methodological SPR approaches in drug-protein interaction studies to avoid biased values of binding constants that are often obtained from the use of too narrow drug concentrations as well as false assumptions made in determining the number of binding sites (88). They also investigated bindings of enantiomeric propranolol to AGP, enantiomeric warfarin to HSA, and stereoisomers of melagatrans to thrombin, AGP, and HSA. Authors found that a single site model could be assumed in the (R,S)-melagatran to thrombin while the (S, R)-form did not bind to thrombin. The data from propranolol to AGP as well as warfarin to HSA system indicated that both systems are heterogeneous comprising high affinity enantioselective binding sites and weak nonselective sites (88). The effect of dimethylsulfoxide (DMSO), a widely used solvent to increase solubility of the compound, was also studied and the results show severe deviations in binding data for the propranolol and AGP system in the presence of 5% DMSO.

Day and Myszka immobilized HSA on the surface of SPR sensor chips using a standard amine coupling protocol to characterize bindings of 12 different drugs to HSA as well as albumin from different species. The binding constants determined from the SPR sensor assay corresponded with binding constants obtained by other methods (89). One particular observation they noticed was a gradual decay in binding capacity of newly immobilized HSA to naproxen and warfarin and the loss of activity was most marked during the first 24 hours of post immobilization suggesting the surfaces stabilize over time (89). They investigated effects of different immobilization strategies, different HSA preparations, alternating binding buffer compositions, regenerating the sensor surface as well dispensing samples into glass instead of plastic vials on the initial stability of the albumin surface and could not find the reason for such loss. They pointed out that the newly immobilized albumins should be stabilized with injecting buffer for 10 hours prior to analysis.

OTHER ISSUES IN PLASMA PROTEIN BINDING STUDIES

Several factors are responsible for different enantioselective binding behaviors of a given drug-plasma protein interactions. Also it should be noted that the majority of drug-plasma protein binding measurements have been studied on controlled environments such as performing drug-protein binding studies on isolated proteins or using samples from healthy donors within the same ethnicity

ENANTIOSELECTIVE BINDING OF DRUGS TO PLASMA PROTEINS   197

in similar age groups. These are important issues in studying drug binding for the patient in the clinic since these patients generally show many different situations (i.e., age, disease state, sex, ethnicity, etc.) than the in vitro experiments in a controlled laboratory. In the remaining parts of this chapter, factors affecting enantioselective bindings of drug in blood will be thoroughly reviewed.

Species Difference

The study of how drugs bind to proteins is crucial in the drug discovery process since how a drug binds to proteins in plasma can limit the overall availability and pharmacology of the drug. In this process, a drug candidate generally undergoes a series of binding experiments in plasma or blood from other species (generally human, monkey, rat, mouse, dog, and rabbit) to assess overall availability of drugs. Extrapolating preclinical protein binding as well as pharmacology/toxicology (PK/TK) studies in different species to the human requires an understanding of the action of the drug from these different species as well as the relationship of the free, pharmacologically active form to total drug concentration in these species. For instance, the binding of an immunomodulator, KE-298 (2 acetylthiomethyl-4-(4-methylphenyl)-4-oxobutanoic acid), and its metabolites in all species of plasma was enantioselective. The (+)-(S)-enantiomers of these compounds bound rat, dog, and human plasma proteins to a greater extent than did the –(R)-enantiometers, except that the case of KE-298 was opposite in rat plasma (92). Fitos and coworkers investigated enantioselective bindings of benzodiazepine and coumarin drugs to serum albumin from different mammalian species using chiral chromatographic techniques and found that the stereoselective binding of 2,3-benzodiazepine drug, tofisopam, in human is opposite to that in all other species while the binding of 1,4-benzodiazepines in dog albumin is very similar to HSA and highly preferred binding of (S)-phenprocoumon was observed with dog albumin (93). Another example of species differences in enantioselective binding of drugs to plasma proteins was in the case of 1-methyl-5-phenyl-5-propyl-barbiturate where the (R)-(–)-form of drug preferably (factor of 2) bound to HSA compared to the (S)-(+)-form. In contrast to HSA, bovine as well as rabbit serum albumin bound (S)-(+)-form to a higher degree than the (R)-(–)-form (93). No enantioselective binding difference was observed in the case of 1-methyl-5-phenyl-5-propyl-barbiturate binding to rat serum albumin. These studies clearly indicated that taking into account for each of these mammalian albumins with the respective number of binding classes and sites in each class, the affinity constant, and total binding constant, it becomes evident that they bind (S)-(+)- and (R)-(–)-1-methyl-5-phenyl-5-propyl-barbiturate quantitatively differently in a species-specific manner (93). The same group also studied stereoselective binding of the enantiomers of four closely related N-methyl-barbiturates to human, bovine, and rat serum albumin, and reported that the structural differences between the serum albumins of these mammalian species cause the different pattern of enantioselective binding of the enantiomers of four closely related N-methyl-barbiturates (94).

In the case of MK-571 ((+)-3-(((3-(2-(7-chloro-2-quinolinyl)ethenyl)phenyl) ((3-(dimethylamino)-3-oxopropyl)thio)methyl)thio)propanoic acid), a potent and specific antagonist of leukotriene D4 action, the binding of the enantiomers of this drug to plasma protein was found to be extensively stereoselective and species-dependent. The (R)-(–)-enantiomer was bound to rat plasma to a greater extent than the (S)-(+)-enantiomer, while in dog and monkey plasma the reverse was the case (95).

These examples showed that the dissimilarities in the enantioselective binding behaviors of structurally related serum albumins (i.e., small differences in the genomic sequence) from different species often result from the differences in the microenvironments of binding process between drugs and serum albumins. These include differences in protein conformation, the degree of noncovalent interaction, pH, as well as relative amount of lipids arising from differences in diet.

Age

Several studies have shown that the amount of albumin in blood tends to decrease with age and this may result in reduced plasma protein binding of certain drugs. This phenomenon of reduced plasma-protein binding in hypoalbuminemia is not universal. For instance, warfarin binding in plasma does not decrease with age while the reduction in plasma diazepam binding in elderly subjects has been reported (96). Viani and coworkers have reported that there was a significant negative correlation of albumin concentration with age and this resulted in about a twofold increase in the amount of unbound fraction of diazepam and salicylic acid in two age groups (30s vs. 70s) while the unbound fraction of digitoxin was independent of age (97). In contrast, Davis and coworkers have reported that the plasma protein binding of the basic drug lignocaine predominantly bound to AGP and tended to increase slightly with age (96). The binding of lignocaine to AGP positively correlated with plasma concentration of AGP concentration, which also increased slightly with age. No stereoselective discrimination of drugs to plasma protein in differences in the age was reported in both Viani and Davis' studies. The enantioselective protein binding of mephobarbital (MPB) was investigated from plasma from 8 young (18–25 years) and 8 elderly (more than 60 years) male subjects. The binding of both enantiomers was significantly lower in the young compared with the elderly subjects and (S)-MPB was approximately 59% bound to plasma protein (29% to HSA) while (R)-MPB was approximately 67% bound to plasma protein (41% to HSA) (98). Stereoselective serum protein binding of the NSAID ibuprofen was performed by ultrafiltration, and disposition of ibuprofen in young healthy volunteers and elderly persons with and without renal impairment showed both elderly groups had significantly decreased binding of (S)-ibuprofen compared to the young group resulting in an increased unbound fraction of (S)-ibuprofen in elderly persons with renal impairment that may increase the risk for NSAID-associated adverse effects (99). A similar study was performed by Tan and coworkers (100). They found aging had little effect on the distribution and metabolism of (R)-ibuprofen in both young and elderly subjects while the unbound fraction of (S)-ibuprofen was significantly greater in the elderly further confirming the risk of elderly subjects on increased exposure of NSAID-associated adverse effects.

Disease States

In the previous section, an example of differences in the amount of unbound fraction of (S)-ibuprofen in the elderly subjects with renal impairment compared with healthy young and elderly groups was discussed. This suggests that certain disease states may potentially alter the enantioselective binding of drugs to plasma proteins and possibly cause unfavorable changes in pharmacokinetic profiles of administered drugs.

Knadler and coworker reported the enantioselective binding an arylpropionic NSAID flurbiprofen to human plasma in eight normal subjects and showed the unbound fraction of (R)-flurbiprofen was greater than that of the (S)-form for most subjects with the ratio of percent unbound (R)- to (S)-flurbiprofen ranging from 0.52 to 1.33 (22). Furthermore, they found that the binding of racemic flurbiprofen in elderly and obese volunteers and patients with liver disease was not significantly different from normal subjects, but binding was less in hypoalbuminemic patients and patients with renal impairment. Uremia decreased the binding of (R)-flurbiprofen preferentially (22). The phenomenon of hypoalbuminemia occurs frequently in severe hepatic and renal disorders. Noctor pointed out there may be relatively elevated serum concentrations of free fatty acids, which may compete with drugs for binding sites on HSA (101). The binding of oxazepam and its glucuronide conjugates to HSA were examined by equilibrium dialysis. It was found that the oxazepam and its (S)-glucuronide are specifically bound to site II on HSA with a greater extent than (R)-glucuronide, which bound to HSA in a nonspecific manner. Furthermore, the unbound fraction of oxazepam and its (S)-(+)-glucuronide conjugate are displaced by elevated levels of fatty acids found in renally impaired patients (102). A similar instance was observed in the case of the enantioselective disposition of an NSAID ketoprofen and ketoprofen glucuronide in renally impaired patients. Grubb and coworkers found that the accumulation of (S)-ketoprofen and (S)-ketoprofen glucuronide, in contrast to the (R)-isomers, unexpectedly increased after repeated dosing achieving S:R ratios of 3.3 ± 1.7 and 11.2 ± 5.3, respectively, possibly caused by decreased protein binding due to decreased albumin and displacement from protein binding sites in uremia (103). Hayball and coworkers published a similar report on the influence of renal function on the enantioselective pharmacokinetics and pharmacodynamics of ketoprofen. In that report, the mean percentage unbound fraction of (S)-ketoprofen in plasma increased at higher concentration of (S)-ketoprofen, whereas no variation was observed for (R)-ketoprofen in renally impaired subjects indicating that diminished renal function is associated with an increased exposure to unbound (S)-ketoprofen, presumably due to regeneration of parent aglycone arising from the hydrolysis of accumulated acyl-glucuronide conjugates (104).

Certain liver diseases are also responsible for an elevated level of enantioselective disposition of ketoprofen. Sakai and coworkers investigated the stereoselective binding of ketoprofen to HSA in patients with liver diseases and found that the serum protein binding of racemic ketoprofen was found to be decreased in hepatic patients, whereas, interestingly, the stereoselectivity of ketoprofen was increased (105). The ratio of the free fraction of (R)- and (S)-ketoprofen increased linearly with albumin concentrations and the data suggested that the difference in stereoselective binding of ketoprofen in hepatic patients was mainly caused by the decrease in serum HSA level as well as the stereoselective inhibition induced by the increased level of bile acid and its metabolite. Enantioselectivity in serum flurbiprofen binding was also reported in patients with liver and renal disease. The significantly lower concentration of serum albumin level was observed in subjects with renal disease or liver disease with ascites and this resulted in higher free fractions of both the (R)-(–)- and (S)-(+)-enantiomers of flurbiprofen. The ratio of unbound fraction of (R)- and (S)-flurbiprofen was lower in the subjects with ascites (0.714 ± 0.298) than in those without ascites (0.796 ± 0.090) ($P < 0.05$) (106).

CONCLUSION

As has been set out in the present chapter, the stereoselectivity of drugs to plasma proteins is important and the understanding of this phenomenon has been improved by several new developments in analytical methodologies for the study of protein interactions with chiral compounds. In certain cases, the consequences of stereoselectivity of drugs to plasma protein are significant since the pharmacokinetics and pharmacodynamics of the chiral compounds are directly related to the plasma concentration level of drugs. It is necessary to consider clinical implications of stereoselectivity in cases where a therapeutic range of plasma concentration needs to be identified or significant interpatient variability exists in the relationship between dose and plasma concentration of drugs. Undoubtedly, the understanding of the stereoselective mechanism of drug interaction to plasma proteins further enhanced the design of lead compounds and this has been facilitated by recent advances in chiral analysis. The several different formats for chiral analysis include methods based on affinity chromatography or ACE as well as SPR. The variety of chiral analysis methodologies is useful since this provides a means for studying a wide array of drug-protein systems and the systemized study of stereoselective binding offers a potential to further characterize the molecular processes involved in the binding of drugs to proteins.

REFERENCES

1. Refetoff RaLP. Endocrinology. Philadelphia: Saunders, 1989.
2. Sager G, Bratlid H, Little C. Binding of catecholamines to alpha-1 acid glycoprotein, albumin and lipoproteins in human serum. Biochem Pharmacol 1987; 36(21):3607–3612.
3. Krauss E, Polnaszek CF, Scheeler DA, et al. Interaction between human serum albumin and alpha 1-acid glycoprotein in the binding of lidocaine to purified protein fractions and sera. J Pharmacol Exp Ther 1986; 239(3):754–759.
4. Brunner F, Muller WE. Prazosin binding to human alpha 1-acid glycoprotein (orosomucoid), human serum albumin, and human serum. Further characterization of the "single drug binding site" of orosomucoid. J Pharm Pharmacol 1985; 37(5):305–309.
5. Belpaire FM, Bogaert MG, Rosseneu M. Binding of beta-adrenoceptor blocking drugs to human serum albumin, to alpha 1-acid glycoprotein and to human serum. Eur J Clin Pharmacol 1982; 22(3):253–256.
6. Haughey DB, Steinberg I, Lee MH. Protein binding of disopyramide—displacement by mono-N-dealkyldisopyramide and variation with source of alpha-1-acid glycoprotein. J Pharm Pharmacol 1985; 37(4):285–288.
7. Kwong TC. Free drug measurements: methodology and clinical significance. Clin Chim Acta 1985; 151(3):193–216.
8. Barre J, Didey F, Delion F, et al. Problems in therapeutic drug monitoring: free drug level monitoring. Ther Drug Monit 1988; 10(2):133–143.
9. Menke G, Worner W, Kratzer W, et al. Kinetics of drug binding to human serum albumin: allosteric and competitive inhibition at the benzodiazepine binding site by free fatty acids of various chain lengths. Naunyn Schmiedebergs Arch Pharmacol 1989; 339(1–2):42–47.

10. Hage DS, Tweed SA. Recent advances in chromatographic and electrophoretic methods for the study of drug-protein interactions. J Chromatogr B Biomed Sci Appl 1997; 699(1–2):499–525.
11. Hage DS. Chromatographic and electrophoretic studies of protein binding to chiral solutes. J Chromatogr A 2001; 906(1–2):459–481.
12. Peters T. All About Albumin. New York: Academic Press, 1996.
13. He XM, Carter DC. Atomic structure and chemistry of human serum albumin. Nature 1992; 358(6383):209–215.
14. Sudlow G, Birkett DJ, Wade DN. The characterization of two specific drug binding sites on human serum albumin. Mol Pharmacol 1975; 11(6):824–832.
15. Sudlow G, Birkett DJ, Wade DN. Spectroscopic techniques in the study of protein binding. A fluorescence technique for the evaluation of the albumin binding and displacement of warfarin and warfarin-alcohol. Clin Exp Pharmacol Physiol 1975; 2(2):129–140.
16. Fehske KJ, Muller WE, Wollert U. The location of drug binding sites in human serum albumin. Biochem Pharmacol 1981; 30(7):687–692.
17. Loun B, Hage DS. Chiral separation mechanisms in protein-based HPLC columns. 1. Thermodynamic studies of (R)- and (S)-warfarin binding to immobilized human serum albumin. Anal Chem 1994; 66(21):3814–3822.
18. Hage DS, Noctor TA, Wainer IW. Characterization of the protein binding of chiral drugs by high-performance affinity chromatography. Interactions of R- and S-ibuprofen with human serum albumin. J Chromatogr A 1995; 693(1):23–32.
19. Rahman MH, Maruyama T, Okada T, et al. Study of interaction of carprofen and its enantiomers with human serum albumin–II. Stereoselective site-to-site displacement of carprofen by ibuprofen. Biochem Pharmacol 1993; 46(10):1733–1740.
20. Rahman MH, Maruyama T, Okada T, et al. Study of interaction of carprofen and its enantiomers with human serum albumin–I. Mechanism of binding studied by dialysis and spectroscopic methods. Biochem Pharmacol 1993; 46(10):1721–1731.
21. Liu X, Song Y, Yue Y, et al. Study of interaction between drug enantiomers and human serum albumin by flow injection-capillary electrophoresis frontal analysis. Electrophoresis 2008; 29(13):2876–2883.
22. Knadler MP, Brater DC, Hall SD. Plasma protein binding of flurbiprofen: enantioselectivity and influence of pathophysiological status. J Pharmacol Exp Ther 1989; 249(2):378–385.
23. Visy J, Fitos I, Mady G, et al. Enantioselective plasma protein binding of bimoclomol. Chirality 2002; 14(8):638–642.
24. Hong Y, Tang Y, Zeng S. Enantioselective plasma protein binding of propafenone: mechanism, drug interaction, and species difference. Chirality 2009; 21(7):692–698.
25. Tsuda Y, Tsunoi T, Watanabe N, et al. Stereoselective binding and degradation of sulbenicillin in the presence of human serum albumin. Chirality 2001; 13(5):236–243.
26. Otagiri M, Masuda K, Imai T, et al. Binding of pirprofen to human serum albumin studied by dialysis and spectroscopy techniques. Biochem Pharmacol 1989; 38(1):1–7.
27. Imamura H, Komori T, Ismail A, et al. Stereoselective protein binding of alprenolol in the renal diseased state. Chirality 2002; 14(7):599–603.
28. Tillement JP, Zini R, d' Athis P, et al. Binding of certain acidic drugs to human albumin: theoretical and practical estimation of fundamental parameters. Eur J Clin Pharmacol 1974; 7(4):307–313.
29. Miller JH, Smail GA. Interaction of the enantiomers of warfarin with human serum albumin, peptides and amino acids [proceedings]. J Pharm Pharmacol 1977; 29 Suppl:33P.
30. Lagercrantz C, Larsson T, Denfors I. Stereoselective binding of the enantiomers of warfarin and tryptophan to serum albumin from some different species studied by affinity chromatography on columns of immobilized serum albumin. Comp Biochem Physiol C 1981; 69(2):375–378.
31. Fitos I, Visy J, Magyar A, et al. Inverse stereoselectivity in the binding of acenocoumarol to human serum albumin and to alpha 1-acid glycoprotein. Biochem Pharmacol 1989; 38(14):2259–2262.

32. Noctor TA, Wainer IW, Hage DS. Allosteric and competitive displacement of drugs from human serum albumin by octanoic acid, as revealed by high-performance liquid affinity chromatography, on a human serum albumin-based stationary phase. J Chromatogr 1992; 577(2):305–315.
33. Brown NA, Jahnchen E, Muller WE, et al. Optical studies on the mechanism of the interaction of the enantiomers of the antiocagulant drugs phenprocoumon and warfarin with human serum albumin. Mol Pharmacol 1977; 13(1):70–79.
34. Joseph KS, Moser AC, Basiaga SB, et al. Evaluation of alternatives to warfarin as probes for Sudlow site I of human serum albumin: characterization by high-performance affinity chromatography. J Chromatogr A 2009; 1216(16):3492–3500.
35. Moser AC, Kingsbury C, Hage DS. Stability of warfarin solutions for drug-protein binding measurements: spectroscopic and chromatographic studies. J Pharm Biomed Anal 2006; 41(4):1101–1109.
36. Moravek L, Saber M, Meloun B. Steric accessibility of tyrosine residues in human serum albumin. Collect Czech Chem Commun 1979; 44:1657–1670.
37. Yang J, Hage DS. Characterization of the binding and chiral separation of D- and L-tryptophan on a high-performance immobilized human serum albumin column. J Chromatogr 1993; 645(2):241–250.
38. Yang J, Hage DS. Role of binding capacity versus binding strength in the separation of chiral compounds on protein-based high-performance liquid chromatography columns. Interactions of D- and L-tryptophan with human serum albumin. J Chromatogr A 1996; 725(2):273–285.
39. Seedher N, Bhatia S. Interaction of non-steroidal anti-inflammatory drugs, etoricoxib and parecoxib sodium, with human serum albumin studied by fluorescence spectroscopy. Drug Metabol Drug Interact 2006; 22(1):25–45.
40. Seedher N, Bhatia S. Reversible binding of celecoxib and valdecoxib with human serum albumin using fluorescence spectroscopic technique. Pharmacol Res 2006; 54(2):77–84.
41. Kim HS, Mallik R, Hage DS. Chromatographic analysis of carbamazepine binding to human serum albumin. II. Comparison of the Schiff base and N-hydroxysuccinimide immobilization methods. J Chromatogr B Analyt Technol Biomed Life Sci 2006; 837(1–2):138–146.
42. Kim HS, Hage DS. Chromatographic analysis of carbamazepine binding to human serum albumin. J Chromatogr B Analyt Technol Biomed Life Sci 2005; 816(1–2):57–66.
43. Matsunaga H, Sadakane Y, Haginaka J. Separation of basic drug enantiomers by capillary electrophoresis using chicken alpha1-acid glycoprotein: insight into chiral recognition mechanism. Electrophoresis 2003; 24(15):2442–2447.
44. Shiono H, Shibukawa A, Kuroda Y, et al. Effect of sialic acid residues of human alpha 1-acid glycoprotein on stereoselectivity in basic drug-protein binding. Chirality 1997; 9(3):291–296.
45. Hanada K, Ohta T, Hirai M, et al. Enantioselective binding of propranolol, disopyramide, and verapamil to human alpha(1)-acid glycoprotein. J Pharm Sci 2000; 89(6):751–757.
46. Fernandez C, Gimenez F, Thuillier A, et al. Stereoselective binding of zopiclone to human plasma proteins. Chirality 1999; 11(2):129–132.
47. Hiep BT, Gimenez F, Khanh VU, et al. Binding of chlorpheniramine enantiomers to human plasma proteins. Chirality 1999; 11(5–6):501–504.
48. Kimura T, Shibukawa A, Matsuzaki K. Biantennary glycans as well as genetic variants of alpha(1)-acid glycoprotein control the enantioselectivity and binding affinity of oxybutynin. Pharm Res 2006; 23(5):1038–1042.
49. Eap CB, Cuendet C, Baumann P. Selectivity in the binding of psychotropic drugs to the variants of alpha-1 acid glycoprotein. Naunyn Schmiedebergs Arch Pharmacol 1988; 337(2):220–224.
50. Eap CB, Cuendet C, Baumann P. Binding of d-methadone, l-methadone, and dl-methadone to proteins in plasma of healthy volunteers: role of the variants of alpha 1-acid glycoprotein. Clin Pharmacol Ther 1990; 47(3):338–346.

51. Kim HS, Siluk D, Wainer IW. Quantitative determination of fenoterol and fenoterol derivatives in rat plasma using on-line immunoextraction and liquid chromatography/mass spectrometry. J Chromatogr A 2009; 1216(16):3526–3532.
52. Got PA, Scherrmann JM. Stereoselectivity of antibodies for the bioanalysis of chiral drugs. Pharm Res 1997; 14(11):1516–1523.
53. Franco EJ, Hofstetter H, Hofstetter O. Enantiomer separation of alpha-hydroxy acids in high-performance immunoaffinity chromatography. J Pharm Biomed Anal 2008; 46(5):907–913.
54. Hofstetter H, Cary JR, Eleniste PP, et al. New developments in the production and use of stereoselective antibodies. Chirality 2005; 17(suppl):S9–S18.
55. Brocks DR, Dennis MJ, Schaefer WH. A liquid chromatographic assay for the stereospecific quantitative analysis of halofantrine in human plasma. J Pharm Biomed Anal 1995; 13(7):911–918.
56. Brocks DR, Ramaswamy M, MacInnes AI, et al. The stereoselective distribution of halofantrine enantiomers within human, dog, and rat plasma lipoproteins. Pharm Res 2000; 17(4):427–431.
57. Mohamed NA, Kuroda Y, Shibukawa A, et al. Binding analysis of nilvadipine to plasma lipoproteins by capillary electrophoresis-frontal analysis. J Pharm Biomed Anal 1999; 21(5):1037–1043.
58. Kuroda Y, Cao B, Shibukawa A, et al. Effect of oxidation of low-density lipoprotein on drug binding affinity studied by high performance frontal analysis-capillary electrophoresis. Electrophoresis 2001; 22(16):3401–3407.
59. Mohamed NA, Kuroda Y, Shibukawa A, et al. Enantioselective binding analysis of verapamil to plasma lipoproteins by capillary electrophoresis-frontal analysis. J Chromatogr A 2000; 875(1–2):447–453.
60. Bowers WF, Fulton S, Thompson J. Ultrafiltration vs equilibrium dialysis for determination of free fraction. Clin Pharmacokinet 1984; 9(suppl 1):49–60.
61. Chiiken I. Analytical Affinity Chromatography. Boca Raton: CRC Press, 1987.
62. Hage DS. Affinity Chromatography. New York: Marcel Dekker, 1998.
63. Ali I, Hussain A, Aboul-Enein HY, et al. Supramolecular systems-based HPLC for chiral separation of beta-adrenergics and beta-adrenolytics in drug discovery schemes. Curr Drug Discov Technol 2007; 4(4):255–274.
64. Millot MC. Separation of drug enantiomers by liquid chromatography and capillary electrophoresis, using immobilized proteins as chiral selectors. J Chromatogr B Analyt Technol Biomed Life Sci 2003; 797(1–2):131–159.
65. Bertucci C, Bartolini M, Gotti R, et al. Drug affinity to immobilized target biopolymers by high-performance liquid chromatography and capillary electrophoresis. J Chromatogr B Analyt Technol Biomed Life Sci 2003; 797(1–2):111–129.
66. Hage DS. Affinity chromatography: a review of clinical applications. Clin Chem 1999; 45(5):593–615.
67. Kaliszan R. Retention data from affinity high-performance liquid chromatography in view of chemometrics. J Chromatogr B Biomed Sci Appl 1998; 715(1):229–244.
68. Hage DS, Austin J. High-performance affinity chromatography and immobilized serum albumin as probes for drug- and hormone-protein binding. J Chromatogr B Biomed Sci Appl 2000; 739(1):39–54.
69. Mallik R, Hage DS. Affinity monolith chromatography. J Sep Sci 2006; 29(12):1686–1704.
70. Hage DS, Jackson A, Sobansky MR, et al. Characterization of drug-protein interactions in blood using high-performance affinity chromatography. J Sep Sci 2009; 32(5–6):835–853.
71. Kimura T, Nakanishi K, Nakagawa T, et al. High-performance frontal analysis of the binding of thyroxine enantiomers to human serum albumin. Pharm Res 2005; 22(4):667–675.
72. Zhivkova ZD, Russeva VN. Stereoselective binding of ketoprofen enantiomers to human serum albumin studied by high-performance liquid affinity chromatography. J Chromatogr B Biomed Sci Appl 1998; 714(2):277–283.

73. Fitos I, Visy J, Simonyi M, et al. Stereoselective allosteric binding interaction on human serum albumin between ibuprofen and lorazepam acetate. Chirality 1999; 11(2):115–120.
74. Chen J, Hage DS. Quantitative analysis of allosteric drug-protein binding by biointeraction chromatography. Nat Biotechnol 2004; 22(11):1445–1448.
75. Markuszewski M, Kaliszan R. Quantitative structure-retention relationships in affinity high-performance liquid chromatography. J Chromatogr B Analyt Technol Biomed Life Sci 2002; 768(1):55–66.
76. Mallik R, Hage DS. Development of an affinity silica monolith containing human serum albumin for chiral separations. J Pharm Biomed Anal 2008; 46(5):820–830.
77. Mallik R, Xuan H, Hage DS. Development of an affinity silica monolith containing alpha1-acid glycoprotein for chiral separations. J Chromatogr A 2007; 1149(2): 294–304.
78. Shimura K, Kasai K. Affinity capillary electrophoresis: a sensitive tool for the study of molecular interactions and its use in microscale analyses. Anal Biochem 1997; 251(1):1–16.
79. Haginaka J. Enantiomer separation of drugs by capillary electrophoresis using proteins as chiral selectors. J Chromatogr A 2000; 875(1–2):235–254.
80. Hage DS. Chiral separations in capillary electrophoresis using proteins as stereoselective binding agents. Electrophoresis 1997; 18(12–13):2311–2321.
81. Martinez-Gomez MA, Villanueva-Camanas RM, Sagrado S, et al. Evaluation of enantioselective binding of basic drugs to plasma by ACE. Electrophoresis 2007; 28(17):3056–3063.
82. Zhu X, Ding Y, Lin B, et al. Study of enantioselective interactions between chiral drugs and serum albumin by capillary electrophoresis. Electrophoresis 1999; 20(9):1869–1877.
83. Oravcova J, Sojkova D, Lindner W. Comparison of the Hummel-Dreyer method in high-performance liquid chromatography and capillary electrophoresis conditions for study of the interaction of (RS)-, (R)- and (S)-carvedilol with isolated plasma proteins. J Chromatogr B Biomed Appl 1996; 682(2):349–357.
84. Amini A, Westerlund D. Evaluation of association constants between drug enantiomers and human alpha 1-acid glycoprotein by applying a partial-filling technique in affinity capillary electrophoresis. Anal Chem 1998; 70(7):1425–1430.
85. Kuroda Y, Shibukawa A, Nakagawa T. The role of branching glycan of human alpha1-acid glycoprotein in enantioselective binding to basic drugs as studied by capillary electrophoresis. Anal Biochem 1999; 268(1):9–14.
86. Gagyi L, Gyeresi A, Kilar F. Role of chemical structure in stereoselective recognition of beta-blockers and H1-antihistamines by human serum transferrin in capillary zone electrophoresis. Electrophoresis 2006; 27(8):1510–1516.
87. Kilar F, Visegrady B. Mapping of stereoselective recognition sites on human serum transferrin by capillary electrophoresis and molecular modelling. Electrophoresis 2002; 23(6):964–971.
88. Sandblad P, Arnell R, Samuelsson J, et al. Approach for reliable evaluation of drug proteins interactions using surface plasmon resonance technology. Anal Chem 2009; 81(9):3551–3559.
89. Day YS, Myszka DG. Characterizing a drug's primary binding site on albumin. J Pharm Sci 2003; 92(2):333–343.
90. Ahmad A, Ramakrishnan A, McLean MA, et al. Use of surface plasmon resonance biosensor technology as a possible alternative to detect differences in binding of enantiomeric drug compounds to immobilized albumins. Biosens Bioelectron 2003; 18(4):399–404.
91. Frostell-Karlsson A, Remaeus A, Roos H, et al. Biosensor analysis of the interaction between immobilized human serum albumin and drug compounds for prediction of human serum albumin binding levels. J Med Chem 2000; 43(10):1986–1992.
92. Endo H, Yoshida H, Hasegawa M, et al. Stereoselectivity and species difference in plasma protein binding of KE-298 and its metabolites. Biol Pharm Bull 2001; 24(7):800–805.

93. Fitos I, Visy J, Simonyi M. Species-dependency in chiral-drug recognition of serum albumin studied by chromatographic methods. J Biochem Biophys Methods 2002; 54(1–3):71–84.
94. Buch HP, Krug R, Knabe J. Stereoselective binding of the enantiomers of four closely related N-methyl-barbiturates to human, bovine, and rat serum albumin. Arch Pharm (Weinheim) 1996; 329(8–9):399–402.
95. Tocco DJ, deLuna FA, Duncan AE, et al. Interspecies differences in stereoselective protein binding and clearance of MK-571. Drug Metab Dispos 1990; 18(4):388–392.
96. Davis D, Grossman SH, Kitchell BB, et al. The effects of age and smoking on the plasma protein binding of lignocaine and diazepam. Br J Clin Pharmacol 1985; 19(2):261–265.
97. Viani A, Rizzo G, Carrai M, et al. The effect of ageing on plasma albumin and plasma protein binding of diazepam, salicylic acid and digitoxin in healthy subjects and patients with renal impairment. Br J Clin Pharmacol 1992; 33(3):299–304.
98. O'Shea NJ, Hooper WD. Enantioselective binding of mephobarbital to plasma proteins. Chirality 1990; 2(4):257–262.
99. Rudy AC, Knight PM, Brater DC, et al. Enantioselective disposition of ibuprofen in elderly persons with and without renal impairment. J Pharmacol Exp Ther 1995; 273(1):88–93.
100. Tan SC, Patel BK, Jackson SH, et al. Influence of age on the enantiomeric disposition of ibuprofen in healthy volunteers. Br J Clin Pharmacol 2003; 55(6):579–587.
101. Noctor TA. Enantioselective Binding of Drugs to Plasma Proteins. In: Wainer IW, ed. Drug Stereochemistry: Analytical Methods and Pharmacology. New York: Marcel Dekker, 1993.
102. Boudinot FD, Homon CA, Jusko WJ, et al. Protein binding of oxazepam and its glucuronide conjugates to human albumin. Biochem Pharmacol 1985; 34(12):2115–2121.
103. Grubb NG, Rudy DW, Brater DC, et al. Stereoselective pharmacokinetics of ketoprofen and ketoprofen glucuronide in end-stage renal disease: evidence for a "futile cycle" of elimination. Br J Clin Pharmacol 1999; 48(4):494–500.
104. Hayball PJ, Nation RL, Bochner F, et al. The influence of renal function on the enantioselective pharmacokinetics and pharmacodynamics of ketoprofen in patients with rheumatoid arthritis. Br J Clin Pharmacol 1993; 36(3):185–193.
105. Sakai T, Maruyama T, Sako T, et al. Stereoselective serum protein binding of ketoprofen in liver diseases. Enantiomer 1999; 4(5):477–482.
106. Blouin R, Chaudhary I, Nishihara K, et al. The effects of liver and renal disease on stereoselective serum binding of flurbiprofen. Br J Clin Pharmacol 1993; 35(1):62–64.
107. Mallik Rangan, Yoo Michelle J, Chen Sike, et al. Studies of verapamil binding to human serum albumin by high-performance affinity chromatography. J Chromatogr B Analyt Technol Biomed Life Sci 2008; 876(1):69–75.

12 Clinical pharmacokinetics and pharmacodynamics of stereoisomeric drugs

Scott A. Van Wart and Donald E. Mager

INTRODUCTION

Stereoisomers may exhibit different pharmacological properties, and many drugs used in clinical practice contain one or more chiral centers. These chiral drugs are often used therapeutically either as pure stereoisomers or as a racemic mixture of equal proportions of two complementary enantiomers. The three-dimensional interaction of two enantiomers with a macromolecule, such as an enzyme or receptor, to form diastereomeric complexes may result in chiral recognition and significant differences in processes such as absorption, distribution, metabolism, and excretion (pharmacokinetics or PK) as well as the time course of pharmacological response (pharmacodynamics or PD). Thus, the clinical pharmacology of chiral drugs is based on understanding the nature of these important differences for the optimization of pharmacotherapy.

ABSORPTION

Oral absorption is dependent on a number of factors including the rate of disintegration and dissolution of the drug from its dosing formulation, the rate of gastric emptying, movement of drug through the enterocyte barrier via passive diffusion or other transport mechanisms, and avoidance of intestinal and hepatic first-pass metabolism. Intestinal drug permeability is often first evaluated in vitro using bidirectional transport studies in human cell lines such as human colon carcinoma (Caco-2) cells. The rate of entry of individual enantiomers into the systemic circulation is also assessed in vivo by several approaches including examination of the maximal plasma drug concentration (C_{max}) and the time at which it is achieved (T_{max}), mathematical deconvolution techniques, and the use of compartmental PK analysis. The extent of absorption or the fraction of the dose accessing the systemic circulation (absolute bioavailability) is most often determined using model-independent approaches, such as comparing the ratio of the area under the plasma concentration–time curve (AUC) following an oral dose relative to an intravenous dose. For racemic drugs, the relative bioavailability for each enantiomer is also often reported using the ratio of AUC values for each dosing route. Stereoselective intestinal and hepatic first-pass metabolism can have a significant impact on the oral bioavailability for each enantiomer and is discussed under drug metabolism.

Passive Intestinal Absorption

For the majority of racemic drugs, absorption across the intestinal epethilial membrane appears to be passive in nature (1–3) and it is anticipated that both enantiomers will pass through the intestinal endothelium and enter the portal circulation at roughly the same rate and to the same extent, provided there is no stereoselectivity in presystemic metabolism. This expectation is based on the assumption that the physicochemical properties of both enantiomers are the same (i.e., lipophilicity, solubility, stability, ionization), resulting in comparable passive diffusion. For example, most chiral antimalarial agents are primarily passively absorbed and do not appear to be stereoselective with respect to T_{max}. Differences in C_{max} among these enantiomers are likely attributable to enantioselectivity in other PK properties, such as clearance or volume of distribution, rather than the absorption rate itself (2).

Intestinal Absorptive and Secretory Transporters

Intestinal membrane transporters can be functionally divided into two main groups, namely, influx and efflux transporters. Influx transporters are located at the apical membrane of enterocytes. The organic anion-transporting polypeptides (OATPs) serve as absorptive transporters for drugs such as fexofenadine (4), whereas the peptide transporter (PEPT1) is involved in the transport of small peptides and peptide-like drugs such as β-lactam antibiotics, cephalosporins, angiotensin-converting enzyme (ACE) inhibitors, ester prodrugs, and some anticancer drugs (5). Interestingly, the coupling of active drugs (e.g., L-dopamine) with an amino acid (e.g., valine or phenylalanine) is one strategy investigated for improving oral absorption by PEPT1 recognition and uptake (6). Efflux transporters at both the apical and basolateral membranes of the enterocyte are responsible for drug passage across the enterocyte in either the absorptive or secretory directions. P-glycoprotein (MDR1), breast cancer resistance protein (BCRP), and multidrug resistance–associated protein 2 (MRP2) are examples of secretory transporters located on the apical membrane of enterocytes, which limit entry of many drugs into the body.

Stereoselective plasma drug concentration–time profiles owing to active or carrier-mediated absorption are rare. Although many racemic drugs have been shown to be substrates for P-glycoprotein efflux in vitro, stereoselectivity in the oral bioavailability of racemic drugs that are also highly permeable and soluble is not expected as enantiomer concentrations should conceptually be high enough in the intestine to saturate transport systems and passive diffusion would predominate (7). Most substrates for P-glycoprotein efflux are also substrates for the metabolic enzyme cytochrome P450 3A4 and the coexpression and interplay of these two enzymes in the intestine further complicates the ability to attribute stereoselective drug absorption to transport by intestinal P-glycoprotein.

Fexofenadine is a racemic drug that is transported by both intestinal P-glycoprotein and OATP but is not metabolized by intestinal CYP3A (8). A PK study conducted in healthy subjects to elucidate the PK of each enantiomer after a single oral dose of racemic fexofenadine demonstrated that plasma concentrations of (R)-fexofenadine were significantly higher than (S)-fexofenadine with [R:S] AUC and C_{max} ratios of 1.75 and 1.63 (9). These findings suggest that either the affinity of P-glycoprotein for (S)-fexofenadine is greater than its affinity for the (R)-enantiomer or that the OATP uptake for (R)-fexofenadine is greater than that for the (S)-enantiomer. In healthy subjects, coadministration of fexofenadine and the P-glycoprotein inhibitor itraconazole increased the AUC of (S)-fexofenadine to a greater degree than (R)-fexofenadine (4.0- and 3.1-fold, respectively), further demonstrating the potential for intestinal P-glycoprotein stereoselectivity (10). In contrast, other studies have shown a lack of stereospecificity with respect to intestinal P-glycoprotein. Apical-to-basolateral transport of talinolol, a nonmetabolized racemic compound, in human Caco-2 cells was shown to increase almost threefold for both (R)- and (S)-talinolol in the presence of grapefruit juice (11). Grapefruit juice also accelerated the in vivo absorption rate for talinolol, but did not significantly affect terminal elimination half-lives, to the same extent for both talinolol enantiomers.

The reduced folate carrier (RFC1) is located on both the apical and basolateral membranes of intestinal epithelial cells and is responsible for transport of folate and its analogues such as methotrexate. After oral administration of both enantiomers to humans, concentrations of L-methotrexate in plasma were markedly higher than for D-methotrexate (40-fold higher C_{max} and AUC) (12). Since the elimination rate constant of D-methotrexate was only 1.7-fold greater than the L-isomer following intravenous dosing, the differences in plasma concentrations were likely due to the stereoselective intestinal absorption. A fivefold greater uptake of L-methotrexate was also observed relative to the D-isomer in human Caco-2 cells (13). Both enantiomers inhibit human RFC1-mediated folic acid uptake into Caco-2 cells, but the affinity of L-methotrexate to human RFC1 is 50-fold greater. These findings suggest stereoselective intestinal transporter is the main cause for the marked differences in the oral absorption of methotrexate enantiomers.

DISTRIBUTION

The volume of distribution correlates the amount of drug in the body to measured plasma concentrations and both the rate and extent of distribution are dependent on a number of physiological factors (body size and volumes), physicochemical drug properties (molecular weight, lipophilicity, ionization), and the degree to which the drug binds to plasma and tissue proteins. Protein binding can be a major determinant of both drug distribution and elimination, and thus stereoselective plasma protein binding could influence these processes. Stereoselective recognition by active or carrier-mediated transporters located in different tissues may also result in differences in the extent of drug distribution and tissue penetration.

Protein Binding

Endogenous and exogenous compounds are known to reversibly bind to plasma proteins such as albumin, α_1-acid glycoprotein, globulins, and lipoproteins. Stereoselective plasma protein binding of enantiomers administered in racemic form may result in a disproportionate composition of free concentrations, as well as agonistic (cooperative allosteric modification of the protein conformation) or antagonistic (noncooperative allosteric modification, competitive displacement) binding interactions with each other and/or other drugs (15). Albumin is the most abundant protein found in plasma and preferentially binds acidic drugs, whereas α_1-acid glycoprotein is present to the extent of 3% of albumin and primarily binds to basic compounds. Albumin contains two principal drug-binding sites. Site I is of larger capacity, is more flexible, and possesses a larger number of individual ligand-binding sites, which might explain why stereoselectivity is more often associated with enantiomers that bind to site II (16). Enantiomers may also bind to one or more of these sites on albumin as demonstrated for ibuprofen (3,17).

Stereoselective binding to plasma proteins has been well documented (Table 12.1). In some cases, the difference in protein binding between the enantiomers is a direct result of noncooperative allosteric modification of the protein conformation. For example, (S)-oxazepam hemisuccinate exerts a negative

CLINICAL PHARMACOKINETICS AND PHARMACODYNAMICS

TABLE 12.1 Stereoselectivity in Human Plasma Protein Binding

Drug	Mean unbound enantiomer ratio		
	Albumin	α_1-Acid glycoprotein	Plasma or serum
Bimoclomol (20)	1.04 [(−):(+)]	6.80 [(+):(−)]	6.43 [(+):(−)]
Bupivacaine (21)			1.40 [S:R]
Cetirizine (22)			1.91 [S:R]
Chloroquine (23,24)	1.2–1.5 [R:S]	1.25–1.59 [S:R]	1.6–1.76 [R:S]
Chlorpheniramine (25)	1.06 [R:S]	1.23 [R:S]	1.24 [R:S]
Disopyramide (26,27)		2.2 [R:S]	2.0 [R:S]
Etodolac (28,29)			2.5 [S:R]
Gallopamil (30)	1.32 [S:R]	1.17 [S:R]	1.40 [S:R]
Hydroxychloroquine (31)	1.42 [R:S]	1.20 [S:R]	1.4–1.75 [R:S]
Ibuprofen (3)			1.7 [S:R]
Ketorolac (32)			1.3 [S:R]
Lorazepam (18)	1.43–1.88 [R:S]		
Mexiletine (27)			1.02–1.40 [S:R]
Oxazepam (18)	3.03 [R:S]		
Propafenone (27,33)	1.02 [R:S]	1.83 [R:S]	1.60–2.06 [R:S]
Propranolol (34)	1.07 [S:R]	1.27 [R:S]	1.0–1.5 [R:S]
Thiamylal (35)			1.61–1.76 [R:S]
Verapamil (27,36)	1.43 [S:R]	1.77 [S:R]	1.7–2.0 [S:R]
Warfarin (37,38)	1.22 [R:S]		1.3 [R:S]
Zopiclone (14)	1.19 [(−):(+)]	1.17 [(−):(+)]	1.58 [(+):(−)]

allosteric effect on the binding of the (R)-enantiomer to immobilized human albumin (18). The magnitude of stereoselectivity in plasma protein binding may differ within a particular drug class. For instance, several NSAIDs (e.g., ibuprofen, indoprofen, carprofen, etodolac, ketoprofen, and flurbiprofen) are bound stereoselectively to albumin to different degrees, whereas both enantiomers of pirprofen and fenoprofen are bound to a similar extent (17). Enantiomers may also display different magnitudes of stereoselectivity between the various proteins found in plasma, such as albumin and α_1-acid glycoprotein. The binding of (R)-propranolol to albumin, for example, is greater relative to (S)-propranolol; however, the opposite stereoselectivity in the extent of binding is observed for α_1-acid glycoprotein, which predominates in plasma (34). Following the administration of racemic zopiclone, a chiral hypnotic drug, stereoselective plasma protein binding is observed with (−)-zopiclone being more highly bound than its antipode (14). The binding of (+)-zopiclone to both α_1-acid glycoprotein and albumin has been shown to be greater than for (−)-zopiclone, indicating that (−)-zopiclone may instead preferentially bind to other plasma proteins.

Stereoselective plasma protein binding can potentially translate into differences in the unbound plasma concentration–time profiles. Although clearance and volume of distribution are independent processes, protein binding may influence both PK properties to varying extents. For drugs with low or moderate hepatic extraction, plasma protein binding is a controlling factor for the transport of drug into the cellular membranes of eliminating organs and can control the effective rate of hepatic clearance. As such, a higher plasma unbound fraction of one of the enantiomers may result in an apparent stereoselectivity in clearance, even for enantiomers with similar intrinsic clearance values based on in vitro studies. Several racemic drugs that exhibit stereoselective volumes of distribution and/or

TABLE 12.2 Stereoselectivity in Volume of Distribution or Clearance for Racemic Drugs

		Mean enantiomer ratio	
Drug	Route	Volume of distribution	Clearance
Albuterol (39,40)	Inhalation	1.11 [R:S] (adult)	1.59 [R:S] (adult)
		3.44 [R:S] (child)	4.6 [R:S] (child)
Bupivacaine (41)	IV	1.56 [R:S]	1.28 [R:S]
	IV[a]	1.05 [R:S]	1.19 [S:R]
Chlorpheniramine (42)	Oral	1.95 [R:S]	1.80 [R:S]
Disopyramide (43,44)	IV	1.82 [R:S]	–
	Oral[a]	1.69 [S:R]	1.73 [S:R]
Etodolac (45)	Oral	7.62 [S:R]	13.1 [S:R]
Ketorolac (46)	IV	2.19 [S:R]	2.60 [S:R]
Hexobarbitone (47)	Oral	–	6.48 [R:S]
Methylphenobarbital (48)	Oral	–	27.6 [R:S]
Mepivacaine (49)	IV	1.55 [R:S]	1.75 [R:S]
Methadone (50)	IV and Oral	1.72 [R:S]	1.30 [R:S]
Pimobendan (51,52)	Oral	1.04 [(+):(−)]	1.31 [(+):(−)]
	IV	1.34 [(−):(+)]	1.07 [(−):(+)]
Propranolol (34)	IV	1.18 [R:S]	1.17 [R:S]
	Oral	–	1.5 [R:S]
Reboxetine (53,54)	IV	2.36 [S,S:R,R]	2.63 [S,S:R,R]
	Oral	2.24 [S,S:R,R]	2.39 [S,S:R,R]
Talinolol (55)	Oral and IV	1.31 [S:R]	1.04 [R:S]
Thiopental (56)	IV	1.27 [R:S]	1.30 [R:S]
Tocainide (57)	IV	1.02 [R:S]	1.79 [R:S]
Verapamil (58,59)	IV	2.34 [S:R]	1.77 [S:R]
	Oral	3.76 [S:R]	4.49 [S:R]
Warfarin (60)	Oral	1.83 [R:S]	1.05 [S:R]

[a]Based on unbound rather than total enantiomer concentration.

total body clearances (Table 12.2) also show stereoselectivity in plasma protein binding (e.g., bupivacaine, chlorpheniramine, disopyramide, and verapamil). Racemic bimoclomol, a cytoprotective agent investigated for the treatment of diabetic complications, shows stereoselective PK properties in vivo, with elimination of (+)-bimoclomol occurring much more rapidly than its antipode (20). The in vitro hepatic metabolism of (−)-bimoclomol is much faster than (+)-bimoclomol when incubated with human liver S9 fractions. However, (−)-bimoclomol stereoselectively binds to α_1-acid glycoprotein, with a 6.8-fold higher unbound fraction of (+)-bimoclomol. Stereoselectivity in plasma protein binding, and not metabolism, was thus implicated in the faster elimination of (+)-bimoclomol.

Tissue Distribution

Stereoselective recognition by active or carrier-mediated transporters located in different tissues may result in differences in the extent of drug distribution and tissue penetration. For example, many transport systems play an integral role in regulating the uptake and efflux of drugs into the brain. The large neutral amino acid transporter mediates the uptake of phenylalanine and neutral amino acid analogues into the brain (61). The rate and extent of uptake into brain tissue and the measured decarboxylation products (pharmacologically active dopamine) were shown to be significantly greater for L-dopa than for D-dopa (62).

P-glycoprotein is an efflux transporter that is expressed in brain endothelial cells and restricts access of numerous drugs to the brain. Stereoselective and regioselective brain tissue uptake in rats and humans has been reported for the antimalarial drug mefloquine (63). Enantiomer concentration ratios ranged from 1.5 to 3.5 across species in various anatomical regions of the brain, with the highest concentrations reported in the hippocampus and the lowest in the cerebellum. Studies using rat brain capillary endothelial cells have suggested that mefloquine is both a substrate for and stereoselective inhibitor of P-glycoprotein and the (+)-enantiomer displayed eightfold greater inhibition of vinblastine efflux (64).

METABOLISM

Stereoselective drug metabolism is commonly observed in vitro for racemic drugs (Table 12.3) and can result in substantial differences in the in vivo plasma

TABLE 12.3 Stereoselective Drug Metabolism In Vitro

Drug	Metabolic pathway	Enzyme (stereoselectivity[a])
± Warfarin (65–67)	7-Hydroxylation	CYP2C9 ($S>>R$)
	6-Hydroxylation	CYP1A2 ($R>>S$)
	8-Hydroxylation	CYP1A2 ($R>>S$)
	10-Hydroxylation	CYP3A4 ($R>>S$)
± Acenocoumarol (68)	6-Hydroxylation	CYP2C9 ($S>>R$); CYP2C19 ($R>S$)
	7-Hydroxylation	CYP2C9 ($S>>R$); CYP2C19 ($R>S$)
	8-Hydroxylation	CYP2C9 ($S>>R$); CYP2C19 ($R>S$)
± Ifosfamide (69–72)	4-Hydroxylation	CYP3A4 ($R>S$); CYP2B6 ($S>>R$)
	N^2-dechloroethylation	CYP3A4 ($R>S$); CYP2B6 ($S>>R$)
	N^3-dechloroethylation	CYP3A4 ($R>>S$); CYP2B6 ($S>>R$)
± Felodipine (73)	Oxidation	CYP3A4 ($S>R$)
± Fluoxetine (74)	N-demethylation	CYP2C9 ($R>>S$)
±Omeprazole (75)	Hydroxylation	CYP2C19 ($R>>S$)
	Sulfoxidation	CYP3A4 ($S>>R$)
	5-O-demethylation	CYP2C19 ($R>>S$)
± Mephenytoin (76)	4′-Hydroxylation	CYP2C19 ($S>>R$)
±Disopyramide (77)	N-Dealkylation	CYP3A4 ($S>R$); CYP3A5 ($S>R$)
(R)-Bufuralol (78)	1″-Hydroxylation[b]	CYP2D6 ($S>R$)
± Metoprolol (79–81)	O-Demethylation	CYP2D6 ($R>S$)
± Verapamil (82–84)	N-demethylation	CYP3A4 ($S>R$); CYP3A5 ($R>S$)
	N-dealkylation	CYP3A4 ($R>S$); CYP3A5 ($R>S$)
± Cibenzoline (85)	p-Hydroxylation	CYP2D6 ($R>>S$)
Risperidone (86)	9-Hydroxylation[b]	CYP2D6 ($S>>R$); CYP3A4 ($R>>S$)
± Etodolac (87)	Acyl Glucuronidation	UGT ($S>R$)
± Propranolol (88)	Glucuronidation	UGT1A9 ($S>>R$); UGT1A10 ($R>>S$)
± Formoterol (89,90)	Glucuronidation	UGT ($S>R$)
± Lansoprazol (91)	Sulfation	PST [(−) > (+)]
± Fenoterol (92)	Sulfation	P-PST ($S>R$); M-PST ($R>S$)
± Metaproterenol (90)	Sulfation	PST ($S>R$)
± Salmeterol (90)	Sulfation	PST ($R>S$)
± Terbutaline (90)	Sulfation	PST ($S>R$)
± Isoproterenol (90)	Sulfation	M-PST ($R>>S$)
± Ibuprofen (93)	Conjugation	Acyle-coenzyme A synthetase ($R>>S$)

[a]Stereoselectivity based on intrinsic clearance or rate of metabolite formation for the enantiomeric metabolites; ">>" defined as the ratio >5, otherwise as ">".
[b]Product enantioselectivity.
Abbreviations: UGT, uridine 5′-diphosphate glucuronyltransferase; PST, phenosulfotransferase.

concentration–time profiles between enantiomers due to stereoselective bioavailability or drug disposition (94,95). Selective enzymatic recognition could occur at the binding and/or catalytic steps, resulting in different enantiomer affinities and/or reactivities (96). Chiral discrimination in metabolism can be classified as either substrate stereoselectivity (differential metabolism of isomers under identical conditions) or the less commonly observed product stereoselectivity (the differential formation of two enantiomeric metabolites from a single prochiral substrate) (97). Enantiomers may also differ with respect to hepatic uptake transporter affinity, which may restrict access to certain metabolic enzymes.

Phase I and Phase II Metabolism
Phase I metabolism generally results in the introduction of a new hydrophilic functional group into the substrate or the exposure of new functional groups on the substrate via oxidation, reduction, or hydrolysis reactions. Although the cytochrome P450 (CYP) family of enzymes is the major phase I enzyme responsible for drug metabolism, other enzymes such as flavin-containing monooxygenase and monoamine oxidases also fall into this category. Two dominant structural features, the iron-porphyrin-oxygen complex and the apoprotein associated with a specific enzyme, control the reactions catalyzed by the CYP P450 enzymes (96). The apoprotein governs which part of a given substrate can gain access to the enzyme and defines metabolic stereoselectivity. Phase II metabolism instead includes pathways such as acetylation, amino acid conjugation, glutathione conjugation, glucuronidation, methylation, and sulfation.

Stereoselectivity in drug metabolism by the P450 enzymes has been reported for a number of racemic drugs. The magnitude of selectivity depends on the metabolic pathway involved. Warfarin enantiomers, for example, are eliminated almost entirely by metabolism by different cytochrome P450 enzymes. The predominant route of metabolism for (S)-warfarin is 7-hydroxylation of the coumarin ring by CYP2C9, whereas (R)-warfarin is preferentially metabolized by CYP1A2 to 6- and 8-hydroxywarfarin and by CYP3A4 to 10-hydroxywarfarin (65). Omeprazole, which is metabolized by multiple pathways including sulfoxidation, demethylation, and hydroxylation, shows stereoselectivity with respect to each of these metabolic processes using human liver microsomes (55). Enantiomers may also interact with each other and compete for the same metabolic pathway. For example, propafenone is extensively metabolized to 5-hydroxypropafenone by CYP2D6. The in vitro metabolism of propafenone enantiomers is inhibited by each other in a competitive manner, and (R)-propafenone is a stronger inhibitor toward 5-hydroxylation than its antipode (98). After oral administration of racemic propafenone or the individual enantiomers, the clearance of the (R)-enantiomer was similar whereas clearance of (S)-propafenone was significantly lower after administration of the racemate as compared with administration of the (S)-enantiomer alone (99,100).

Stereoselectivity in the metabolism of racemic drugs by acetylation, glucuronidation (propranolol), and sulfation (isoproterenol and albuterol) pathways has been reported in the literature (101). For the β_2-receptor agonists isoproterenol and albuterol, a 6.1-fold higher K_m was reported for (R)-isoproterenol while the [R:S] ratio for the K_m of albuterol is reported to range from 0.1 to 0.2 (19,90). Propanolol is metabolized by two separate uridine 5'-diphosphate glucuronyl transferase enzymes (UGT1A9 and UGT1A10) as well as by sidechain/ring oxidation. Hepatic UGT1A9 metabolizes (S)-propranolol faster than (R)-propranolol, whereas extrahepatic UGT1A10 metabolizes (R)-propranolol faster than (S)-propranolol (88).

First-Pass Metabolism and Bioavailability

Differences in oral bioavailability observed between enantiomers are relatively common as a consequence of differences in intestinal or hepatic metabolism. For low-extraction drugs (hepatic blood flow >> intrinsic clearance), stereoselectivity in the intrinsic clearance will directly impact hepatic clearance and result in differences in the enantiomeric plasma concentrations after both oral and intravenous dosing. For highly extracted chiral drugs (hepatic blood flow << intrinsic clearance), stereoselective intrinsic clearance may not alter enantiomer plasma concentrations following intravenous dosing as systemic clearance is mainly governed by organ blood flow. However, stereoselective intrinsic clearance can lead to differences in enantiomer plasma concentrations and oral bioavailability.

Stereoselective first-pass metabolism following oral administration of the racemate has been reported for several racemic drugs such as albuterol, propranolol, and verapamil (90,102–104). Verapamil is administered as a racemic mixture, despite (S)-verapamil being 10- to 20-times more potent than (R)-verapamil. Verapamil exhibits stereoselective first-pass metabolism, with preferential metabolism of (S)-verapamil following oral administration. The differences between the plasma concentrations of verapamil enantiomers are more pronounced after oral administration, with twofold higher C_{max} and AUC values for (R)-verapamil because of increased presystemic metabolism of (S)-verapamil (33,83,104). Oral bioavailability can often be enhanced by administering high doses to saturate the enzymes responsible for presystemic metabolism.

Chiral Inversion

Several drug stereoisomers may switch from one enantiomeric configuration to the other (chiral inversion), and this may be spontaneous and bidirectional or facilitated by the presence of proteins or enzymes, in which case it is often unidirectional in nature. Thalidomide is known to undergo a bidirectional chiral inversion in human serum (105). Similarly, the NSAID oxindanac appears to undergo bidirectional chiral inversion when spiked in plasma and after in vivo administration, although at equilibrium the enantiomer ratio appears to favor (S)-oxindanac (106).

More specific biochemically mediated reactions are possible that are primarily unidirectional. The 2-arylpropionic acid NSAIDs, such as ibuprofen and fenoprofen, possess a chiral center α to the carboxyl group (93,107–109). The (S)-enantiomer is primarily responsible for the inhibition of COX enzyme and the (R)-enantiomer is relatively inactive in vitro. However, enantioselective inversion from the (R)-enantiomer to its antipode for this class of drugs has been observed in vivo (93). The anatomical site of chiral inversion for NSAIDs is controversial, but there is evidence that the inversion site resides in the liver and the intestinal tract (110–112). The mechanism of this inversion reaction involves initial enantioselective formation of a coenzyme A (CoA) thioester followed by epimerization and hydrolytic cleavage of the formed antipode-CoA thioester by acyl CoA ligase to regenerate free acids. Competing metabolic pathways may influence the potential and extent of chiral inversion for this drug class (113). This may explain in part why unidirectional (R)- to (S)-enantiomer chiral inversion appears to be species-dependent and sometimes absent in humans, as reported for some NSAIDs such as ketoprofen, tiaprofenic acid, and flurbiprofen (114–116).

The therapeutic consequence of both unidirectional or bidirectional chiral inversion is that the concentration-time profile of the active enantiomer(s) may be substantially different than expected based on the administered racemic dose. Chiral inversion may be detected based on recognition of several features, primarily the appearance of one enantiomer in the plasma after dosing of the other isomer (this can be used to determine if conversion is unidirectional or bidirectional). In addition, the concentration-time profiles for each enantiomer should appear to be polyexponential with parallel terminal elimination phases, even when the disposition of each enantiomer may appear to be monoexponential after administered alone (117). Characterization of the disposition of racemic drugs undergoing chiral conversion is complex and resolving system parameters often require PK data after dosing each enantiomer separately. A PK model describing the bidirectional chiral conversion of two enantiomers with monoexponential disposition is shown in Figure 12.1. This structural model can characterize plasma concentrations of both enantiomers after administration of each enantiomer alone or as a racemic mixture (117,118).

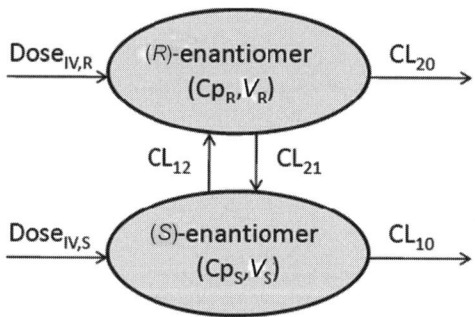

Where:
   $Dose_{IV,S}$ = Intravenous dose of the (S)-enantiomer alone, or 50% of a racemic dose
   $Dose_{IV,R}$ = Intravenous dose of the (R)-enantiomer alone, or 50% of a racemic dose
   Cp_S = Plasma concentration of the (S)-enantiomer
   Cp_R = Plasma concentration of the (R)-enantiomer
   V_S = Volume of distribution of the (S)-enantiomer
   V_R = Volume of distribution of the (R)-enantiomer
   CL_{10} = Clearance of the (S)-enantiomer
   CL_{20} = Clearance of the (R)-enantiomer
   CL_{12} = Interconversion clearance of the (S)-enantiomer to the (R)-enantiomer
   CL_{21} = Interconversion clearance of the (R)-enantiomer to the (S)-enantiomer
   $CL_{T,S}$ = Total clearance of the (S)-enantiomer, equal to $CL_{10} + CL_{12}$
   $CL_{T,R}$ = Total clearance of the (R)-enantiomer, equal to $CL_{20} + CL_{21}$

FIGURE 12.1 Pharmacokinetic model diagram for a drug undergoing bidirectional chiral inversion following intravenous administration of the racemate or each enantiomer alone.

TABLE 12.4 Influence of Polymorphic Drug Metabolism on Stereoselectivity

Drug	Metabolic enzyme	Mean AUC ratio	
		Extensive metabolizers	Poor metabolizers
Chlorpheniramine (124)	CYP2D6	2.4 [S:R]	3.5 [S:R]
Flecainide (125)	CYP2D6	1.03 [R:S]	1.29 [R:S]
Fluoxetine (120)	CYP2D6	1.45 [S:R]	5.59 [S:R]
Lansoprazole (126)	CYP2C19	8.54 [R:S]	5.71 [R:S]
Metoprolol (127)	CYP2D6	0.60 [(+):(−)]	1.11 [(+):(−)]
Mexiletine (128)	CYP2D6	0.96 [R:S]	0.92 [R:S]
Omeprazole (129)	CYP2C19	0.62 [(+):(−)]	1.5 [(+):(−)]
Pantoprazole (130)	CYP2C19	0.84 [(+):(−)]	3.26 [(+):(−)]
Phenprocoumon (123)	CYP2D6	1.05 [S:R]	1.54 [S:R]
Propafenone (95)	CYP2D6	1.71 [S:R]	1.48 [S:R]
Propranolol (131)	CYP2D6	0.70 [(+):(−)]	0.93 [(+):(−)]
Trimipramine (122)	CYP2D6	1.15^a [L:D]	0.60^a [L:D]
	CYP2C19		1.66^a [L:D]

^{a}Cssmin

Genetic Polymorphisms in Metabolism

Polymorphic cytochrome P450 enzymes have been shown to influence stereoselectivity in plasma concentrations of racemic drugs (Table 12.4). Human CYP2D6 is involved in the oxidative metabolism of approximately 25% of all commonly prescribed drugs, and genotypes can be classified into four subgroups: extensive metabolizers (CYP2D6*1, 70–80% of Caucasians); intermediate metabolizers (CYP2D6*10/*17, 10–15% of Caucasians); poor metabolizers (CYP2D6*3/*4/*5, 7–10% of Caucasians and 1% of Asians); and ultrarapid metabolizers (3–5% of Caucasians). For human CYP2C9 and CYP2C19, the *1 alleles are considered to be extensive metabolizers, while the CYP2C19*2/*3 alleles (3% of Caucasians and 15–20% of Asians) and the CYP2C9*2/*3 alleles (8–12% of Caucasians) are poor metabolizers (95).

Debrisoquine is a prochiral drug used as a CYP2D6 phenotyping probe (119). In extensive metabolizers, hydroxylation generates the (S)-metabolite almost exclusively, whereas in poor metabolizers, the (R)-metabolite represents a significant fraction (5–36%) of the total metabolite excreted. Poor metabolizers of the antidepressant fluoxetine, which is metabolized by CYP2D6, exhibit a 3.8-fold higher enantiomeric [S:R] AUC ratio as compared to extensive metabolizers (120,121). Stereoselectivity in the metabolism of trimipramine has also been reported, with a preferential N-demethylation of D-trimipramine and hydroxylation of L-trimipramine. It was suggested that CYP2C19 is involved in the demethylation pathway with preferential metabolism of D-trimipramine (122). As a result, poor metabolizers of CYP2C19 exhibit lower exposures to D-trimipramine with [L:D] trimipramine AUC ratios of 1.7 versus 1.2 reported for extensive metabolizers. Phenprocoumon, an inhibitor of vitamin K reductase, is stereoselectively metabolized by CYP2C9 to inactive hydroxylated metabolites. A genetic polymorphism of CYP2C9 reduces phenprocoumon (S)-7-hydroxylation and increases concentrations of (S)-phenprocoumon, thereby increasing risk of excessive anticoagulation in poor metabolizers (123).

EXCRETION

Biliary and urinary excretory processes have mechanistic components that may show stereoselectivity. Transport proteins such as P-glycoprotein, MRP, and organic anion and cation transporters are all implicated in drug transport across canalicular and renal tubular cells, and binding site specificity may lead to enantioselectivity in substrate transport. Both plasma drug concentrations and excretion data should be collected so that the biliary or renal clearance can be determined as the ratio of the cumulative amount excreted to the total plasma AUC. Only when a drug is eliminated solely by one route of elimination can the cumulative excretion data alone be used to judge the presence of stereoselective secretion in that pathway. For example, the [S:R] ratio of the cumulative urinary excretion of bisoprolol in beagle dogs is about 1.29, whereas renal clearance is essentially nonstereoselective (S:R ratio = 0:96) (132).

Renal Clearance

Glomerular filtration, tubular secretion, and tubular reabsorption are the major processes governing renal clearance. Both glomerular filtration and tubular reabsorption are generally considered to be passive processes and are not expected to contribute to stereoselectivity in renal clearance. However, transport systems responsible for the renal tubular secretion of organic anions and cations, which move ions from the blood across the basolateral membrane and into the proximal tubular fluid, may display stereoselectivity and differences in renal clearance (133). The organic anion transporters (OATs) and OATPs are responsible for entry of organic anions into the proximal tubule cells from the blood and urine and have been implicated in the accumulation of NSAIDs in proximal tubule cells (134–136). The organic cation transporters (OCTs) and P-glycoprotein are also involved in renal elimination of cationic drugs such as quinidine and verapamil (137).

Despite the fact that many of these transporters are involved in the renal secretion and reabsorption of racemic drugs, there is currently little evidence of stereoselective recognition by these transporters in humans. For the antibiotic ofloxacin, (R)-ofloxacin appears to have a greater affinity for OCTs in humans than its antipode resulting in a slightly faster [R:S] renal clearance ratio of 1.05 (138). The tubular secretion of (S)-cetirizine is 1.9-fold higher than that of the more active (R)-cetirizine, which may have consequences for drug interactions at the renal level (22). The greater tubular secretion of (S)-cetirizine is likely due to the higher free fraction available for secretion or due to a greater affinity for renal secretory transporters. There is also evidence implicating stereoselective binding of drug enantiomers to OCTs, as a number of compounds such as disopyramide, pindolol, and verapamil stereoselectively inhibit transport in vitro (139,140).

Biliary Clearance

Stereoselective recognition by transport proteins could also potentially lead to differences in biliary clearance. Several canalicular transport proteins have been identified and are known to be involved in the biliary clearance of racemic drugs, with the major factors being P-glycoprotein, MRP2, and BCRP (141). P-glycoprotein is involved in the transport of cationic substrates such as quinidine, talinolol, and verapamil while MRP2 plays a key role in the biliary excretion of organic anions such as methotrexate (141–143). There is little information available regarding stereoselective biliary clearance, with the exception of quinine and quinidine. Clinically relevant drug-drug interactions for digoxin (~40% reduction in biliary clearance) have been reported following coadministration of verapamil, quinine, and quinidine, which was attributed to inhibition of hepatic P-glycoprotein by these racemic drugs (144). For example, quinidine reduces biliary clearance to a greater extent (42%) than quinine (35%).

OTHER FACTORS AFFECTING STEREOISOMER DISPOSITION
Age
Age-related differences in physiological processes may contribute to the PK variability for many racemic drugs (145,146). Age-related differences in hexobarbital enantiomers have been reported, with elderly males having a significantly lower clearance of L-hexobarbital compared to younger adults despite the fact that there was no difference in clearance for D-hexobarbital (147). Elderly patients had increased exposure to active (S)-ibuprofen following oral administration of racemic ibuprofen ([S:R] AUC ratio of 1.8 and 1.4 in young and elderly groups), which potentially may increase efficacy in the elderly relative to younger adults (148). Other age-specific differences in stereoselectivity have been reported for amlodipine, verapamil, and citalopram (149–151).

Gender
Gender-related differences in stereoselective drug disposition are more likely to be associated with drug distribution as opposed to elimination, given the typical body size differences between men and women. A population PK analysis of methadone indicated stereoselectivity in the volume of distribution between the (R)- and (S)-enantiomers, which decreased with increasing α_1-glycoprotein concentration and was also lower in females relative to males (152). None of the patient covariates tested, including gender, were significant predictors of the total clearance of either enantiomer. Gender-related differences in the PK of *trans*-tramadol have also been reported (19). The ratio of (+):(−) *trans*-tramadol for the apparent volume of distribution but not total clearance was higher in males, whereas no difference was observed for either parameter in females, suggesting that body weight may be a confounding factor.

Diet and Lifestyle
Lifestyle factors such as smoking and dietary intake may influence drug enantiomer concentrations. Smoking is associated with the induction of a variety of drug metabolizing enzymes, most notably CYP1A1. For mexiletine, which is metabolized mainly by CYP2D6 and to a lesser extent by CYP1A2, the total clearance of each enantiomer was 40% to 60% higher in smokers than in nonsmoking subjects (153). Consumption of cruciferous vegetables such as broccoli and brussel sprouts, as well as charcoal-cooked meats, has been shown to induce CYP1A2, and alcohol consumption induces CYP2E1 (154). Drinking grapefruit juice or use of St. John's Wart can potentially inhibit CYP3A4. Grapefruit juice may also inhibit intestinal transporter function and may either decrease (e.g., OATP inhibition) or increase (e.g., P-glycoprotein inhibition) oral bioavailability (4). However, whether ingestion of grapefruit juice can lead to stereoselective PK is controversial. For example, both intestinal

and hepatic CYP3A4-mediated first-pass metabolism of (R)- and (S)-lansoprazole were not influenced by grapefruit juice in healthy subjects (155). Ingestion of grapefruit juice also failed to influence the enantiomeric ratio of AUC values for nitrendipine, whereas coadministration of the CYP450 inhibitor cimetidine increased the [S:R] ratio for AUC by 20% compared to nitrendipine alone (156).

Drug-Drug Interactions

Enantiomers present in a racemic mixture may interact in a stereospecific manner on the disposition of other concurrently administered drugs. In other cases, an interaction may take the form of one drug interacting preferentially on the PK of a single enantiomer. Some examples of PK drug-drug interactions involving one or both enantiomers are summarized in Table 12.5 and discussed further in this section.

Protein Binding Displacement

In addition to enantiomer-enantiomer protein binding interactions, concomitant administration of enantiomers and other drugs may result in competitive displacement from plasma proteins. Salicylate, tolbutamide, valproic acid, ibuprofen, and probenecid represent drugs that could potentially alter plasma protein binding in a stereoselective manner, given their high binding affinities for sites on albumin and may achieve relatively high concentrations in plasma following therapeutic doses. Clinically important examples of protein binding drug-drug interactions resulting in high plasma concentrations of unbound drug are rare (for drugs that are greater than 90% to 95% protein bound) but include phenytoin displacement by valproic acid and warfarin displacement by

TABLE 12.5 Pharmacokinetic Drug-Drug Interactions in Humans Involving Drug Enantiomers

Chiral drug	Interacting drug	Change in plasma concentrations	Probable mechanism
Carvedilol (157)	Amiodarone	$S\uparrow, R\leftrightarrow$	Apparent selective inhibition of CYP2C9
Chlorpheniramine (124)	Quinidine	R and $S\uparrow$	Inhibition of CYP 2D6
Etodalac (158)	Phenobarbital	R and $S\downarrow$	Hepatic enzyme induction
Hexobarbital (159)	Rifampin	$R\uparrow, S\leftrightarrow$	Induction of CYP2C19
Mexiletine (128)	Ciprofloxacin	R and $S\uparrow$	Inhibition of CYP1A2
Nicardipine (160)	Grapefruit juice	$(+)$ and $(-)\uparrow$	Inhibition of CYP3A4
Nitrendipine (156)	Grapefruit juice Cimetidine	R and $S\uparrow$	Inhibition of CYP3A4
Pindolol (161)	Cimetidine	R and $S\uparrow$	Inhibition of renal tubular secretion and hepatic metabolism
Propranolol (162–165)	Nicardipine, verapamil, cimetidine, or quinidine	R and $S\uparrow$	Inhibition of first-pass metabolism
Verapamil (166)	Rifampin	R and $S\uparrow$	Induction of hepatic and gastrointestinal P450
Verapamil (167)	Grapefruit juice	R and $S\uparrow$	Inhibition of CYP3A4
Warfarin (168,169)	Cimetidine	$R\uparrow, S\leftrightarrow$	Inhibition of CYP
	Bucolome	$R\leftrightarrow, S\uparrow$	

$\uparrow$, increase; $\downarrow$, decrease; $\leftrightarrow$, unchanged.

phenylbutazone (170). As outlined by Benet and Hoener (171), changes in protein binding are rarely clinically important, with the exception of a drug with a high extraction ratio and narrow therapeutic index given parenterally or a narrow therapeutic index drug given orally with a rapid onset of effect.

Altered Hepatic Clearance

For racemic drugs with low hepatic extraction, coadministration with drugs that inhibit or induce intrinsic clearance may alter both total clearance and the elimination half-life, but not oral bioavailability. In contrast, racemic drugs with a high hepatic extraction are expected to have altered total clearance, elimination half-life, and oral bioavailability when coadministered with medications that alter hepatic blood flow (e.g., hydralazine), and drugs that affect intrinsic clearance under such conditions will primarily affect oral bioavailability. Propranolol is an example of a high-extraction drug that exhibits stereoselective increases in plasma concentrations and oral bioavailability following coadministration of cimetidine, resulting in significant increases in an (R)-propranolol concentration ratio in plasma relative to its antipode (165). Similarly, plasma concentrations of (R)-warfarin were increased by cimetidine, although (S)-warfarin concentrations are not significantly affected (168). Some inhibitors themselves are chiral and subject to stereoselective drug disposition and in fact a greater degree of in vitro inhibition of CYP3A4 has been demonstrated for (–)-ketoconazole than (+)-ketoconazole (172). Induction of drug metabolic enzymes can also lead to stereoselectivity in plasma concentrations as is demonstrated for oral verapamil. After coadministration of rifampin, a known inducer of a number of CYP450 enzymes, induction of both intestinal and hepatic metabolism disproportionately reduces (R)-verapamil exposure relative to (S)-verapamil (166).

PHARMACODYNAMICS

Similar to drug disposition, differences in pharmacological or toxicological activities may be observed between individual enantiomers owing to stereoselective recognition at receptor binding sites and subsequent activation pathways. The inability of an enantiomer to bind with the three-dimensional targeted receptor binding site may lead to reduced efficacy for that enantiomer. Different receptor isoforms may also result in stereoselective drug-target binding. Thus, administration of the racemate may impact efficacy or the side effect profile compared to giving the same dose of the more active enantiomer. These caveats will be discussed in this section, along with rationale for chiral switching and improving the target selectivity and/or specificity.

Greater Activity with Racemic Mixture

Dobutamine is an inotropic agent that increases the force of myocardial contraction without increasing heart rate or blood pressure. Although both enantiomers are active, (–)-dobutamine is 33% as effective as (+)-dobutamine in increasing cardiac contractility while possessing much greater pressor activity, indicating different selectivity for α- and β-adrenoreceptors (173). (+)-Dobutamine acts as a potent agonist at the β_1-receptor and exhibits weak α-antagonist properties, whereas (–)-dobutamine possesses weaker β_1-receptor agonist activity and is a potent α_1-receptor agonist (174). Both enantiomers contribute to the positive inotropic effects, while the peripheral vasoconstrictor and vasodilator effects likely cancel out. The clinical use of the racemate thus provides advantages that could not be achieved by dosing the individual enantiomers. This example also illustrates the distinction between stereoselective receptor binding (affinity) and activation of receptors (efficacy). Both stereoisomers bind to α-receptors with equal affinity; however, (–)-dobutamine is an agonist and the (+)-isomer is a potent competitive antagonist (174). Thus, the point of asymmetry on the drug may be related to the ability of an agonist such as dobutamine to activate α-receptors, yet may have a limited role in the binding process itself. Such is not the case for the β_1-receptor, where the asymmetric center influences affinity but not efficacy, highlighting receptor-specific structural requirements for activation.

Enantiomer Differential Effects and Indications
Enantiomers of propoxyphene are pharmacologically active and exert distinct effects, with D-propoxyphene producing analgesic effects and L-propoxyphene serving as an antitussive (175). In fact, these molecules are marketed under separate trade names that are mirror images of each other, Darvon and Novrad. Differential pharmacological effects between enantiomers have also been reported for related opiate derivatives. D-Methorphan is an antitussive agent that is devoid of substantial analgesic or sedative effects, whereas L-methorphan is a potent opioid analgesic.

Primary Activity from a Single Enantiomer
For a number of chiral drugs, pharmacologic activity is restricted to a single enantiomer, commonly referred to as the so-called eutomer, while the other enantiomer is inactive or weakly active and is referred to as the distomer. It is common for the eutomer to exhibit several fold greater pharmacologic activity than its distomer.

β-Adrenoreceptor Antagonists
The β-adrenoreceptor antagonists (β-blockers) are indicated for the management of various conditions such as cardiac arrhythmias, angina, and hypertension. This class includes nonselective β-blocking agents (propranolol, pindolol, and sotolol) and β_1-selective drugs (acebutolol, atenolol, alprenolol, betaxolol, and metoprolol). Some racemic β-blockers (e.g., labetalol and carvedilol) exhibit mixed antagonism of both β- and α_1-receptors, which provides additional arteriolar vasodilating action relative to some other β-blockers. Structurally, the β-blockers may be classified as either arylethanolamine or aryloxypropanolamine derivatives. Many β-blockers are marketed in racemic form even though the L-isomer is more potent at blocking β-receptors than the D-isomer (activity typically resides in the (R)-enantiomers of the arylethanolamine series and the (S)-enantiomers of the aryloxypropanolamine series). Differences in enantiomer binding affinity ratios exist between receptor subtypes, with higher ratios typically observed for β_1-receptors than β_2-receptors. Despite stereoselective activity, only three of the β-blockers (timolol, penbutolol, and levobutanol) are marketed as the single (S)-enantiomer. There appears to be little advantage of using single enantiomers since adverse effects are primarily related to their pharmacologic action and thus a significant reduction in side effects is unlikely.

Stereoselective binding affinity for the β-blockers is highly variable, with differences in the [R:S] enantiomer potency ratio ranging from 10-fold for atenolol to 1000-fold for pindolol. Sotalol exhibits both β-receptor and potassium-channel blocking activity and is used clinically as an antiarrhythmic agent. Although sotalol enantiomers are equipotent in their ability to lengthen the cardiac action potential duration, (–)-sotalol shows 30- to 60-fold greater β-blocking potency than (+)-sotalol (176,177). The β-blocking potency of (S)-propranolol is 40- to 100-fold greater than that of (R)-propranolol. Alprenolol has equal affinity for $β_1$- and $β_2$-receptors, but the (–)-enantiomer is 100 times more potent at $β_1$-antagonism (176). Timolol is marketed as (S)-enantiomer rather than racemate due to the fact that (S)-enantiomer is 6- to 18-fold more potent than (R)-timolol.

Carvedilol and labetalol are nonselective β-blockers used to treat heart failure. (S)-Carvedilol is a 100-fold more potent $β_1$-receptor antagonist than the (R)-isomer in isolated guinea pig atrium and both isomers are equipotent $α_1$-receptor blockers (178,179). In addition, (S)-carvedilol is sixfold more potent as an antihypertensive agent in rats. Two of the four marketed labetalol isomers are completely inactive ((S,S)- and (R,S)-labetalol) and another is an $α_1$-antagonist ((S,R)-labetalol). The nonselective β-blocker dilevalol is the (R,R)-enantiomer of labetalol, possessing slightly more potent $β_2$- than $β_1$-adrenoreceptor antagonism, contributing to its vasodilator effect, and exhibits negligible affinity for α-receptors. Dilevalol is a fourfold more potent β-blocker, but is a three- to sixfold less potent α-blocker as compared to labetalol (180,181).

Calcium Channel Antagonists

The calcium channel antagonists prevent entry of calcium into cardiac muscle and vascular smooth muscle by preferentially binding to voltage-gated calcium slow channels (L-type) (182). There are three main classes of calcium channel blockers. Dihydropyridines (e.g., amlodipine, nicardipine, nimodipine, nisoldipine, felodipine, prandipine, and mandipine) are selective primarily for smooth muscle leading to reduced systemic vascular resistance and vasodilation, but to a lesser degree cause coronary vasodilation and other negative inotropic effects and are used clinically in the treatment of hypertension and angina. Phenylalkylamines (verapamil and gallopamil) are more cardioselective and cause reductions in myocardial oxygen demand, contractility, and atrio-ventricular (AV) conduction, and produce a small degree of vasodilatory effects making these agents effective in the treatment of cardiac arrhythmias. Benzothiazepines (diltiazem) represent an intermediate class that possesses both cardiac depressant and vasodilator actions.

All three classes of calcium channel blockers have shown stereoselective PD effects, with the L-isomer generally producing greater activity than the D-isomer (183). For example, the potency of (S)-verapamil on reducing AV conduction and producing vasodilation in humans is 10- to 20-fold greater than the (R)-isomer (103). Amlodipine also shows marked stereoselectivity, with the (S)-enantiomer exhibiting 1000-fold greater calcium channel blocker activity than the (R)-isomer. The calcium channel blocking action of the (S)-prandipine is 50 times more potent than that of (R)-prandipine (184). (S)-gallopamil and D-cis-diltiazem show stereoselective PD effects in animal models, with these agents producing 63- and 10-fold greater negative inotropic effects and 13- and 79-fold greater vasodilating activity than their distomers (185). Although plasma concentrations of (R)- and (S)-gallopamil are higher in healthy subjects following administration of the racemate as compared to individual enantiomers(saturable first-pass metabolism), a comparable concentration-effect relationship for the active (S)-gallopamil was observed after dosing either the single enantiomer or the racemate (186).

Other Notable Examples
Hypnotics and sedatives such as hexobarbital, secobarbital, mephobarbital, and thiopental are racemic drugs in which only the L-isomer is active, whereas the D-isomer is either inactive or mildly excitative (183). β_2-Receptor agonists (e.g., albuterol, salmeterol, and terbutaline) are administered as racemic mixtures for the treatment of asthma, although bronchodilator properties primarily reside only in the L-isomer. ACE inhibitors (e.g., captopril, benazepril, enalapril) are typically marketed only as the (S)-isomer, which accounts for nearly all of the cardiovascular activity. The analgesic and anti-inflammatory effects of NSAIDs reside in the D-isomer (i.e., (S)-ibuprofen and (S)-ketoprofen exhibit greater potency than their (R)-isomers) (187–190). (S)-citalopram is 100-fold more potent as a selective serotonin reuptake inhibitor than (R)-citalopram for the treatment of depression (191). (R)-methadone is 25- to 50-fold more potent on μ-opiod receptors for use as a centrally acting analgesic than the (S)-isomer in humans (192).

Stereoselective Activity and Toxicity
The development of racemates with only one eutomer is common; however, the toxicological properties of the distomer in some cases have prompted a change in clinical use or a postapproval switch from the racemate to an individual enantiomer. For example, the initial use of racemic dopa for the treatment of Parkinson's disease resulted in a number of adverse effects including nausea, vomiting, anorexia, involuntary movements, and granulocytopenia (193). The subsequent use of L-dopa enabled use of half the dose and resulted in mitigation of many of these adverse events (194). Penicillamine, a metabolite of penicillin with no antibiotic activity, was shown to increase urinary copper excretion via chelation and was used in the 1950s for the treatment of a genetic disorder of copper metabolism (Wilson's disease). Preclinical toxicology was linked to L-penicillamine and the greater mutagenicity of the L-isomer relative to the D-isomer was confirmed (195). Initial clinical studies conducted in the United States using the racemate resulted in a high incidence of optic neuritis and the drug was withdrawn. In the United Kingdom, D-penicillamine was later evaluated and this adverse effect was no longer observed (196).

Ethambutol, an antimycobacterial drug used to treat tuberculosis, acts by inhibiting the enzyme arabinosyltransferase, thereby disrupting the synthesis of cell wall complexes and leading to increased permeability of the bacterial cell wall. Ethambutol contains two chiral centers and exists primarily in three stereoisomeric forms, the enantiomeric pair (+)-(S,S)- and (–)-(R,R)-ethambutol, together with the optically inactive diasteromeric *meso* form. Ethambutol activity resides with the (+)-enantiomer, which is 500- and 12-fold more potent than (–)-ethambutol and the *meso* form (197). This drug was initially introduced for clinical use as the racemate, but was switched to (+)-ethambutol due to reports of dose and treatment-duration related incidences of ocular neuropathy. All three stereoisomers appear to be equipotent with respect to the adverse effect, theorized to be related to chelation of copper in the retinal ganglion cells and their axons in the optic nerve, and use of the single enantiomer provided an improved safety margin (198).

Local anesthetic agents such as prilocaine, mepivacaine, and bupivacaine are marketed as the racemate; however, the enantiomers differ in duration of action, disposition, and acute toxicity. These drugs act by inhibition of nerve impulse in the peripheral nervous system via blockade of sodium and potassium ion channels. Voltage-gated sodium channels exist in three conformational states: resting (closed), open (activated), and inactivated. The affinity of the open and inactivated channel states for the local anesthetics are greater than that of the resting state and compounds that bind with higher affinity, or dissociate more slowly, exhibit greater blockade potency. (R)-Bupivacaine exhibits stereoselective action with [R:S] in vitro activity ratios ranging from one to threefold (199). Resting sodium channels appear to show a slightly greater affinity for (S)-bupivacaine with an [S:R] ratio of 1.2. In contrast, the inactivated channel binds both enantiomers with greater affinity (>10-fold) and opposite stereoselectivity ([S:R] ratio of 1.2), confirming the greater potency of the (R)-isomer for inhibition of both neuronal action potential and sodium currents. The rates of dissociation of the enantiomers from inactivated sodium channels also differ, with (R)-bupivacaine dissociating slower than (S)-bupivacaine. Bupivacaine exhibits stereoselectivity on the flicker potassium ion channel with an [R:S] ratio of 73, which appears to be related to binding site dissociation kinetics ([S:R] ratio of 64) (200). Cardiotoxicity has been reported for bupivacaine and attributed to the (R)-enantiomer, which has slightly greater affinity for cardiac sodium channels and longer dissociation times than (S)-bupivacaine (201).

PROSPECTUS
Chiral Switching

Knowledge of the PK-PD differences between stereoisomers, coupled with cost-effective advances in industrial chemistry, has provided a clear rationale for the development of individual isomers. The potential advantages of administering specific enantiomers rather than racemates include reduction of total dose, reduction of variability in exposure-response relationships, and minimization of toxicity for enantiomers that are either therapeutically inactive or exhibit unwanted pharmacological effects (202–204). Development of specific enantiomers also avoids the risk of administering racemates before establishing the safety of each isomer.

The United States Food and Drug Administration (FDA) issued guidelines in 1992 governing stereoisomerism in drug development (205). Although these guidelines strongly encourage the development of single isomers, there is no requirement to develop single enantiomer drugs, and development decisions are left to the sponsor. It is important to consider a review of appropriate pharmacological and toxicological data, together with a risk-benefit analysis, when assessing whether to develop a single enantiomer versus a racemate (206). In addition to new agents, a number of established drugs originally marketed in racemic form have been reevaluated for use as individual enantiomers (Table 12.6). In many cases, the FDA has allowed established or generic drugs to be patented and marketed under another name for a period of three or five years based on market exclusivity provisions provided by the Hatch-Waxman Act (207).

TABLE 12.6 Clinical Examples of Chiral Switching

Racemate	Single enantiomer	Class/Indication	Reasons for switch
Albuterol (39, 214,215)	(R)-Albuterol	β_2-Agonist	(R)-Enantiomer is a more potent β-agonist; (S)-enantiomer is associated with loss of bronchodilator potency
Amlodipine (224)	(S)-Amlodipine	Calcium channel blocker	(S)-Amlodipine is eutomer
Amphetamine (225)	D-Amphetamine	CNS stimulant used to treat narcolepsy, ADD, and ADHD	D-Amphetamine acts primarily on the dopaminergic systems and is responsible for behavioral-stimulant effects; L-amphetamine is comparatively noradrenergic
Atracurium (226,227)	Cisatricurium	Neuromuscular blocker	Racemic mixture of atracurium contains only 15% of this stereoisomer, yet it is threefold more potent than the racemate and results in reduced formation of a toxic metabolite
Bupivacaine (228,229)	(S)-Bupivacaine	Local analgesic	Despite being more potent, cardiotoxicity is associated with (R)-bupivacaine; (S)-enantiomer shown to produce less negative inotropic effects
Cetirizine (216,230)	(S)-Cetirizine	Antihistamine	(S)-Enantiomer is devoid of activity; exclusion of (S)-enantiomer produces less sedation
Citalopram (231–233)	(S)-Citalopram	Selective serotonin reuptake inhibitor	(S)-Enantiomer is 130- to 160-fold more potent, has reduced side effects, and improved tolerability profile
Formoterol (225)	(R,R)-Formoterol	β_2-agonist	(R)-Enantiomer is more potent β-agonist; (S)-enantiomer is associated with loss of bronchodilator potency
Ibuprofen (187–189)	(S)-Ibuprofen	NSAID	Activity resides only in (S)-enantiomer
Ketamine (212, 213)	(S)-Ketamine	Analgesic	(S)-Enantiomer has greater analgesic effect and reduced adverse reactions (hallucinations and agitation)
Ketoprofen (234)	(S)-Ketoprofen	NSAID	Activity resides only in (S)-enantiomer; (R)-enantiomer has more risk for ulcer
Methylphenidate (235,236)	(R,R)-Methylphenidate	Attention-deficit hyperactivity disorder	(R,R)-Enantiomer has higher oral bioavailability due to less first-pass metabolism than (S,S)-enantiomer and is tenfold more potent inhibitor of dopamine and noradrenaline

(Continued)

CLINICAL PHARMACOKINETICS AND PHARMACODYNAMICS 225

Racemate	Single enantiomer	Class/Indication	Reasons for switch
Modafinil (237)	(R)-Modafinil	Sleep apnea, narcolepsy	(R)-Enantiomer has a longer half-life and enhanced efficacy
Ofloxacin (238, 239)	(S)-Ofloxacin	Antibiotic	(R)-Enantiomer is devoid of activity
Omeprazole (219–220, 240, 241)	(S)-Omeprazole	Proton pump inhibitor	(S)-Enantiomer has greater oral bioavailability and is more effective
Zopiclone (242)	(S)-Zopiclone	Hypnotic, sedative	(R)-Zopiclone is inactive

Clinical Examples

Ketamine is a general anesthetic agent with limited use owing to postanesthesia emergence reactions such as hallucinations, vivid dreams, and agitation. Ketamine pharmacological activity is stereoselective, with (S)-ketamine producing a greater analgesic effect than its antipode (208–210). The effects of ketamine are mediated by N-methyl D-aspartate (NMDA), nicotinic, muscarinic, monoaminergic, and opioid receptors. Ketamine has been shown to stereoselectively bind to the NMDA receptor, with (S)-ketamine exhibiting threefold greater affinity compared to (R)-ketamine (211). Similarly, enantiomeric binding to the μ- and κ-opioid receptors at high doses shows a two- to fourfold selectivity for (S)-ketamine. However, binding of (S)-ketamine to opiod receptors shows a reduced affinity (10- to 20-fold) compared to the NMDA receptor. In surgical patients given racemic ketamine or individual enantiomers, (S)-ketamine produced a more effective analgesia (3.4-fold greater potency for (S)-ketamine), as well as fewer emergence reactions and incidences of agitated behavior as compared to equivalent doses of (R)-ketamine or the racemate (212). Ketamine has undergone the chiral switch process and is now marketed in Germany as the (S)-enantiomer (213).

Chronic albuterol use is associated with loss of bronchodilator potency, decreased protection against bronchoconstriction, and increased sensitivity to allergen challenge (214). Albuterol was initially marketed as the racemate even though (R)-albuterol is 68-fold more active than (S)-albuterol as a β_2-agonist. (R)-albuterol exhibits 90- to 100-fold greater in vitro binding affinity for β_1- and β_2-receptors relative to (S)-albuterol. In addition, (R)-albuterol increases intracellular cyclic adenosine monophosphate (cAMP) and intrinsic activity equivalent to twice the dose of the racemate and inhibits activation of mast cells and eosinophils. Conversely, (S)-albuterol intensifies bronchoconstrictor responses, induces hypersensitivty of asthmatic airways, and promotes the activation of eosinophils. (R)-albuterol results in prolonged protection in the metacholine-induced bronchoconstrictor challenge test compared to the racemate in humans, whereas the (S)-enantiomer significantly increases metacholine sensitivity (215). Stereoselective disposition of inhaled albuterol is also evident, with faster clearance of (R)-albuterol (1.6- to 3-fold compared to (S)-albuterol) resulting in a longer plasma half-life and greater exposure to the counterproductive enantiomer. Thus, (R)-albuterol should have advantages over the racemate in the treatment of asthma (39).

Cetirizine is a second generation H_1 antihistamine administered as a racemate for the treatment of symptoms associated with seasonal allergic rhinitis. Competitive-binding experiments revealed that (S)-cetirizine has a 30-fold higher affinity for the human H_1-receptor than (R)-cetirizine and 2-fold higher affinity than the racemate (K_i values of 6, 3, and 100 nM, respectively) (216). (S)- and (R)-cetirizine dissociate from H_1-receptors with a half time of 142 and 6 minutes. Thus, (R)-cetirizine could act as a pseudo-irreversible antagonist in vivo. (S)-cetirizine also has the potential for producing fewer sedative side effects due to having a smaller volume of distribution, reduced transport across the blood-brain barrier, and enhanced peripheral receptor binding than the (R)-isomer. Although (S)-cetirizine is approved for use in the United States, there is debate as to whether the single enantiomer produces less sedation than the racemate (217,218).

The proton pump inhibitors omeprazole and pantoprazole are used to treat disorders such as gastroesophageal reflux disease and *Helicobacter pylori* infection and associated duodenal ulcer disease (219). The (S)-enantiomers of both drugs are more potent H^+/K^+-ATPase inhibitors and reduce the secretion of hydrochloric acid by gastric parietal cells to a greater degree than the (R)-enantiomer (220). The (S)-enantiomers also exhibit more favorable metabolic profiles than the (R)-enantiomers. The (R)-enantiomers are preferentially metabolized by CYP2C19, whereas the (S)-enantiomers are additionally metabolized by CYP3A4 and sulfotransferases (75). This results in 1.5-fold higher concentrations of the less active (R)-enantiomer in poor metabolizers of CYP2C19, yielding greater interpatient variability in response and increased risk of adverse effects (129,130). Chiral switching for these two racemic drugs exemplifies how the safety profile may be improved by removing an isomer metabolized by a polymorphic enzyme (221).

Other Approaches for Improving Chiral Drug Safety and Efficacy

Physicochemical modification of racemic drugs could influence the rate and extent of their absorption, transport across biological membranes, as well as their ability to interact with various receptors and enzyme systems. Lisdexamfetamine is an inactive prodrug designed for oral administration and consists of the stimulant D-amphetamine covalently bonded to the essential amino acid L-lysine. Lisdexamfetamine is currently approved in the United States for the treatment of attention-deficit hyperactivity disorder in children. The prodrug is converted to active D-amphetamine via rate-limited hydrolysis in the systemic circulation, essentially serving as an extended-release formulation of D-amphetamine, which may further lead to a reduced potential for abuse as compared to an immediate-release formulation. Adult stimulant drug abusers reported less favorability for the 50 or 100 mg doses of oral lisdexamfetamine as compared to a 40 mg dose of D-amphetamine, despite comparable efficacy and safety profiles (222). Another strategy for improving the therapeutic index of racemic drugs is to explore the selectivity or specificity for individual enantiomers to a particular biological target or signaling network (223).

Pharmacokinetic-Pharmacodynamic Modeling

PK-PD modeling can be used to help identify factors that determine the relationship between the dose of a racemic drug and the extent and time course of response. Such approaches can provide insight into mechanisms of interactions between enantiomers and processes controlling differences in both PK and PD for each enantiomer when administered individually or as the racemate. In many instances, it is not appropriate to simply relate drug response to plasma concentrations of the racemate (sum of enantiomer concentrations). For racemic drugs in which both enantiomers are active, modeling represents one approach to potentially identify an optimal enantiomer ratio associated with a safety or efficacy outcome.

The Hill function is commonly used to describe the exposure-response relationship of rapid (i.e., lack of hysteresis) reversibly acting drugs (243) as follows:

$$E = \frac{E_{max} \cdot C_p^n}{EC_{50}^n + C_p^n} \quad (1)$$

where E is the effect, E_{max} is the maximal effect, EC_{50} is the plasma drug concentration (C_p) corresponding with 50% maximal effect, and n is the Hill coefficient. If two enantiomers are pharmacologically active, but act on completely different receptor types, then agonist and/or antagonist activity could be simply determined as the sum of two Hill functions. However, when two enantiomers competitively act at the same receptor, then characterization of response is more complex. The case of two enantiomers that are reversibly binding agonists with affinity for the same receptor, but with different efficacy and potency, has been described in detail elsewhere (244). A general equation to describe the relationship between the drug effect and plasma enantiomer concentrations (C_R and C_S) acting competitively at the same receptor (245) is as follows:

$$E_{R+S} = \frac{E_{max R} \cdot (C_R/EC_{50R})^n + E_{max S} \cdot (C_S/EC_{50S})^n}{1 + (C_R/EC_{50R})^n + (C_S/EC_{50S})^n} \quad (2)$$

For effects of a racemic drug controlled by only one active enantiomer, the apparent EC_{50} will be higher than that of the pure active enantiomer. A general equation to describe such a relationship can be defined as shown in Eq. 3, assuming that the effect is controlled by the (R)-enantiomer (246).

$$E_{R+S} = \frac{E_{max R} \cdot C_R^n}{EC_{50R}^n \cdot (1 + C_s/EC_{50S}) + C_R^n} \quad (3)$$

The PD modeling of ketamine effects on an electroencephalogram (EEG)-based biomarker is a classic example of quantifying pharmacological effects in the presence of competitively acting stereoisomers (247). Five healthy males were given short-term constant rate intravenous infusions of ketamine and each enantiomer in a three-way crossover study design. Power-spectral analysis was used to calculate median frequency of continuous EEG recordings. The ketamine concentration-effect relationship for a representative subject is shown in Figure 12.2. The solid lines are model-fitted curves defined by subtracting the empirical Hill function (Eq. 1) from a baseline value for each chemical species. The efficacy (E_{max}) for the (R)-enantiomer is significantly less than racemic and (S)-ketamine, whereas the greater potency (i.e., lower IC_{50}) of the (S)-isomer was significantly different from the racemic

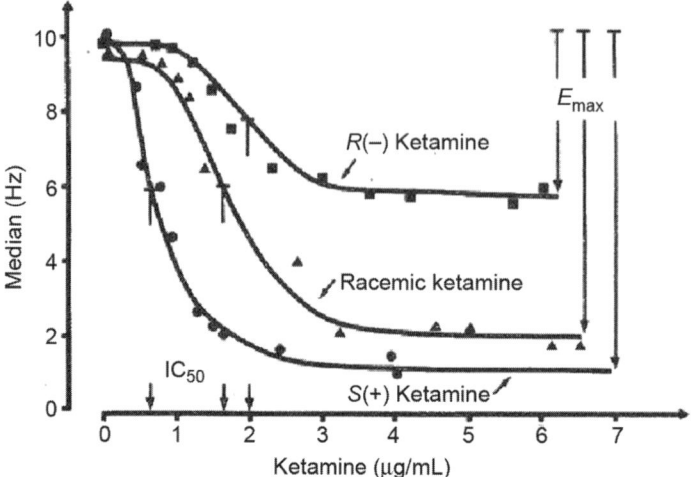

FIGURE 12.2 Pharmacodynamic characterization of the concentration-effect relationship for a racemic mixture and individual enantiomers of ketamine. *Source*: From Ref. 247.

and (R)-ketamine. Although differences in potency may result from stereoselective changes in PK or PD properties, measurements of serum drug concentrations over time excluded PK causes. Furthermore, simulations of a theoretical partial antagonism model (Eq. 2) suggested that the effects of the ketamine racemic mixture can be explained (almost entirely) by a classical competitive interaction.

The complexities of stereoselective interactions in biological systems require careful experimentation, and mathematical modeling provides the best platform for analyzing and interpreting the clinical pharmacology of single enantiomers and racemic mixtures. The simple pharmacodynamic models in this section (Eqs. 1–3) are valid for rapid reversibly acting drugs (i.e., no temporal delays between drug exposure and response); however, diverse structural models are also available for more complex mechanisms of action (248).

ACKNOWLEDGMENTS

The authors wish to thank partial support received from the Intramural Research Program of the National Institute on Aging of the National Institutes of Health.

REFERENCES

1. Brocks DR. Drug disposition in three dimensions: an update on stereoselectivity in pharmacokinetics. Biopharm Drug Dispos 2006; 27(8):387–406.
2. Brocks DR, Mehvar R. Stereoselectivity in the pharmacodynamics and pharmacokinetics of the chiral antimalarial drugs. Clin Pharmacokinet 2003; 42(15):1359–1382.
3. Davies NM. Clinical pharmacokinetics of ibuprofen. The first 30 years. Clin Pharmacokinet 1998; 34(2):101–154.
4. Dresser GK, Bailey DG, Leake BF, et al. Fruit juices inhibit organic anion transporting polypeptide-mediated drug uptake to decrease the oral availability of fexofenadine. Clin Pharmacol Ther 2002; 71(1):11–20.

5. Terada T, Inui K. Peptide transporters: structure, function, regulation and application for drug delivery. Curr Drug Metab 2004; 5 (1):85–94.
6. Tamai I, Nakanishi T, Nakahara H, et al. Improvement of L-dopa absorption by dipeptidyl derivation, utilizing peptide transporter PepT1. J Pharm Sci 1998; 87 (12):1542–1546.
7. Wu CY, Benet LZ. Predicting drug disposition via application of BCS: transport/absorption/elimination interplay and development of a biopharmaceutics drug disposition classification system. Pharm Res 2005; 22(1):11–23.
8. Cvetkovic M, Leake B, Fromm MF, et al. OATP and P-glycoprotein transporters mediate the cellular uptake and excretion of fexofenadine. Drug Metab Dispos 1999; 27(8):866–871.
9. Miura M, Uno T, Tateishi T, et al. Pharmacokinetics of fexofenadine enantiomers in healthy subjects. Chirality 2007; 19(3):223–227.
10. Tateishi T, Miura M, Suzuki T, et al. The different effects of itraconazole on the pharmacokinetics of fexofenadine enantiomers. Br J Clin Pharmacol 2008; 65(5): 693–700.
11. Spahn-Langguth H, Langguth P. Grapefruit juice enhances intestinal absorption of the P-glycoprotein substrate talinolol. Eur J Pharm Sci 2001; 12(4):361–367.
12. Hendel J, Brodthagen H. Entero-hepatic cycling of methotrexate estimated by use of the D-isomer as a reference marker. Eur J Clin Pharmacol 1984; 26(1):103–107.
13. Narawa T, Shimizu R, Takano S, et al. Stereoselectivity of the reduced folate carrier in Caco-2 cells. Chirality 2005; 17(8):444–449.
14. Fernandez C, Gimenez F, Thuillier A, et al. Stereoselective binding of zopiclone to human plasma proteins. Chirality 1999; 11(2):129–132.
15. Chuang VT, Otagiri M. Stereoselective binding of human serum albumin. Chirality 2006; 18(3):159–166.
16. Rahman MH, Maruyama T, Okada T, et al. Study of interaction of carprofen and its enantiomers with human serum albumin–II. Stereoselective site-to-site displacement of carprofen by ibuprofen. Biochem Pharmacol 1993; 46(10):1733–1740.
17. Whitlam JB, Crooks MJ, Brown KF, et al. Binding of nonsteroidal anti-inflammatory agents to proteins–I. Ibuprofen-serum albumin interaction. Biochem Pharmacol 1979; 28(5):675–678.
18. Fitos I, Visy J, Simonyi M, et al. Stereoselective allosteric binding interaction on human serum albumin between ibuprofen and lorazepam acetate. Chirality 1999; 11(2):115–120.
19. Hui-Chen L, Yang Y, Na W, et al. Pharmacokinetics of the enantiomers of trans-tramadol and its active metabolite, trans-O-demethyltramadol, in healthy male and female Chinese volunteers. Chirality 2004; 16(2):112–118.
20. Visy J, Fitos I, Mady G, et al. Enantioselective plasma protein binding of bimoclomol. Chirality 2002; 14(8):638–642.
21. Mazoit JX, Cao LS, Samii K. Binding of bupivacaine to human serum proteins, isolated albumin and isolated alpha-1-acid glycoprotein. Differences between the two enantiomers are partly due to cooperativity. J Pharmacol Exp Ther 1996; 276(1):109–115.
22. Strolin BM, Whomsley R, Mathy FX, et al. Stereoselective renal tubular secretion of levocetirizine and dextrocetirizine, the two enantiomers of the H1-antihistamine cetirizine. Fundam Clin Pharmacol 2008; 22(1):19–23.
23. Ofori-Adjei D, Ericsson O, Lindstrom B, et al. Protein binding of chloroquine enantiomers and desethylchloroquine. Br J Clin Pharmacol 1986; 22(3):356–358.
24. Augustijns P, Verbeke N. Stereoselective pharmacokinetic properties of chloroquine and de-ethyl-chloroquine in humans. Clin Pharmacokinet 1993; 24(3):259–269.
25. Hiep BT, Gimenez F, Khanh VU, et al. Binding of chlorpheniramine enantiomers to human plasma proteins. Chirality 1999; 11(5–6):501–504.
26. Lima JJ, Boudoulas H. Stereoselective effects of disopyramide enantiomers in humans. J Cardiovasc Pharmacol 1987; 9(5):594–600.
27. Mehvar R, Brocks DR, Vakily M. Impact of stereoselectivity on the pharmacokinetics and pharmacodynamics of antiarrhythmic drugs. Clin Pharmacokinet 2002; 41(8):533–558.

28. Muller N, Lapicque F, Monot C, et al. Stereoselective binding of etodolac to human serum albumin. Chirality 1992; 4(4):240–246.
29. Brocks DR, Jamali F. Enantioselective pharmacokinetics of etodolac in the rat: tissue distribution, tissue binding, and in vitro metabolism. J Pharm Sci 1991; 80(11): 1058–1061.
30. Gross AS, Eser C, Mikus G, et al. Enantioselective gallopamil protein binding. Chirality 1993; 5(6):414–418.
31. McLachlan AJ, Cutler DJ, Tett SE. Plasma protein binding of the enantiomers of hydroxychloroquine and metabolites. Eur J Clin Pharmacol 1993; 44(5):481–484.
32. Vakily M, Corrigan B, Jamali F. The problem of racemization in the stereospecific assay and pharmacokinetic evaluation of ketorolac in human and rats. Pharm Res 1995; 12(11):1652–1657.
33. Hong Y, Tang Y, Zeng S. Enantioselective plasma protein binding of propafenone: mechanism, drug interaction, and species difference. Chirality 2009; 21:692–698.
34. Mehvar R, Brocks DR. Stereospecific pharmacokinetics and pharmacodynamics of beta-adrenergic blockers in humans. J Pharm Pharm Sci 2001; 4(2):185–200.
35. Sueyasu M, Fujito K, Shuto H, et al. Protein binding and the metabolism of thiamylal enantiomers in vitro. Anesth Analg 2000; 91(3):736–740.
36. Gross AS, Heuer B, Eichelbaum M. Stereoselective protein binding of verapamil enantiomers. Biochem Pharmacol 1988; 37(24):4623–4627.
37. He J, Shibukawa A, Tokunaga S, et al. Protein-binding high-performance frontal analysis of (R)- and (S)-warfarin on HSA with and without phenylbutazone. J Pharm Sci 1997; 86(1):120–125.
38. Yacobi A, Levy G. Protein binding of warfarin enantiomers in serum of humans and rats. J Pharmacokinet Biopharm 1977; 5(2):123–131.
39. Maier G, Rubino C, Hsu R, et al. Population pharmacokinetics of (R)-albuterol and (S)-albuterol in pediatric patients aged 4–11 years with asthma. Pulm Pharmacol Ther 2007; 20(5):534–542.
40. Boulton DW, Fawcett JP. Enantioselective disposition of salbutamol in man following oral and intravenous administration. Br J Clin Pharmacol 1996; 41(1):35–40.
41. Mather LE, McCall P, McNicol PL. Bupivacaine enantiomer pharmacokinetics after intercostal neural blockade in liver transplantation patients. Anesth Analg 1995; 80(2):328–335.
42. Bui TH, Fernandez C, Vu K, et al. Stereospecific versus nonstereospecific assessments for the bioequivalence of two formulations of racemic chlorpheniramine. Chirality 2000; 12 (8):599–605.
43. Giacomini KM, Nelson WL, Pershe RA, et al. In vivo interaction of the enantiomers of disopyramide in human subjects. J Pharmacokinet Biopharm 1986; 14(4):335–356.
44. Le Corre P, Gibassier D, Sado P, et al. Stereoselective metabolism and pharmacokinetics of disopyramide enantiomers in humans. Drug Metab Dispos 1988; 16 (6):858–864.
45. Brocks DR, Jamali F. Etodolac clinical pharmacokinetics. Clin Pharmacokinet 1994; 26(4):259–274.
46. Hamunen K, Maunuksela EL, Sarvela J, et al. Stereoselective pharmacokinetics of ketorolac in children, adolescents and adults. Acta Anaesthesiol Scand 1999; 43 (10):1041–1046.
47. Adedoyin A, Prakash C, O'Shea D, et al. Stereoselective disposition of hexobarbital and its metabolites: relationship to the S-mephenytoin polymorphism in Caucasian and Chinese subjects. Pharmacogenetics 1994; 4(1):27–38.
48. Lim WH, Hooper WD. Stereoselective metabolism and pharmacokinetics of racemic methylphenobarbital in humans. Drug Metab Dispos 1989; 17(2):212–217.
49. Vree TB, Beumer EM, Lagerwerf AJ, et al. Clinical pharmacokinetics of R(+)- and S(–)-mepivacaine after high doses of racemic mepivacaine with epinephrine in the combined psoas compartment/sciatic nerve block. Anesth Analg 1992; 75(1):75–80.
50. Kristensen K, Blemmer T, Angelo HR, et al. Stereoselective pharmacokinetics of methadone in chronic pain patients. Ther Drug Monit 1996; 18(3):221–227.

51. Chu KM, Shieh SM, Hu OY. Plasma and red blood cell pharmacokinetics of pimobendan enantiomers in healthy Chinese. Eur J Clin Pharmacol 1995; 47(6):537–542.
52. Chu KM, Shieh SM, Hu OY. Pharmacokinetics and pharmacodynamics of enantiomers of pimobendan in patients with dilated cardiomyopathy and congestive heart failure after single and repeated oral dosing. Clin Pharmacol Ther 1995; 57(6):610–621.
53. Fleishaker JC, Mucci M, Pellizzoni C, et al. Absolute bioavailability of reboxetine enantiomers and effect of gender on pharmacokinetics. Biopharm Drug Dispos 1999; 20(1):53–57.
54. Fleishaker JC. Clinical pharmacokinetics of reboxetine, a selective norepinephrine reuptake inhibitor for the treatment of patients with depression. Clin Pharmacokinet 2000; 39(6):413–427.
55. Zschiesche M, Lemma GL, Klebingat KJ, et al. Stereoselective disposition of talinolol in man. J Pharm Sci 2002; 91(2):303–311.
56. Nguyen KT, Stephens DP, McLeish MJ, et al. Pharmacokinetics of thiopental and pentobarbital enantiomers after intravenous administration of racemic thiopental. Anesth Analg 1996; 83(3):552–558.
57. Thomson AH, Murdoch G, Pottage A, et al. The pharmacokinetics of R- and S-tocainide in patients with acute ventricular arrhythmias. Br J Clin Pharmacol 1986; 21(2):149–154.
58. Gupta S, Modi NB, Sathyan G, et al. Pharmacokinetics of controlled-release verapamil in healthy volunteers and patients with hypertension or angina. Biopharm Drug Dispos 2002; 23(1):17–31.
59. Eichelbaum M, Mikus G, Vogelgesang B. Pharmacokinetics of (+)-, (–)- and (+/–)-verapamil after intravenous administration. Br J Clin Pharmacol 1984; 17(4):453–458.
60. Breckenridge A, Orme M, Wesseling H, et al. Pharmacokinetics and pharmacodynamics of the enantiomers of warfarin in man. Clin Pharmacol Ther 1974; 15(4):424–430.
61. Egleton RD, Davis TP. Bioavailability and transport of peptides and peptide drugs into the brain. Peptides 1997; 18(9):1431–1439.
62. Wiese C, Cogoli-Greuter M, Argentini M, et al. Metabolism of 5-fluoro-dopa and 6-fluoro-dopa enantiomers in aggregating cell cultures of fetal rat brain. Biochem Pharmacol 1992; 44(1):99–105.
63. Baudry S, PhamYT, Baune B, et al. Stereoselective passage of mefloquine through the blood-brain barrier in the rat. J Pharm Pharmacol 1997; 49(11):1086–1090.
64. Pham YT, Regina A, Farinotti R, et al. Interactions of racemic mefloquine and its enantiomers with P-glycoprotein in an immortalised rat brain capillary endothelial cell line, GPNT. Biochim Biophys Acta 2000; 1524(2–3):212–219.
65. Kaminsky LS, Zhang ZY. Human P450 metabolism of warfarin. Pharmacol Ther 1997; 73(1):67–74.
66. Rettie AE, Korzekwa KR, Kunze KL, et al. Hydroxylation of warfarin by human cDNA-expressed cytochrome P-450: a role for P-4502C9 in the etiology of (S)-warfarin-drug interactions. Chem Res Toxicol 1992; 5(1):54–59.
67. Yamazaki H, Shimada T. Human liver cytochrome P450 enzymes involved in the 7-hydroxylation of R- and S-warfarin enantiomers. Biochem Pharmacol 1997; 54(11):1195–1203.
68. Thijssen HH, Flinois JP, Beaune PH. Cytochrome P4502C9 is the principal catalyst of racemic acenocoumarol hydroxylation reactions in human liver microsomes. Drug Metab Dispos 2000; 28(11):1284–1290.
69. Lu H, Wang JJ, Chan KK, et al. Stereoselectivity in metabolism of ifosfamide by CYP3A4 and CYP2B6. Xenobiotica 2006; 36(5):367–385.
70. Roy P, Tretyakov O, Wright J, et al. Stereoselective metabolism of ifosfamide by human P-450s 3A4 and 2B6. Favorable metabolic properties of R-enantiomer. Drug Metab Dispos 1999; 27(11):1309–1318.
71. Granvil CP, Madan A, Sharkawi M, et al. Role of CYP2B6 and CYP3A4 in the in vitro N-dechloroethylation of (R)- and (S)-ifosfamide in human liver microsomes. Drug Metab Dispos 1999; 27(4):533–541.

72. Chen CS, Jounaidi Y, Waxman DJ. Enantioselective metabolism and cytotoxicity of R-ifosfamide and S-ifosfamide by tumor cell-expressed cytochromes P450. Drug Metab Dispos 2005; 33(9):1261–1267.
73. Eriksson UG, Lundahl J, Baarnhielm C, et al. Stereoselective metabolism of felodipine in liver microsomes from rat, dog, and human. Drug Metab Dispos 1991; 19(5):889–894.
74. Margolis JM, O'Donnell JP, Mankowski DC, et al. (R)-, (S)-, and racemic fluoxetine N-demethylation by human cytochrome P450 enzymes. Drug Metab Dispos 2000; 28(10):1187–1191.
75. Abelo A, Andersson TB, Antonsson M, et al. Stereoselective metabolism of omeprazole by human cytochrome P450 enzymes. Drug Metab Dispos 2000; 28(8):966–972.
76. de Morais SM, Wilkinson GR, Blaisdell J, et al. The major genetic defect responsible for the polymorphism of S-mephenytoin metabolism in humans. J Biol Chem 1994; 269(22):15419–15422.
77. Echizen H, Tanizaki M, Tatsuno J, et al. Identification of CYP3A4 as the enzyme involved in the mono-N-dealkylation of disopyramide enantiomers in humans. Drug Metab Dispos 2000; 28(8):937–944.
78. Narimatsu S, Takemi C, Kuramoto S, et al. Stereoselectivity in the oxidation of bufuralol, a chiral substrate, by human cytochrome P450s. Chirality 2003; 15(4):333–339.
79. Ellis SW, Rowland K, Ackland MJ, et al. Influence of amino acid residue 374 of cytochrome P-450 2D6 (CYP2D6) on the regio- and enantio-selective metabolism of metoprolol. Biochem J 1996; 316(pt 2):647–654.
80. Kim M, Shen DD, Eddy AC, et al. Inhibition of the enantioselective oxidative metabolism of metoprolol by verapamil in human liver microsomes. Drug Metab Dispos 1993; 21(2):309–317.
81. Mautz DS, Shen DD, Nelson WL. Regioselectivity and enantioselectivity of metoprolol oxidation by two variants of cDNA-expressed P4502D6. Pharm Res 1995; 12(12):2053–2056.
82. Ha PT, Sluyts I, Van Dyck S, et al. Chiral capillary electrophoretic analysis of verapamil metabolism by cytochrome P450 3A4. J Chromatogr A 2006; 1120(1–2):94–101.
83. Shen L, Fitzloff JF, Cook CS. Differential enantioselectivity and product-dependent activation and inhibition in metabolism of verapamil by human CYP3As. Drug Metab Dispos 2004; 32(2):186–196.
84. Tracy TS, Korzekwa KR, Gonzalez FJ, et al. Cytochrome P450 isoforms involved in metabolism of the enantiomers of verapamil and norverapamil. Br J Clin Pharmacol 1999; 47(5):545–552.
85. Niwa T, Shiraga T, Mitani Y, et al. Stereoselective metabolism of cibenzoline, an antiarrhythmic drug, by human and rat liver microsomes: possible involvement of CYP2D and CYP3A. Drug Metab Dispos 2000; 28(9):1128–1134.
86. Yasui-Furukori N, Hidestrand M, Spina E, et al. Different enantioselective 9-hydroxylation of risperidone by the two human CYP2D6 and CYP3A4 enzymes. Drug Metab Dispos 2001; 29(10):1263–1268.
87. Brocks DR, Jamali F, Russell AS. Stereoselective disposition of etodolac enantiomers in synovial fluid. J Clin Pharmacol 1991; 31(8):741–746.
88. Sten T, Qvisen S, Uutela P, et al. Prominent but reverse stereoselectivity in propranolol glucuronidation by human UDP-glucuronosyltransferases 1A9 and 1A10. Drug Metab Dispos 2006; 34(9):1488–1494.
89. Zhang M, Fawcett JP, Kennedy JM, et al. Stereoselective glucuronidation of formoterol by human liver microsomes. Br J Clin Pharmacol 2000; 49(2):152–157.
90. Hartman AP, Wilson AA, Wilson HM, et al. Enantioselective sulfation of beta 2-receptor agonists by the human intestine and the recombinant M-form phenolsulfotransferase. Chirality 1998; 10(9):800–803.
91. Katsuki H, Hamada A, Nakamura C, et al. Role of CYP3A4 and CYP2C19 in the stereoselective metabolism of lansoprazole by human liver microsomes. Eur J Clin Pharmacol 2001; 57(10):709–715.

92. Wilson AA, Wang J, Koch P, et al. Stereoselective sulphate conjugation of fenoterol by human phenolsulphotransferases. Xenobiotica 1997; 27(11):1147–1154.
93. Hall SD, Quan X. The role of coenzyme A in the biotransformation of 2-arylpropionic acids. Chem Biol Interact 1994; 90(3):235–251.
94. Testa B, Mayer JM. Stereoselective drug metabolism and its significance in drug research. Prog Drug Res 1988; 2:249–303.
95. Tucker GT, Lennard MS. Enantiomer specific pharmacokinetics. Pharmacol Ther 1990; 45(3):309–329.
96. Trager WF. Stereochemistry of cytochrome P-450 reactions. Drug Metab Rev 1989; 20(2–4):489–496.
97. Lu H. Stereoselectivity in drug metabolism. Expert Opin Drug Metab Toxicol 2007; 3(2):149–158.
98. Kroemer HK, Fischer C, Meese CO, et al. Enantiomer/enantiomer interaction of (S)- and (R)-propafenone for cytochrome P450IID6-catalyzed 5-hydroxylation: in vitro evaluation of the mechanism. Mol Pharmacol 1991; 40(1):135–142.
99. Kroemer HK, Funck-Brentano C, Silberstein DJ, et al. Stereoselective disposition and pharmacologic activity of propafenone enantiomers. Circulation 1989; 79(5):1068–1076.
100. Kroemer HK, Fromm MF, Buhl K, et al. An enantiomer-enantiomer interaction of (S)- and (R)-propafenone modifies the effect of racemic drug therapy. Circulation 1994; 89(5):2396–2400.
101. Walle T, Walle UK, Thornburg KR, et al. Stereoselective sulfation of albuterol in humans. Biosynthesis of the sulfate conjugate by HEP G2 cells. Drug Metab Dispos 1993; 21(1):76–80.
102. Vakily M, Mehvar R, Brocks D. Stereoselective pharmacokinetics and pharmacodynamics of anti-asthma agents. Ann Pharmacother 2002; 36(4):693–701.
103. Echizen H, Manz M, Eichelbaum M. Electrophysiologic effects of dextro- and levo-verapamil on sinus node and AV node function in humans. J Cardiovasc Pharmacol 1988; 12(5):543–546.
104. Echizen H, Vogelgesang B, Eichelbaum M. Effects of d,l-verapamil on atrioventricular conduction in relation to its stereoselective first-pass metabolism. Clin Pharmacol Ther 1985; 38(1):71–76.
105. Eriksson T, Bjorkman S, Roth B, et al. Enantiomers of thalidomide: blood distribution and the influence of serum albumin on chiral inversion and hydrolysis. Chirality 1998; 10(3):223–228.
106. King JN, Mauron C, LeGoff C, et al. Bidirectional chiral inversion of the enantiomers of the nonsteroidal antiinflammatory drug oxindanac in dogs. Chirality 1994; 6(6):460–466.
107. Davies NM. Chiral inversion. In: Reddy IK, Mehvar R, eds. Chirality in Drug Design and Development. New York: Marcel Dekker, 2004:351–392.
108. Abas A, Meffin PJ. Enantioselective disposition of 2-arylpropionic acid nonsteroidal anti-inflammatory drugs. IV. Ketoprofen disposition. J Pharmacol Exp Ther 1987; 240(2):637–641.
109. Barbanoj MJ, Antonijoan RM, Gich I. Clinical pharmacokinetics of dexketoprofen. Clin Pharmacokinet 2001; 40(4):245–262.
110. Jamali F, Mehvar R, Russell AS, et al. Human pharmacokinetics of ibuprofen enantiomers following different doses and formulations: intestinal chiral inversion. J Pharm Sci 1992; 81(3):221–225.
111. Kaiser DG, Vangiessen GJ, Reischer RJ, et al. Isomeric inversion of ibuprofen (R)-enantiomer in humans. J Pharm Sci 1976; 65(2):269–273.
112. Sattari S, Jamali F. Evidence of absorption rate dependency of ibuprofen inversion in the rat. Chirality 1994; 6(5):435–439.
113. Mohri K, Okada K, Benet LZ. Stereoselective taurine conjugation of (R)-benoxaprofen enantiomer in rats: in vivo and in vitro studies using rat hepatic mitochondria and microsomes. Pharm Res 2005; 22(1):79–85.
114. Jamali F, Russell AS, Foster RT, et al. Ketoprofen pharmacokinetics in humans: evidence of enantiomeric inversion and lack of interaction. J Pharm Sci 1990; 79(5):460–461.

115. Erb K, Brugger R, Williams K, et al. Stereoselective disposition of tiaprofenic acid enantiomers in rats. Chirality 1999; 11(2):103–108.
116. Singh NN, Jamali F, Pasutto FM, et al. Pharmacokinetics of the enantiomers of tiaprofenic acid in humans. J Pharm Sci 1986; 75(5):439–442.
117. Cheng H, Jusko WJ. Pharmacokinetics of reversible metabolic systems. Biopharm Drug Dispos 1993; 14(9):721–766.
118. Ebling WF, Jusko WJ. The determination of essential clearance, volume, and residence time parameters of recirculating metabolic systems: the reversible metabolism of methylprednisolone and methylprednisone in rabbits. J Pharmacokinet Biopharm 1986; 14(6):557–599.
119. Marzo A, Balant LP. Investigation of xenobiotic metabolism by CYP2D6 and CYP2C19: importance of enantioselective analytical methods. J Chromatogr B Biomed Appl 1996; 678(1):73–92.
120. Fjordside L, Jeppesen U, Eap CB, et al. The stereoselective metabolism of fluoxetine in poor and extensive metabolizers of sparteine. Pharmacogenetics 1999; 9(1):55–60.
121. Scordo MG, Spina E, Dahl ML, et al. Influence of CYP2C9, 2C19 and 2D6 genetic polymorphisms on the steady-state plasma concentrations of the enantiomers of fluoxetine and norfluoxetine. Basic Clin Pharmacol Toxicol 2005; 97(5):296–301.
122. Eap CB, Bender S, Gastpar M, et al. Steady state plasma levels of the enantiomers of trimipramine and of its metabolites in CYP2D6-, CYP. Ther Drug Monit 2000; 22(2):209–214.
123. Kirchheiner J, Ufer M, Walter EC, et al. Effects of CYP2C9 polymorphisms on the pharmacokinetics of R- and S-phenprocoumon in healthy volunteers. Pharmacogenetics 2004; 14(1):19–26.
124. Yasuda SU, Zannikos P, Young AE, et al. The roles of CYP2D6 and stereoselectivity in the clinical pharmacokinetics of chlorpheniramine. Br J Clin Pharmacol 2002; 53(5):519–525.
125. Gross AS, Mikus G, Fischer C, et al. Stereoselective disposition of flecainide in relation to the sparteine/debrisoquine metaboliser phenotype. Br J Clin Pharmacol 1989; 28(5):555–566.
126. Kim JS, Nafziger AN, Tsunoda SM, et al. Limited sampling strategy to predict AUC of the CYP3A phenotyping probe midazolam in adults: application to various assay techniques. J Clin Pharmacol 2002; 42(4):376–382.
127. Lennard MS, Tucker GT, Silas JH, et al. Differential stereoselective metabolism of metoprolol in extensive and poor debrisoquin metabolizers. Clin Pharmacol Ther 1983; 34(6):732–737.
128. Abolfathi Z, Fiset C, Gilbert M, et al. Role of polymorphic debrisoquin 4-hydroxylase activity in the stereoselective disposition of mexiletine in humans. J Pharmacol Exp Ther 1993; 266(3):1196–1201.
129. Tybring G, Böttiger Y, Widén J, et al. Enantioselective hydroxylation of omeprazole catalyzed by CYP2C19 in Swedish white subjects. Clin Pharmacol Ther 1997; 62(2):129–137.
130. Tanaka M, Ohkubo T, Otani K, et al. Stereoselective pharmacokinetics of pantoprazole, a proton pump inhibitor, in extensive and poor metabolizers of S-mephenytoin. Clin Pharmacol Ther 2001; 69(3):108–113.
131. Ward SA, Walle T, Walle UK, et al. Propranolol's metabolism is determined by both mephenytoin and debrisoquin hydroxylase activities. Clin Pharmacol Ther 1989; 45(1):72–79.
132. Horikiri Y, Suzuki T, Mizobe M. Stereoselective pharmacokinetics of bisoprolol after intravenous and oral administration in beagle dogs. J Pharm Sci 1997; 86(5):560–564.
133. Inui KI, Masuda S, Saito H. Cellular and molecular aspects of drug transport in the kidney. Kidney Int 2000; 58(3):944–958.
134. Sekine T, Cha SH, Endou H. The multispecific organic anion transporter (OAT) family. Pflugers Arch 2000; 440(3):337–350.
135. Khamdang S, Takeda M, Noshiro R, et al. Interactions of human organic anion transporters and human organic cation transporters with nonsteroidal anti-inflammatory drugs. J Pharmacol Exp Ther 2002; 303(2):534–539.

136. Izzedine H, Launay-Vacher V, Deray G. Renal tubular transporters and antiviral drugs: an update. AIDS 2005; 19(5):455–462.
137. Yabuuchi H, Tamai I, Nezu J, et al. Novel membrane transporter OCTN1 mediates multispecific, bidirectional, and pH-dependent transport of organic cations. J Pharmacol Exp Ther 1999; 289(2):768–773.
138. Okazaki O, Kojima C, Hakusui H, et al. Enantioselective disposition of ofloxacin in humans. Antimicrob Agents Chemother 1991; 35(10):2106–2109.
139. Ott RJ, Giacomini KM. Stereoselective interactions of organic cations with the organic cation transporter in OK cells. Pharm Res 1993; 10(8):1169–1173.
140. Zhang L, Schaner ME, Giacomini KM. Functional characterization of an organic cation transporter (hOCT1) in a transiently transfected human cell line (HeLa). J Pharmacol Exp Ther 1998; 286(1):354–361.
141. Morris ME, You G, eds. Drug Transporters: Molecular Characterization and Role in Drug Disposition. Hoboken, NJ: Wiley and Sons, 2007:709–745.
142. Spahn-Langguth H, Baktir G, Radschuweit A, et al. P-glycoprotein transporters and the gastrointestinal tract: evaluation of the potential in vivo relevance of in vitro data employing talinolol as model compound. Int J Clin Pharmacol Ther 1998; 36(1):16–24.
143. Pauli-Magnus C, von Richter O, Burk O, et al. Characterization of the major metabolites of verapamil as substrates and inhibitors of P-glycoprotein. J Pharmacol Exp Ther 2000; 293(2):376–382.
144. Hedman A, Angelin B, Arvidsson A, et al. Interactions in the renal and biliary elimination of digoxin: stereoselective difference between quinine and quinidine. Clin Pharmacol Ther 1990; 47(1):20–26.
145. Mayersohn MB. Special pharmacokinetic considerations in the elderly. In: Evans WE, Schentag JJ, Jusko WJ, eds. Applied Pharmacokinetics: Principles of Therapeutic Drug Monitoring. 3rd ed. Vancouver, WA: Applied Therapeutics, 1992:9-1–9-43.
146. Abernethy DR. Drug therapy in the elderly. In: Atkinson AJ Jr., Daniels CE, Dedrick RL, et al., eds. Principles of Clinical Pharmacology. New York: Academic Press, 2001:307–317.
147. Chandler MH, Scott SR, Blouin RA. Age-associated stereoselective alterations in hexobarbital metabolism. Clin Pharmacol Ther 1988; 43(4):436–441.
148. Tan SC, Patel BK, Jackson SH, et al. Influence of age on the enantiomeric disposition of ibuprofen in healthy volunteers. Br J Clin Pharmacol 2003; 55(6):579–587.
149. Ohmori M, Arakawa M, Harada K, et al. Stereoselective pharmacokinetics of amlodipine in elderly hypertensive patients. Am J Ther 2003; 10(1):29–31.
150. Foglia JP, Pollock BG, Kirshner MA, et al. Plasma levels of citalopram enantiomers and metabolites in elderly patients. Psychopharmacol Bull 1997; 33(1):109–112.
151. Sasaki M, Tateishi T, Ebihara A. The effects of age and gender on the stereoselective pharmacokinetics of verapamil. Clin Pharmacol Ther 1993; 54(3):278–285.
152. Foster DJ, Somogyi AA, White JM, et al. Population pharmacokinetics of (R)-, (S)- and rac-methadone in methadone maintenance patients. Br J Clin Pharmacol 2004; 57(6):742–755.
153. Labbe L, Robitaille NM, Lefez C, et al. Effects of ciprofloxacin on the stereoselective disposition of mexiletine in man. Ther Drug Monit 2004; 6(5):492–498.
154. Sweeney BP, Bromilow J. Liver enzyme induction and inhibition: implications for anaesthesia. Anaesthesia 2006; 61(2):159–177.
155. Miura M, Kagaya H, Tada H, et al. Intestinal CYP3A4 is not involved in the enantioselective disposition of lansoprazole. Xenobiotica 2006; 36(1):95–102.
156. Soons PA, Vogels BA, Roosemalen MC, et al. Grapefruit juice and cimetidine inhibit stereoselective metabolism of nitrendipine in humans. Clin Pharmacol Ther 1991; 50(4):394–403.
157. Fukumoto K, Kobayashi T, Komamura K, et al. Stereoselective effect of amiodarone on the pharmacokinetics of racemic carvedilol. Drug Metab Pharmacokinet 2005; 20(6):423–427.
158. Brocks DR, Jamali F. Pharmacokinetics of etodolac enantiomers in the rat after administration of phenobarbital or cimetidine. Eur J Drug Metab Pharmacokinet 1992; 17(4):293–299.

159. Smith DA, Chandler MH, Shedlofsky SI, et al. Age-dependent stereoselective increase in the oral clearance of hexobarbitone isomers caused by rifampicin. Br J Clin Pharmacol 1991; 32(6):735–739.
160. Uno T, Ohkubo T, Sugawara K, et al. Effects of grapefruit juice on the stereoselective disposition of nicardipine in humans: evidence for dominant presystemic elimination at the gut site. Eur J Clin Pharmacol 2000; 56(9–10):643–649.
161. Somogyi AA, Bochner F, Sallustio BC. Stereoselective inhibition of pindolol renal clearance by cimetidine in humans. Clin Pharmacol Ther 1992; 51(4):379–387.
162. Vercruysse I, Belpaire F, Wynant P, et al. Enantioselective inhibitory effect of nicardipine on the hepatic clearance of propranolol in man. Chirality 1994; 6(1):5–10.
163. Hunt BA, Bottorff MB, Herring VL, et al. Effects of calcium channel blockers on the pharmacokinetics of propranolol stereoisomers. Clin Pharmacol Ther 1990; 47(5):584–591.
164. Zhou HH, Anthony LB, Roden DM, et al. Quinidine reduces clearance of (+)-propranolol more than (−)-propranolol through marked reduction in 4-hydroxylation. Clin Pharmacol Ther 1990; 47(6):686–693.
165. Donn KH, Powell JR, Rogers JF, et al. The influence of H2-receptor antagonists on steady-state concentrations of propranolol and 4-hydroxypropranolol. J Clin Pharmacol 1984; 24(11–12):500–508.
166. Fromm MF, Busse D, Kroemer HK, et al. Differential induction of prehepatic and hepatic metabolism of verapamil by rifampin. Hepatology 1996; 24(4):796–801.
167. Ho PC, Ghose K, Saville D, et al. Effect of grapefruit juice on pharmacokinetics and pharmacodynamics of verapamil enantiomers in healthy volunteers. Eur J Clin Pharmacol 2000; 56(9–10):693–698.
168. Choonara IA, Cholerton S, Haynes BP, et al. Stereoselective interaction between the R enantiomer of warfarin and cimetidine. Br J Clin Pharmacol 1986; 21(3):271–277.
169. Matsumoto K, Ishida S, Ueno K, et al. The stereoselective effects of bucolome on the pharmacokinetics and pharmacodynamics of racemic warfarin. J Clin Pharmacol 2001; 41(4):459–464.
170. Mackichan JJ. Protein binding drug displacement interactions fact or fiction? Clin Pharmacokinet 1989; 16(2):65–73.
171. Benet LZ, Hoener BA. Changes in plasma protein binding have little clinical relevance. Clin Pharmacol Ther 2002; 71(3):115–121.
172. Dilmaghanian S, Gerber JG, Filler SG, et al. Enantioselectivity of inhibition of cytochrome P450 3A4 (CYP3A4) by ketoconazole: testosterone and methadone as substrates. Chirality 2004; 16(2):79–85.
173. Tuttle RR, Mills J. Dobutamine: development of a new catecholamine to selectively increase cardiac contractility. Circ Res 1975; 36(1):185–196.
174. Ruffolo RR Jr., Spradlin TA, Pollock GD, et al. Alpha and beta adrenergic effects of the stereoisomers of dobutamine. J Pharmacol Exp Ther 1981; 219(2):447–452.
175. Wainer IW. Three-dimensional view of pharmacology. Am J Hosp Pharm 1992; 49 (9 suppl 1):S4–S8.
176. Ranade VV, Somberg JC. Chiral cardiovascular drugs: an overview. Am J Ther 2005; 12(5):439–459.
177. Funck-Brentano C. Pharmacokinetic and pharmacodynamic profiles of d-sotalol and d,l-sotalol. Eur Heart J 1993; 14(suppl H):30–35.
178. Nichols AJ, Sulpizio AC, Ashton DJ, et al. The interaction of the enantiomers of carvedilol with alpha 1- and beta 1-adrenoceptors. Chirality 1989; 1(4):265–270.
179. Bartsch W, Sponer G, Strein K, et al. Pharmacological characteristics of the stereoisomers of carvedilol. Eur J Clin Pharmacol 1990; 38(suppl 2):S104–S107.
180. Monopoli A, Bamonte F, Forlani A, et al. Effects of the R, R-isomer of labetalol, SCH 19927, in isolated tissues and in spontaneously hypertensive rats during a repeated treatment. Arch Int Pharmacodyn Ther 1984; 272(2):256–263.
181. Sybertz EJ, Sabin CS, Pula KK, et al. Alpha and beta adrenoceptor blocking properties of labetalol and its R,R-isomer, SCH 19927. J Pharmacol Exp Ther 1981; 218(2):435–443.

182. Triggle DJ. Calcium-channel drugs: structure-function relationships and selectivity of action. J Cardiovasc Pharmacol 1991; 18(suppl 10):S1-S6.
183. Nguyen LA, He H, Pham-Huy C. Chiral drugs. An overview. Inter J Biomed Sci 2006; 2(2):85-100.
184. Hirano T, Mori T, Kido M, et al. Differential properties of the optical-isomers of pranidipine, a 1,4-dihydropyridine calcium channel modulator. Fundam Clin Pharmacol 1999; 13(6):650-655.
185. van Amsterdam FT, Punt NC, Haas M, et al. Stereoisomers of calcium antagonists distinguish a myocardial and vascular mode of protection against cardiac ischemic injury. J Cardiovasc Pharmacol 1990; 15(2):198-204.
186. Gross AS, Mikus G, Ratge D, et al. Pharmacokinetics and pharmacodynamics of the enantiomers of gallopamil. J Pharmacol Exp Ther 1997; 281(3):1102-1112.
187. Adams SS, Bresloff P, Mason CG. Pharmacological differences between the optical isomers of ibuprofen: evidence for metabolic inversion of the (–)-isomer. J Pharm Pharmacol 1976; 28(3):256-257.
188. Caldwell J, Hutt AJ, Fournel-Gigleux S. The metabolic chiral inversion and dispositional enantioselectivity of the 2-arylpropionic acids and their biological consequences. Biochem Pharmacol 1988; 37(1):105-114.
189. Rahlfs VW, Stat C. Reevaluation of some double-blind, randomized studies of dexibuprofen (Seractil): a state-of-the-art overview. Studies in patients with lumbar vertebral column syndrome, rheumatoid arthritis, distortion of the ankle joint, gonarthrosis, ankylosing spondylitis, and activated coxarthrosis. J Clin Pharmacol 1996; 36(12 suppl):33S-40S.
190. Mauleon D, Artigas R, Garcia ML, et al. Preclinical and clinical development of dexketoprofen. Drugs 1996; 52(suppl 5):24-45.
191. Rentsch KM. The importance of stereoselective determination of drugs in the clinical laboratory. J Biochem Biophys Methods 2002; 54(1-3):1-9.
192. Olsen GD, Wendel HA, Livermore JD, et al. Clinical effects and pharmacokinetics of racemic methadone and its optical isomers. Clin Pharmacol Ther 1977; 21(2): 147-157.
193. Cotzias GC, Van Woert MH, Schiffer LM. Aromatic amino acids and modification of Parkinsonism. N Engl J Med 1967; 276(7):374-379.
194. Cotzias GC, Papavasiliou PS, Gellene R. Modification of Parkinsonism—chronic treatment with L-dopa. N Engl J Med 1969; 280(7):337-345.
195. Glatt H, Oesch F. Mutagenicity of cysteine and penicillamine and its enantiomeric selectivity. Biochem Pharmacol 1985; 34(20):3725-3728.
196. Walshe JM. Chirality of penicillamine. Lancet 1992; 339(8787):254.
197. Lee RE, Protopopova M, Crooks E, et al. Combinatorial lead optimization of [1,2]-diamines based on ethambutol as potential antituberculosis preclinical candidates. J Comb Chem 2003; 5(2):172-187.
198. Li J, Tripathi RC, Tripathi BJ. Drug-induced ocular disorders. Drug Saf 2008; 31 (2):127-141.
199. Foster RH, Markham A. Levobupivacaine: a review of its pharmacology and use as a local anaesthetic. Drugs 2000; 59(3):551-579.
200. Valenzuela C, Delpon E, Tamkun MM, et al. Stereoselective block of a human cardiac potassium channel (Kv1.5) by bupivacaine enantiomers. Biophys J 1995; 69 (2):418-427.
201. Nau C, Vogel W, Hempelmann G, et al. Stereoselectivity of bupivacaine in local anesthetic-sensitive ion channels of peripheral nerve. Anesthesiology 1999; 91 (3):786-795.
202. Hutt AJ. The development of single-isomer molecules: why and how. CNS Spectr 2002; 7(4 suppl 1):14-22.
203. Hutt AJ, Valentová J. The chiral switch: the development of single enantiomer drugs from racemates. Acta Facultatis Pharmaceuticae Universitatis Comen 2003; 50:7-23.
204. Caldwell J. Do single enantiomers have something special to offer? Hum Psychopharmacol 2001; 16(S2):S67-S71.

205. FDA's policy statement for the development of new stereoisomeric drugs. Chirality 1992; 4(5):338–340.
206. Strong M. FDA policy and regulation of stereoisomers: paradigm shift and the future of safer, more effective drugs. Food Drug Law J 1999; 54(3):463–487.
207. Food and Drug Administration, U.S. Department of Health and Human Services. Policy on period of marketing exclusivity for newly approved drug products with enantiomer active ingredients; Request for comments. Food and Drug Administration, HHS. Notice. Fed Regist 1997; 62(10):2167–2169.
208. White PF, Ham J, Way WL, et al. Pharmacology of ketamine isomers in surgical patients. Anesthesiology 1980; 52(3):231–239.
209. Marietta MP, Way WL, Castagnoli N Jr., et al. On the pharmacology of the ketamine enantiomorphs in the rat. J Pharmacol Exp Ther 1977; 202(1):157–165.
210. Kohrs R, Durieux ME. Ketamine: teaching an old drug new tricks. Anesth Analg 1998; 87(5):1186–1193.
211. Sinner B, Graf BM. Ketamine. Handb Exp Pharmacol 2008; 182:313–333.
212. Himmelseher S, Pfenninger E. [The clinical use of S-(+)-ketamine—a determination of its place]. Anasthesiol Intensivmed Notfallmed Schmerzther 1998; 33(12):764–770.
213. Adams HA, Werner C. [From the racemate to the eutomer: (S)-ketamine. Renaissance of a substance?] Anaesthesist 1997; 46(12):1026–1042.
214. Nelson HS. Clinical experience with levalbuterol. J Allergy Clin Immunol 1999; 104 (2 Pt 2):S77–S84.
215. Page CP, Morley J. Contrasting properties of albuterol stereoisomers. J Allergy Clin Immunol 1999; 104(2 pt 2):S31–S41.
216. Tillement JP, Testa B, Bree F. Compared pharmacological characteristics in humans of racemic cetirizine and levocetirizine, two histamine H1-receptor antagonists. Biochem Pharmacol 2003; 6(7):1123–1126.
217. Hindmarch I, Johnson S, Meadows R, et al. The acute and sub-chronic effects of levocetirizine, cetirizine, loratadine, promethazine and placebo on cognitive function, psychomotor performance, and weal and flare. Curr Med Res Opin 2001; 17 (4):241–255.
218. Gandon JM, Allain H. Lack of effect of single and repeated doses of levocetirizine, a new antihistamine drug, on cognitive and psychomotor functions in healthy volunteers. Br J Clin Pharmacol 2002; 54(1):51–58.
219. Spencer CM, Faulds D. Esomeprazole. Drugs 2000; 60(2):321–329.
220. Lind T, Rydberg L, Kyleback A, et al. Esomeprazole provides improved acid control vs. omeprazole in patients with symptoms of gastro-oesophageal reflux disease. Aliment Pharmacol Ther 2000; 14(7):861–867.
221. Andersson T, Hassan-Alin M, Hasselgren G, et al. Pharmacokinetic studies with esomeprazole, the (S)-isomer of omeprazole. Clin Pharmacokinet 2001; 40(6):411–426.
222. Blick SK, Keating GM. Lisdexamfetamine. Paediatr Drugs 2007; 9(2):129–135.
223. Koh JT. Engineering selectivity and discrimination into ligand-receptor interfaces. Chem Biol 2002; 9(1):17–23.
224. Kim SA, Park S, Chung N, et al. Efficacy and safety profiles of a new S(-)-amlodipine nicotinate formulation versus racemic amlodipine besylate in adult Korean patients with mild to moderate hypertension: an 8-week, multicenter, randomized, double-blind, double-dummy, parallel-group, Phase III, noninferiority clinical trial. Clin Ther 2008; 30(5):845–857.
225. Op't Holt TB. Inhaled beta agonists. Respir Care 2007; 52(7):820–832.
226. Bryson HM, Faulds D. Cisatracurium besilate. A review of its pharmacology and clinical potential in anaesthetic practice. Drugs 1997; 53(5):848–866.
227. Sun Wai WYS, Flynn PJ. 51W89, the 1R cis – 1'R cis isomer of atracurium. Anaesth Pharmacol Rev 1995; 3:218–221.
228. Graf BM, Martin E, Bosnjak ZJ, et al. Stereospecific effect of bupivacaine isomers on atrioventricular conduction in the isolated perfused guinea pig heart. Anesthesiology 1997; 86(2):410–419.

229. Bardsley H, Gristwood R, Baker H, et al. A comparison of the cardiovascular effects of levobupivacaine and rac-bupivacaine following intravenous administration to healthy volunteers. Br J Clin Pharmacol 1998; 46(3):245–249.
230. Wang DY, Hanotte F, De Vos C, et al. Effect of cetirizine, levocetirizine, and dextrocetirizine on histamine-induced nasal response in healthy adult volunteers. Allergy 2001; 56(4):339–343.
231. Hyttel J, Bogeso KP, Perregaard J, et al. The pharmacological effect of citalopram residues in the (S)-(+)-enantiomer. J Neural Transm Gen Sect 1992; 88(2):157–160.
232. Montgomery SA, Loft H, Sanchez C, et al. Escitalopram (S-enantiomer of citalopram): clinical efficacy and onset of action predicted from a rat model. Pharmacol Toxicol 2001; 88(5):282–286.
233. Moore N, Verdoux H, Fantino B. Prospective, multicentre, randomized, double-blind study of the efficacy of escitalopram versus citalopram in outpatient treatment of major depressive disorder. Int Clin Psychopharmacol 2005; 20(3):131–137.
234. Moore RA, Barden J. Systematic review of dexketoprofen in acute and chronic pain. BMC Clin Pharmacol 2008; 8:11.
235. Patrick KS, Caldwell RW, Ferris RM, et al. Pharmacology of the enantiomers of threo-methylphenidate. J Pharmacol Exp Ther 1987; 241(1):152–158.
236. Srinivas NR, Hubbard JW, Korchinski ED, et al. Enantioselective pharmacokinetics of dl-threo-methylphenidate in humans. Pharm Res 1993; 10(1):14–21.
237. Nishino S, Okuro M. Armodafinil for excessive daytime sleepiness. Drugs Today (Barc) 2008; 44(6):395–414.
238. Davis R, Bryson HM. Levofloxacin. A review of its antibacterial activity, pharmacokinetics and therapeutic efficacy. Drugs 1994; 47(4):677–700.
239. Hayakawa I, Atarashi S, Yokohama S, et al. Synthesis and antibacterial activities of optically active ofloxacin. Antimicrob Agents Chemother 1986; 29(1):163–164.
240. McKeage K, Blick SK, Croxtall JD, et al. Esomeprazole: a review of its use in the management of gastric acid-related diseases in adults. Drugs 2008; 68(11):1571–1607.
241. Wagner JG. Kinetics of pharmacologic response. I. Proposed relationships between response and drug concentration in the intact animal and man. J Theor Biol 1968; 20(2):173–201.
242. Hair PI, McCormack PL, Curran MP. Spotlight on eszopiclone in insomnia. CNS Drugs 2008, 22(11).975–978.
243. Oosterhuis B, van Boxtel CJ. Kinetics of drug effects in man. Ther Drug Monit 1988; 10(2):121–132.
244. Ariens EJ, Simonis AM. A molecular basis for drug action. The interaction of one or more drugs with different receptors. J Pharm Pharmacol 1964; 16:289–312.
245. Gaddum JH. Theories of drug antagonism. Pharmacol Rev 1957; 9(2):211–218.
246. Scott LJ, Dunn CJ, Mallarkey G, et al. Esomeprazole: a review of its use in the management of acid-related disorders in the US. Drugs 2002; 62(7):1091–1118.
247. Schüttler J, Stanski DR, White PF, et al. Pharmacodynamic modeling of the EEG effects of ketamine and its enantiomers in man. J Pharmacokinet Biopharm 1987; 15(3):241–253.
248. Mager DE, Wyska E, Jusko WJ. Diversity of mechanism-based pharmacodynamic models. Drug Metab Dispos 2003; 31(5):510–518.

13 Regulatory perspective on the development of new stereoisomeric drugs

Sarah K. Branch[a] and Andrew J. Hutt

INTRODUCTION

The regulation of chiral drugs is, in principle, no different from the regulation of any other drug substance. Satisfactory standards of quality, safety, and efficacy need to be demonstrated for medicines containing chiral active ingredients before they can receive an authorization to place them on the market. Of course, there are some special considerations or even challenges to take into account, particularly regarding the similar physicochemical properties of potential impurities in the traditional sense. Significant advances in synthetic, preparative, and analytical chemistry in the 1980s and 1990s allowed the development of synthetic, as opposed to the natural or semisynthetic, drug substances, which had dominated the area of stereochemically pure medicinal products up to then (e.g., Gal) (1). At the same time, there was increasing awareness of the different biological activities of stereoisomers and indeed the means to study these differences.

It is now recognized that stereoisomers may differ in their pharmacodynamic activity at their biological target, for example, receptor, enzyme active site, etc., or in their pharmacokinetic properties (absorption, distribution, and clearance by metabolism and excretion). Regarding pharmacodynamic activity, differences between stereoisomers may be quantitative, in that the resulting action is the same but one isomer is more potent than the other. Alternatively, the difference may be qualitative in nature, the "unrequired" isomer may be biologically inert, that is, have no detectable biological activity, may act on a different biological target possibly giving an unwanted biological response, or may even have an opposing effect reducing the required pharmacodynamic activity if a racemate was to be administered.

In terms of pharmacokinetics, differences in the way chiral substances are handled by the body can have far-reaching consequences for individual patients depending on their disease state, age, gender, and genetic profile. In particular, differences in liver enzymes as a result of these factors can lead to fast/extensive or slow/poor metabolism of individual enantiomers leading to differential duration, or extent, of activity. Stereoselectivity in protein binding resulting in stereoselective drug distribution may also contribute to differential pharmacodynamic activity. All these factors need to be taken into account during modern drug development to ensure patients are not subject to unnecessary chemical exposure or pharmacological effects.

The full potential of technological advances coupled with the greater recognition of the significance of the differential pharmacological activity of stereoisomers has not necessarily been completely realized. New drugs may be designed deliberately to be achiral to avoid the complications presented by the

---

[a]The views expressed in this chapter are those of the author and do not necessarily represent the views or opinions of the Medicines and Healthcare products Regulatory Agency (MHRA) or other regulatory agencies or their advisory committees.

production of single stereoisomers on a commercial scale. The area of "chiral switches," whereby an enantiomerically pure form of a drug previously available as a racemate, has also perhaps been more challenging than had been anticipated. This is because it has not always proved easy to translate the predicted differential activity into meaningful differences of therapeutic value during clinical trials or differences in side effect profiles of significance to the patient. Indeed, the relative merits of single enantiomers versus their corresponding racemates of a number of "blockbuster drugs" have been the subject of considerable debate, for example, citalopram/escitalopram, omeprazole/esomeprazole, cetirizine/levocetirizine (2,3). The major criticism is that comparative clinical and pharmacoeconomic studies clearly indicating the benefits of the single enantiomers to patients, prescribing physicians, and health funding agencies are not readily available (2–6). Nonetheless, even though the number of new drug introductions has decreased in recent years the number of chiral compounds approved and marketed as single stereoisomers rather than racemates has increased (see section "Impact of Regulatory Guidelines on Chiral Drug Development").

Typically, regulatory authorities require the production of stereochemically pure drug substances unless there is sound scientific justification as to why this is either unnecessary or not possible on a commercial scale. Accordingly, the regulatory requirements for investigation of chiral drugs in Europe, the United States, and Japan are outlined in this chapter. These regions sponsor the International Conference on Harmonisation (ICH), which aims to unify the process of drug registration globally (see section "International Conference on Harmonization"). There will be a particular focus on synthetic chemical drug substances and the quality aspects of their development, that is, the chemical and pharmaceutical components that form part of the technical dossier needed to support an application for a marketing authorization. The requirements for this technical dossier itself have been harmonized through ICH and therefore apply in the three sponsoring regions.

REQUIREMENTS IN THE EUROPEAN UNION
Introduction
The regulatory requirements for pharmaceuticals in the European Union (EU) are set out by European Community (EC) law. Within certain parameters, companies have a choice of procedures through which they may apply for an authorization to market a medicinal product. This choice allows a company to take into consideration likely future use and acceptance of their product in different member states, although it should be borne in mind that it is a core European principle that patients across the community should have equal access to medicines.

The available application procedures may be summarized as follows:

- *Centralized procedure* leading to a community-wide authorization directly applicable in all member states:
    - Compulsory for certain products specified by law for example, *biotechnology or gene technology products deemed to be of high value to public health*;
    - Optional for products that can demonstrate *novelty, community interest*, etc.

- *Mutual recognition procedure* whereby a company has obtained a license in one member state first and then requests its recognition by all or a selected number of other member states.
- *Decentralized procedure* where an application is made to all or some member states simultaneously but one member state leads the assessment of the dossier.
- *National procedure* where a company makes an application in one member state only.

Whatever the procedural route chosen, or used, an application to place a medicinal product on the market is assessed according to the criteria of quality, safety, and efficacy. The quality of a product relates to its pharmaceutical characteristics including the synthesis and control of the active ingredient, quality of excipients, manufacture of the dosage form, specifications, and stability. Safety aspects concern both nonclinical (toxicological) and clinical use of the active ingredient and product to determine both safe quantitative levels of the active substance, excipients, and any impurities and their qualitative effects or safety profile. Clinical efficacy must be demonstrated either through appropriate clinical trials or adequate supporting published data. The outcome of assessment and approval of an application represents an agreement on, among other things, the way the product is to be manufactured, its shelf life, and how it will be used clinically, which is recorded together with safety information and other details in a document known as the *Summary of Product Characteristics* (SmPC), and then reflected in associated information for patients.

The procedural requirements for marketing authorization applications and the content of the regulatory dossier of evidence needed to support such submissions is set out in common European legislation. There is also an extensive range of technical guidance, some of which has been developed internationally. Those aspects of European law and guidance that relate to the authorization of chiral drugs will be summarized in this chapter. However, it should be borne in mind that there has been a subtle shift in approach in the EU since the pharmaceutical legislation was updated in 2005. Before that time, new single enantiomer versions of compounds of previously marketed racemates were almost universally treated as new drugs and the supporting data were evaluated as such, taking into account relevant information from studies with the racemate. Guidance existed to the effect that enantiomers were considered to be the same active ingredient unless they had a different safety and efficacy profile but it was generally accepted that it was necessary to treat the single enantiomer as a new drug evaluation before this fact could be established. However, the revised legislation in 2005 transferred this concept into law since when it has been applied much more rigorously leading to assessment of single enantiomers as new drugs or otherwise on a case-by-case basis. While the outcome of this shift has not altered the fundamental principles of the scientific evaluation and terms of marketing authorization with respect to clinical use of the product, it has led to changes in interpretation of data and marketing exclusivity periods for single enantiomers, and indeed to on-going legal debate.

Current EU Legislation
Common legislation governs the criteria for the approval of human medicines throughout the countries of the EU whatever the procedural route chosen for the

marketing authorization application. The legislation takes the form of regulations, which are directly binding in member states, and directives, which have to be transposed and implemented nationally. Regulation EC 726/2004 governs the centralized procedure and sets out the structure and function of the European Medicines Agency (EMA) and pharmaceutical advisory bodies such as the Committee for Human Medicinal Products (CHMP). Directive 2001/83/EC describes how authorizations are obtained for noncentralized products and sets out the requirements for the supporting information to be included in the dossier that must be submitted with marketing authorization applications. The law is accompanied by various forms of guidance, which help to interpret the pharmaceutical legislation and both are set out in the *Rules Governing Medicinal Products in the European Union,* which are available on the European Commission Health and Consumers website (7). Such guidelines are not legally binding but the applicant would be expected to provide a satisfactory (scientifically based) justification in cases where the recommendations had not been followed. Recommendations on the studies to be conducted in support of an application for a marketing authorization are made in *Scientific Guidelines on Medicinal Products for Human Use* contained in volume 3 of the *Rules.* These guidelines are available on the European Commission Health and Consumers website or directly on the EMA website (8). These notes for guidance begin life under the auspices of the various working parties of CHMP [or its predecessor Committee for Proprietary Medicinal Products (CPMP)] and include guidelines that are the result of international harmonization (see section "International Guidance"). The guidelines are available at their draft consultation stages and in their final form on the EMA website.

Note for Guidance on Investigation of Chiral Active Substances

The note for guidance of primary interest for drugs that may exist as optical isomers is entitled *Investigation of Chiral Active Substances* (originally CPMP/III/3501/91, current reference 3CC29A) (9). It came into operation in 1994 so is now somewhat dated and probably in need of revision in order to reflect current practice. Its requirements are discussed below. Its contents must be considered additional to other guidelines relating to the quality, safety, and efficacy of medicinal products authorized in the EU. It is up to manufacturers to decide whether to market a drug as a single enantiomer or a racemate. They should provide sufficient justification of their decision so that the competent authority can assess the benefit-risk ratio. The key to a successful outcome for an application for a marketing authorization is proper justification of the decisions made concerning the product during development. The chiral substances guideline sets out requirements for studies to justify the chosen strategy in the areas corresponding to the three technical parts of the dossier accompanying the application, that is, quality, safety, and efficacy.

Chemistry and Pharmacy Aspects

At the time the *Chiral Active Substances* guideline came into effect, chemical aspects of the information required to support an application for a medicinal product were set out in the guideline *Chemistry of Active Substances* (Eudra/Q/87/011, current reference 3AQ5a) (10) published in volume 3 of the *Rules.* Further advice has since been provided in the note for guidance on the *Chemistry* of *New Active Substances* (CPMP/QWP/130/96 Rev 1) (11), which came into operation in 2004. The guideline on Summary of Requirements for Active Substances in the Quality Part of the Dossier (CHMP/QWP/297/97 Rev. 1) (12) came into effect a year later. It explains how the requirements may be met for existing drug substances through cross-reference to master files or pharmacopoeial monographs and gives directions to relevant guidance for new active substances.

Synthesis of the active substance. The information required concerning the synthesis of a single stereoisomer of a chiral drug substance is the same as for any other agent. However, particular attention must be paid to the step where the chiral center is introduced, which must be described in detail, and maintenance of the desired configuration during subsequent stages of synthesis must also be demonstrated. Validated analytical methods for determination of chiral compounds are therefore of key importance. The end product must be fully characterized with respect to identity, related substances, and other impurities as for any other drug substance, but with the additional requirement of establishing stereochemical purity. There is an increasing trend toward the use of "bought-in" intermediates in the preparation of active substances. However, it is usually expected that sufficient synthetic steps prior to achieving the final drug substance will be described so that potential by-products carrying through to the end material can be evaluated. Sometimes more than one intermediate supplier is used (perhaps less so in the case of chiral active substances) but it should be remembered that full characterization of such intermediates may be required and that information should be presented on their synthesis to allow evaluation of potential impurities in the final active ingredient.

An increasing number of synthetic strategies are possible and the information that should be presented in a regulatory dossier will depend on the route chosen. Full characterization is needed for starting materials that already contain the required chiral center, whether a racemate or single enantiomer, and should include assessment of stereochemical purity using validated stereospecific analytical procedures. Evidence should be provided that the required enantiomeric ratio is consistently achieved in cases where a racemate, or other mixture of isomers, is intended as the resulting product, unless obvious from the synthetic route employed. Where separation methods are used to obtain the preferred enantiomer, for example, simulated moving bed (SMB) technology, the resolution step is considered part of the overall manufacturing process and the usual details of the procedure should be given together with the number of cycles used.

All relevant information should be provided in the dossier in cases where the known active enantiomer cannot be obtained on a scale suitable for commercial manufacture, for example, because of difficulties with scale-up or failure to obtain material in a suitable physical form for pharmaceutical manufacture. Advances in preparative techniques should eventually make this scenario less common but if necessary, all the experimental results available should be described and the reason for the failure given to justify use of the racemate over the active enantiomer. Likewise, if sufficient enantiomeric material could not be provided for preclinical and clinical studies (see below), this should also be discussed.

Quality of the active substance. A specification drawn up by the applicant and agreed by the regulatory authority is the basis of assuring the quality of a drug substance. More recent guidance on specifications and tests has been published by ICH and is discussed later in this chapter. The guideline on chiral active substances states that particular attention should be paid to identity and stereochemical purity. It states that specifications for a racemate should include a test to show that the substance is indeed a racemate and this is a position supported by the requirements of the European Pharmacopoeia in its *Technical Guide for Elaboration of Monographs*, 5th edition, 2010 (13).

The chiral drugs guideline lists examples of methods that may be used for the control of drug substances, ranging from the simpler ones, such as optical rotation, melting point, stereospecific high-performance liquid chromatography (HPLC), to the more sophisticated techniques including optical rotatory dispersion, circular dichroism, or nuclear magnetic resonance (NMR) spectroscopy with chiral shift reagents. The advances in analytical methodology since 1994 have seen the introduction of newer, more sensitive, and alternative techniques, including capillary zone electrophoresis (CZE), with chiral selectors added to the background electrolyte, and related electrophoretic techniques. Use of newer methods is not precluded by the guideline but it is the responsibility of applicants to decide on the techniques that are most appropriate for the satisfactory control of their drug substance and to ensure that they are fully validated. The guidelines on analytical validation have been internationally harmonized and are discussed below (see section "ICH Guidelines on Analytical Validation").

In situations where stereoisomeric reference substances are required for test procedures, the stereochemical purity of the reference material must be stated. Obviously the characterization of such materials must be carried out with considerable care particularly when they are required to support the stereochemical identification of the drug substance. Past experience has shown that applicants can fall into circular arguments when trying to establish the absolute configuration of a compound argued on mechanistic grounds and/or the chirality of starting materials. Proof is required that the synthetic process has actually delivered the required stereoisomer! Single crystal X-ray diffraction studies of the final drug substance with methods appropriate for the determination of absolute configuration provide the greatest confidence.

The enantiomer of the required stereoisomer may arise during synthesis of a drug substance or on degradation during storage. The unwanted enantiomer is considered to be an impurity as stated in the guideline on chiral active substances. Further guidance is available in the ICH guideline on impurities in drug substances described below (see section "ICH Guidelines on Impurities") but it is recognized that the usual limits may not apply to chiral impurities due to the lack of sufficient sensitivity of some of the available test methods. However, the principles of the guidance for achiral active substances should be followed, in particular the process of "qualification" of impurities whereby the biological safety of an individual impurity, or an impurity profile, is established at a specified level. Qualification is achieved by linking the use of development and commercial scale batches to particular toxicological or clinical studies. The impurity profile of these batches must be recorded and, assuming that no adverse effects were observed during use of the development batches, then the levels of impurities are considered to be qualified. Together with the batch analysis data, the qualification studies should be used to justify the specification limits for individual known, unknown, and total impurities.

Impurities that are also significant human metabolites do not need further qualification as exposure to them would be automatic on administration of the drug in clinical trials. Thus, enantiomers formed by chiral inversion of the drug substance, either enzymatically or chemically mediated in the biological environment, would not need further qualification. This is one aspect of the importance of establishing the metabolic, particularly the stereochemical, profile of chiral drugs.

Having established the specifications for the stereochemical purity, or enantiomeric composition, at the end of the synthetic process (batch release), the applicant should demonstrate that no unacceptable change in these parameters occurs on storage of the active ingredient.

Chemical development. The discussion of chemical development requires proof of structure and configuration. The physicochemical properties of the drug substance should be characterized, for example, crystallinity, polymorphism, and rate of dissolution. Physical parameters such as melting point, solubility, and crystal properties should be investigated to establish whether a drug produced as a racemate is a true racemate or a conglomerate. The validation of the analytical methods used at batch release to guarantee the identity and purity of the substance should be described.

Finished product. The applicant should show during the manufacture of the dosage form, or finished medicinal product, that there are no unacceptable changes in the stereochemical purity of the active ingredient or during storage for the proposed shelf life. Specifications for the finished product on release and during shelf life will be required although it may be possible to justify the omission of certain tests if, for example, it can be shown that racemization does not occur.

Preclinical and Clinical Studies

Single enantiomer. The development of a single enantiomer as a new active substance should be described in the same manner as for any other new chemical entity. Pivotal studies should be carried out with the single enantiomer but if development began with the racemate then these studies may also be taken into account. Chiral inversion, either chemically or biologically mediated, should be considered early, ideally during the preclinical evaluation of the drug, so that enantiospecific bioanalytical methods may be developed if required. These methods should be presented in the part of the regulatory dossier describing the clinical studies. If following administration of a single stereoisomeric drug the enantiomer is formed in vivo, then it should be evaluated in the same way as other metabolites. For endogenous human chiral compounds, enantiospecific analysis may not be necessary. The enantiomeric purity of the active ingredient used in preclinical and clinical studies should be defined (for the purposes of qualification as described above).

Racemate. The applicant should provide justification for using the racemate. Where interconversion of enantiomers occurs in vivo and the interconversion is more rapid than the distribution and elimination rates, then use of the racemate is justified. In cases where interconversion does not occur, or the rate of interconversion is slow compared to other pharmacokinetic processes, then differential pharmacological effects and fate of the enantiomers may be apparent. Use of the racemate may also be justified if any toxicity is associated with the pharmacological action and the therapeutic index is the same for both isomers. For preclinical assessment, pharmacodynamic, pharmacokinetic (using stereospecific analytical methods), and appropriate toxicological studies of the individual enantiomers and the racemate will be required. Clinical studies on human pharmacodynamics and tolerance, human pharmacokinetics, and pharmacotherapeutics will be required for the racemate and for the enantiomers as appropriate.

New single enantiomer from approved racemate or new racemate from approved single enantiomer. The guideline states that in principle these situations concern the development of a new active substance requiring a completely new application. Since the introduction of Directive 2001/83/EC, however, the legal basis of such applications may not actually be that of a new drug, although the information provided may have to be equivalent. This is discussed further in the section "Regulatory Aspects."

Whatever the case, the application should include an explanation of the decision to develop the enantiomer or racemate. Data on the existing racemate or enantiomer may be included, where appropriate, with bridging studies as necessary, assuming that the applicant can provide the contents of the full dossier on the previously marketed material to the regulatory authority. Obviously this can be problematic if the applicant is not the originator of the initially marketed material. Justification for the extrapolations should be provided. Some pharmacodynamic or pharmacokinetic studies with the opposite enantiomer may be required to account for any racemate interactions.

Nonracemic mixture from approved racemate or single enantiomer. This can be viewed as optimization of the pharmacotherapeutic profile and therefore is treated as a fixed combination product, for which a separate note for guidance applies. Data on stereoisomers not previously approved should be provided together with justification for the fixed combination.

Abridged applications. Abridged applications are those for an existing drug substance that has passed the periods of data and market exclusivity afforded to it when its initial authorization as a new drug was granted. This period may be 6, 10, or 11 years depending on when the authorization was granted, in which country, and whether an extension has been awarded for a significant new clinical use of the product. The validity of abridged applications is further discussed below (see section "Regulatory Aspects"). The guideline states that applications for generic chiral medicinal products should be supported by bioequivalence studies using enantiospecific bioanalytical methods unless both they and their reference products contain the same, stable, single enantiomer or both products contain a racemate where both enantiomers exhibit linear pharmacokinetics.

Concluding remarks. The guideline concludes with a note that there is no intention to require further data on older medicinal products containing racemates where use was already established at the time it came into force, unless new evidence emerges concerning the safety or efficacy of one enantiomer. If new claims related to the chiral nature of the active substance are made, then supporting studies on the individual enantiomers will be required.

Regulatory Aspects
Prior to the introduction of Directive 2001/83/EC in October 2005, new active substances were defined in guidance and single enantiomers derived from existing racemates—the so-called chiral switches—were generally treated as new chemical entities in line with the *Chiral Active Substance* guideline described above. However, the new legislation states firmly that different salts, esters, isomers, mixtures of isomers, complexes, or derivatives shall be considered to be the same active substance *unless they differ significantly in their safety and efficacy profiles*. Proof of safety and efficacy must be provided by the applicant, which raises challenges in determination of the legal classification of single enantiomer products at the outset of a procedure before detailed assessment of safety and efficacy has occurred.

Furthermore, the concept of a global marketing authorization applies for the purposes of determining the validity of generic applications. A global marketing authorization includes all subsequently approved strengths and dosage forms containing the same active ingredient and any changes to those individual authorizations. Following on from the argument that isomers with the same safety and efficacy properties are considered to be the same active substance, this means that single enantiomers may, in certain circumstances, be included in the global marketing authorization for the original racemate. This leads to some interesting questions about the validity of applications for generic versions of existing single enantiomer products. If the single enantiomer is indeed part of the global marketing authorization for the racemate, then a new period of data protection would not apply to the single enantiomer product and generic applications for either the racemate or the enantiomer could be made once the original exclusivity period for the racemic drug has expired. The question currently being debated is whether such considerations can be applied retrospectively even though the single enantiomer was first authorized as a new chemical entity. In some jurisdictions, the matter has progressed to the courts.

If it is established that a generic application is possible for a single enantiomer by cross-referring to racemate data held by a brand leader, there is also the question of how much additional information needs to be provided by the generic applicant to establish the necessary safety and efficacy for approval of the authorization.

These regulatory debates are taking place at the same time as legal challenges in the courts as to validity of patents for single enantiomer versions of existing racemic drugs (see chap. 15).

INTERNATIONAL GUIDANCE
International Conference on Harmonisation
Over the past two decades considerable efforts have been made in order to develop common guidance for worldwide registration of medicinal products with the aim of reducing the regulatory burden for the pharmaceutical industry and avoiding duplication of tests and studies. Significant progress has been made by the International Conference on Harmonisation of Technical Requirements for the

Registration of Pharmaceuticals for Human Use (ICH) since its inauguration in 1990. Its aim is to promote harmonization of regulatory requirements and interpretation of technical guidance and thus to eliminate unnecessary delay in the global development and availability of new medicines. ICH is a tripartite body sponsored by the regulatory authorities, and research-based industry representatives, from the EC, Japan, and the United States, which holds biennial conferences and workshops to promote its work. The Secretariat of the ICH is provided by the International Federation of Pharmaceutical Manufacturers Associations (IFPMA).

Development of Guidelines

The ICH Steering Committee is responsible for identifying topics for which harmonized guidelines are then developed through its Expert Working Groups (EWGs), which have a membership nominated by regulatory authorities and pharmaceutical industry organizations in the three participating regions. There are five stages in the ICH process for developing a guideline, represented in Figure 13.1, which starts with consideration of the topic and development of a consensus by the relevant EWG. Both the consensus and the draft guideline are subject to consultation process. The topics span the three main areas of scientific evaluation of a regulatory dossier (quality, safety, and efficacy) plus a group of multidisciplinary subjects. There is also a revision procedure for updating guidance when necessary. When guidelines are finally agreed, they are adopted in each of the three sponsoring regions. A number of observers are invited to the Steering Committee: World Health Organisation (WHO), Canada (represented by Health Canada), and the European Free Trade Association (EFTA). Therefore, ICH guidelines have a much wider global impact than Europe, United States, and Japan. ICH has developed a number of guidelines of relevance to the development of chiral drugs, in particular regarding specifications and tests, impurities and validation of analytical methods, which are discussed below. These guidelines can be obtained from the ICH website (14).

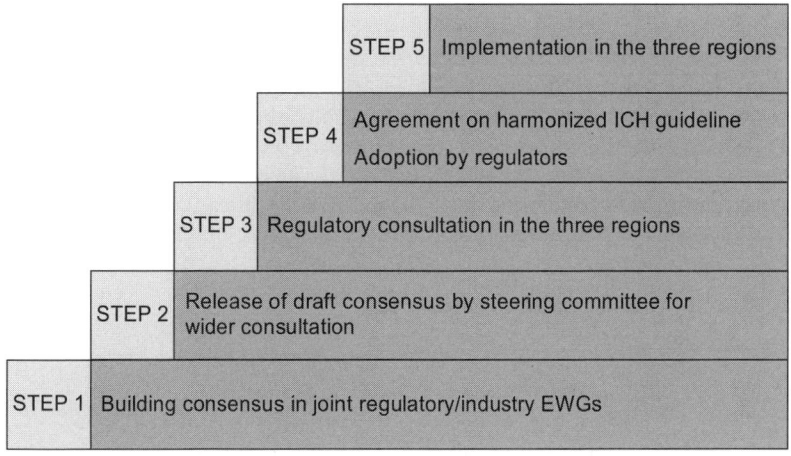

FIGURE 13.1 ICH process for developing harmonized guidelines.

Implementation in the Three ICH Regions
Europe
In the EU, the CHMP endorses the consensus at step 2 and the final guideline at step 4. The CHMP, together with the EC, decides on the duration for consultation with interested parties, which may be up to six months. The EMA publishes and distributes the step 2 guidelines for comment. At step 4 the guidelines are endorsed by the CHMP and a timeframe for implementation is established (usually six months). The step 2 and step 4 guidelines are published on the EMA website (15) and included in the Rules Governing Medicinal Products.

United States
In the United States, when step 2 or step 4 has been reached, the Food and Drug Administration (FDA) publishes a notice in the Federal Register with the full text of the guidance. Notices for step 2 guidelines include a date for receipt of written comment; step 4 guidelines are available for use on the date they are published in the Federal Register. FDA guidances for chemical drug substances are available on the Center for Drug Evaluation and Research (CDER) website (16).

Japan
When step 2 or step 4 has been reached, the ICH texts are translated into Japanese. Subsequently a Pharmaceutical and Medical Safety Bureau (PMSB) notification for the promulgation or consultation of guidelines, written in Japanese, is issued with a deadline for comments in the case of consultation drafts, or an implementation date for finalized guidelines. The notifications on guidelines in Japanese, and also English attachments (ICH Texts), are available from PMSB, which is part of the Ministry of Health, Labour and Welfare (MHLW) through the Pharmaceutical and Medical Devices Agency (PMDA) website (17).

Common Technical Document
Another significant piece of work on the part of ICH has been the development of the common technical document (CTD), introduced in 2000, and standards for its electronic counterpart, which followed in 2002. The EWG for the CTD was extended to include the observers to ICH, representatives of the generics industry, and manufacturers of products for self-medication, reflecting the wider applicability of this topic. CTD provides a common structure for the main body of data supporting applications to market medicinal products in the three sponsoring ICH regions. It has thus led to a significant reduction in the administrative aspects of submitting regulatory dossiers globally.

ICH Guidelines on Specifications and Tests
ICH Topic Q6A *Specifications: Test procedures and acceptance criteria for new drug substances and new drug products: chemical substances* (18) reached step 4 in October 1999 and was approved by the CPMP in November 1999 with a date of May 2000 (step 5) for coming into operation in the EU [CPMP/ICH/367/96]. It provides guidance on the selection of test procedures and the setting and justification of acceptance criteria for new drug substances of synthetic chemical

origin, and drug products made from them, that have not been previously registered in the EU, Japan, or the United States. Detailed recommendations are made regarding the specifications for active ingredients and different types of dosage forms, and reference is made to chiral drugs. Thus, this ICH guideline may supersede, or at least provide additional guidance to, the recommendations in regional guidelines. It should be noted that this ICH guideline does not apply to drugs of natural origin.

Guidance on specifications is divided into universal tests/criteria, which are considered generally applicable to all new substances/products, and specific tests/criteria. The latter may need to be addressed on a case-by-case basis when they have an impact on the quality for batch control. Tests are expected to follow the ICH guideline on analytical validation (see section "ICH Guidelines on Analytical Validation"). Identification of the drug substance is included in the universal category and such a test must be able to discriminate between compounds of closely related structure that are likely to be present. It is acknowledged here that optically active substances may need specific identification testing or performance of a stereospecific assay in addition to this requirement.

Tests for chiral drug substances are included in the category of specific tests/criteria. A decision tree (Fig. 13.2) summarizes when and if chiral identity tests, impurity tests, and assays may be needed in the drug substance and finished product specifications. For a drug substance developed as a single stereoisomer, an identity test should be capable of distinguishing between the enantiomers and the racemate. A stereospecific assay, or enantiomeric impurity procedure, may also serve to provide a chiral identity test. When the active ingredient is a racemate, a stereospecific test is appropriate where there is a significant possibility that substitution of an enantiomer for a racemate may occur, or when preferential crystallization may lead to unintentional production

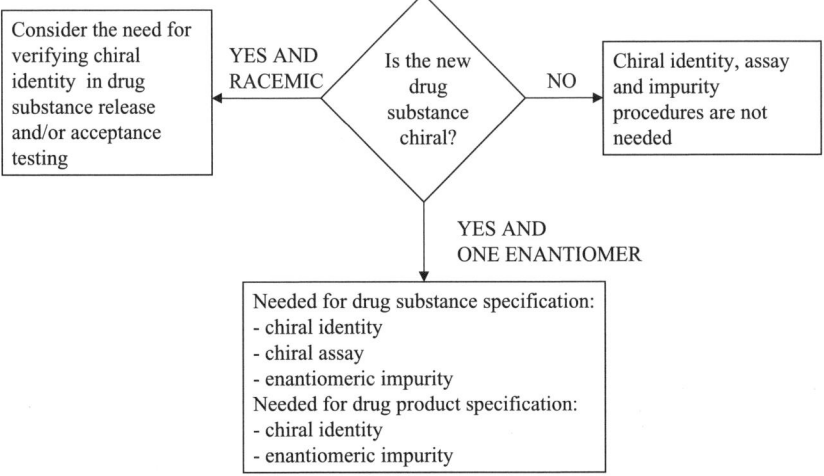

FIGURE 13.2 Establishing procedures for chiral new drug substances and new medicinal products containing chiral drug substances.

of a nonracemic mixture. Such a test is generally not needed in the finished product specification if there is insignificant racemization during manufacture of the dosage form or on storage and a test is included in the drug substance specification. If the unrequired enantiomer is formed on storage then a stereospecific assay for enantiomeric impurity testing will also serve to identify the active substance.

With respect to impurities, it is acknowledged that, where the substance is predominantly a single stereoisomer, the enantiomer is excluded from the qualification and identification thresholds given in the ICH guideline on impurities (see section "ICH Guidelines on Impurities") because of potential difficulties in quantification at the recommended levels. Otherwise, it is expected that the principles of that guidance apply. The guideline allows that appropriate testing of a starting material or intermediate, with suitable justification from studies conducted during development, could give assurance of control. This approach may be necessary, for example, when there are multiple chiral centers present in the drug molecule. Control of the unrequired enantiomer in the finished product is needed unless racemization during manufacture of the dosage form or on storage is insignificant. The procedure used may be the one used for the assay or it may be a separate methodology.

Determination of the drug substance is expected to be enantiospecific and this may be achieved by including a stereospecific assay in the specification or an achiral assay together with appropriate methods of controlling the enantiomeric impurity. For a drug product where racemization does not occur during manufacture or on storage, an achiral assay may suffice. If racemization does occur, then a chiral assay should be used or an achiral method combined with a validated procedure to control the presence of the other enantiomer.

ICH Guidelines on Impurities

There are two ICH guidelines on impurities: Topic Q3A makes recommendations on *Impurities in new drug substances* and Topic Q3B on *Impurities in new medicinal products* (19) with effective dates in Europe of August 2002 and August 2003, respectively. The two should be read in conjunction. The documents were revised in 2006 to provide further information on the rounding of analytical results in relation to the limits for impurities denoted in the texts and are now published as CPMP/ICH/2737/99 and CPMP/ICH/2738/99. As previously mentioned, enantiomeric impurities are excluded from the guideline but the principles expressed are expected to apply. There are two aspects of control of impurities: first, their chemical classification and identification, and second, assessment of their safety at the level imposed by the drug substance specification. The latter is the process of qualification already mentioned in section "Note for Guidance on Investigation of Chiral Active Substances" under quality of the active substance.

The guidance on impurities in new drug substances state that the sources of actual and potential impurities, whether arising from synthesis, purification, or degradation, should be discussed. Analytical data are required that show the level of individual and total impurities in development and commercial-scale batches. The impurity profiles, for example chromatograms, must be available if requested. Samples should be intentionally degraded so that potential impurities arising from storage can be identified. Such studies would reveal whether racemization of single enantiomers was likely to occur. In normal application of the guideline, identification of organic impurities is required above certain specified thresholds, usually by isolation and spectroscopic characterization, or if this has not been possible, the unsuccessful laboratory studies described. Below these thresholds, identification is not required but it is useful to present this data if available and identification should be attempted in any case for compounds expected to be unusually potent or toxic. Stereospecific analysis would enable the identification of enantiomeric impurities.

The guideline gives thresholds depending on the maximum daily dose of the drug above which qualification studies are required. Lower or higher qualification thresholds may be appropriate for certain classes of drugs. Where the qualification threshold is exceeded, additional safety studies may be required according to a decision tree provided in the guideline. Similar qualification of enantiomeric impurities by their presence in batches of drug substance used in safety and/or clinical studies would be expected, although they are not strictly covered by the guideline. The guideline was revised in 2006 to include designation of reporting thresholds for impurities.

The guideline on impurities in new medicinal products parallels the drug substance text but the designated thresholds concern only degradation products. The thresholds should be applied to the product at the end of its shelf life, as that is when the greatest level of degradation is expected to have occurred.

ICH Guidelines on Analytical Validation

Two guidelines were developed on analytical validation but have now been incorporated into a single document (ICH Topic Q2) (20) published in the EU as CPMP/ICH/381/95. The first guideline ICH Topic Q2A *Validation of analytical procedures: Definitions and terminology* came into operation in Europe in June 1995 and provided a glossary of terms. It sought only to present a collection of terms and definitions and not to provide direction on how to accomplish validation. The guideline was intended to bridge the differences that could exist between the various compendia and regulators in the three regions of the ICH at that time.

The guideline states that the objective of validation is to demonstrate that an analytical method is fit for its purpose and summarizes the characteristics required of tests for identification, control of impurities, and assay procedures (Table 13.1). As such, it applies to chiral drug substances as to any other active ingredients. Requirements for other analytical procedures may be added in due course.

TABLE 13.1 Characteristics of Analytical Procedures Requiring Validation (indicated by a tick)

	Identity	Control of impurities		Assay
		Quantification	Limit test	
Accuracy and precision		√		√
Specificity	√	√	√	√
Limit of detection		(√)	√	
Limit of quantitation		√		
Linearity and range		√		√

Assays may be applied to the active moiety in the drug substance or drug product or to other selected components of the product. They are used for content/potency determinations and for measurement of dissolution. Precision includes repeatability (intra-assay precision) and intermediate precision (within laboratory) except the latter is not required where reproducibility (inter-laboratory) has been performed. If there is lack of specificity in one analytical procedure, compensation by other supporting methods is allowed. The characteristics listed in Table 13.1 are considered typical but allowance is made for dealing with exceptions on a case-by-case basis. Robustness is not listed but should be considered at an appropriate stage in development. Revalidation of analytical procedures is required following changes in the synthesis of a drug substance, composition of the finished product, or in the analytical procedure.

The second guideline, ICH Topic Q2B, *Validation of analytical procedures: Methodology* came into operation in Europe in June 1997. It is complementary to the first guideline and provides some guidance and recommendations on acceptable methods for validating the characteristics of an analytical procedure. An indication of the data that should be provided in an application for a marketing authorization is given. The following characteristics are separately discussed: specificity; linearity; range; accuracy; precision; detection limit; quantitation limit; robustness; and system suitability testing.

REQUIREMENTS IN THE UNITED STATES
Introduction

The United States FDA is responsible for the authorization of human medicinal products in the United States through CDER. Policy and guidance relating to drug registration for chemical substances is published in the Federal Register and is available on the FDA website (21).

The first mention of stereochemical requirements to be applied in drug development in the United States may be found in the New Drug Application (NDA) guideline of 1987 concerning manufacturing documentation (22). The document suggested that the FDA may consider enantiomers to be an impurity if present in small quantities or when present in a racemate, that is, 50%. Additionally, when the new drug substance was asymmetric the guideline indicated that sponsors should ideally separate, or synthesize, the various potential stereoisomers, provide physical and chemical information, and that they may need to be examined both pharmacologically and toxicologically (23). The wording of the guideline implied that the FDA may require physical, chemical, and possibly biological data on the enantiomers present in a racemate prior to clinical examination. The concerns raised by the document resulted in the formation of the Pharmaceutical Manufacturers Association (PMA) Ad Hoc Committee of Racemic Mixtures (23). The PMA provided an industrial perspective on the regulatory requirements, including examples where a racemate could be developed on a case-by-case basis, and approaches where a single enantiomer drug could be derived from an approved racemate. In 1989 the CDER established a Stereoisomeric Committee with the remit of determining the requirements, if any, which should be imposed on an applicant developing a stereoisomeric drug (24). The FDA's *Policy statement for the development of new stereoisomeric drugs* was first published in January 1992 (25) with correction and modification in January 1997.

Policy Statement for the Development of New Stereoisomeric Drugs

The FDA has taken essentially the same view as the EU with respect to the development of chiral drugs but emphasizes different aspects in its guidance. The FDA policy statement was produced in response to the technological advances that permitted production of single stereoisomers on a commercial scale. The policy relates only to enantiomers and not to geometric isomers, or diastereoisomers, which have chemically and pharmacologically distinct properties. Except in rare cases where biotransformation occurs, such compounds are treated as separate drugs and mixtures are not developed unless fortuitously as a fixed dose combination. The guideline acknowledges that the development of racemates may continue to be appropriate but identifies two areas that should be considered in product development.

The first is the manufacture and control of a product to assure its stereoisomeric composition with respect to identity, strength, quality, and purity. The quantitative composition of the material used in the pharmacological, toxicological, and clinical studies conducted during development must be known.

The second point of consideration is the pharmacokinetic evaluation of a chiral drug. Results from such studies will be misleading if the disposition of the enantiomers is different, unless stereospecific analytical methodology is employed, a point emphasized on a number of occasions by several authors (26,27). Such stereospecific methodology would need to be established for in vivo use early during the drug development process as results from initial pharmacokinetic investigations, including information concerning in vivo enantiomeric interconversion, will inform the decision to develop either a single enantiomer or racemate. If the drug product is to contain a racemate and the pharmacokinetic profiles of the individual isomers are different, appropriate studies should be conducted to measure characteristics such as the dose linearity, the effects of altered metabolism and excretion, and drug-drug interactions for the individual enantiomers. An achiral assay or monitoring of only one enantiomer is acceptable if the pharmacokinetics of the isomers is the same or in a fixed ratio in the target population. The in vivo measurement of individual enantiomers would be of assistance in assessing the results of toxicological studies. However, if this is not possible then human pharmacokinetic studies would be sufficient.

The pharmacological activities of the isomers should be compared in vitro and in vivo in animals and in vivo in humans. Separate toxicological evaluation of the enantiomers would not usually be required when the profile of the racemate was relatively benign but unexpected effects, especially if unusual or near effective doses in animals or near planned human exposure, would warrant further studies with the individual isomers.

The guideline notes that the FDA invites discussion with sponsors on whether to pursue development of the racemate or single enantiomer. This reflects the somewhat different regulatory approach in the United States where there is greater interaction between the FDA and compound sponsor during the drug development process than occurs in Europe (where such discussion is more formalized through Scientific Advice procedures; for example at EMA http://www.ema.europa.eu/ema/index.jsp?curl=pages/regulation/general/general_content_000050.jsp&mid=WC0b01ac05800229bb or MHRA http://www.mhra.gov.uk/Howweregulate/Medicines/

Licensingofmedicines/Informationforlicenceapplicants/Otherusefulservicesandinformation/Scientificadviceforlicenceapplicants/index.htm). All information obtained by the sponsor, or available in the published literature, relating to the chemistry, pharmacology, toxicology, or clinical actions of the stereoisomers should be included in the investigational new drug (IND) or new drug (NDA) submissions.

Chemistry, Manufacturing, and Controls
The policy gives further recommendations on the information that should be provided on the chemistry, manufacturing, and controls (CMC) in addition to that found in other guidance (see section "Other Relevant FDA Guidance").

Methods, specifications, and impurity limits. For drug substances and products, applications for single enantiomers and racemates should include a stereochemically specific test for identity and/or a stereospecific assay. The selection of appropriate control tests should be based on the method of manufacture, stability, and, for the product, its composition. Similarly, methodology for the assessment of the stereochemical integrity of single isomer drugs and products should be included in stability studies. However, such stereospecific analysis may not be required provided that racemization can be shown not to occur.

Stereochemical impurities, together with other related manufacturing impurities and contaminants, in a drug substance are required to be defined prior to use of material in clinical investigations. In addition, the maximal level of impurities, be they isomeric impurities or not, should not exceed those present in the drug substance examined in the preclinical toxicological studies. These issues are addressed in the ICH impurities guideline (see section "ICH Guidelines on Impurities").

Pharmacology and toxicology. The biological activity of the individual enantiomers should be characterized with respect to the principal, and other significant, pharmacological effects in terms of the usual parameters, including potency, specificity, and maximum effect, etc. The pharmacokinetic profile of each enantiomer determined following administration to animals should subsequently be compared to data obtained following administration to healthy volunteers in phase I studies. It is normally sufficient to carry out toxicity studies on the racemate. However, if toxicity is observed, which could not be predicted from the pharmacological evaluation, at low doses/exposure compared to that intended for use in the clinical trials, then the studies should be repeated with the individual stereoisomers to determine if a single enantiomer is responsible, or predominantly responsible, for the adverse effect(s). If this is the case, then it would be desirable to eliminate the toxicity by developing the appropriate single enantiomer with only the desired effect.

Developing a Single Enantiomer After a Racemate Is Studied
Following the preclinical evaluation of a racemic mixture an abbreviated examination of both the pharmacology and toxicology could be carried out to allow the sponsor to apply previously available racemate data to a single pure enantiomer. Additional investigations would not be required if both the single enantiomer and the racemate had the same toxicological profile. However, if

the single enantiomer appeared to be more toxic, then further investigation would be required to produce an explanation and consider the implications for dosing to humans.

Clinical and biopharmaceutical. Where the individual enantiomers and the racemate show little difference in terms of activity and pharmacokinetic profile, the development of the racemate is justifiable. In other instances, the development of the single enantiomer is especially desirable, for example, where one isomer is toxic and the other is not. Cases where unexpected toxicity, or pharmacological effects, occur with clinically relevant doses of the racemate then the mixture should be investigated further with respect to the properties of the individual enantiomers and their potential active metabolites. Such investigations may be undertaken in animals but studies in humans may be required. The unexpected effects may not be directly associated with the parent stereoisomer but may be associated with an enantiomer-specific metabolite. Generally, it is not as important to consider the development of a single stereoisomer if the enantiomer is biologically inert. However, clinical evaluation of both enantiomers and the potential development of a single stereoisomer is of greater significance when both enantiomers are pharmacologically active but differ significantly in their potency, specificity, or maximum effect. If both enantiomers exhibit desirable but qualitatively different biological properties then development of a mixture, not necessarily a racemate, as a fixed combination might be reasonable.

Phase I clinical studies of a racemate should include an examination of the pharmacokinetic profile of the individual enantiomers, including the possibility of in vivo interconversion. Phase I or II investigations in the target population should indicate if an achiral assay, or alternatively, monitoring of only one isomer where a fixed ratio is confirmed, will be adequate for pharmacokinetic evaluation. If the racemate has already been marketed and the sponsor wishes to develop a single stereoisomer, additional studies should include determination of possible enantiomeric interconversion and whether differences in the pharmacokinetic profile occur following administration of the single enantiomer alone or as the racemate.

Other Relevant FDA Guidance

The FDA's *Guideline for submitting supporting documentation in drug applications for the manufacture of drug substances* makes some specific references to chiral drug substances, the requirements of which are similar to those in the EU. Elucidation of the structure of a chiral drug molecule should include determination of its configuration, information supported by knowledge of the synthetic route employed, particularly in relation to the introduction of, and methodology involved in, the introduction of the chiral center (e.g., from the chiral pool, asymmetric synthesis and/or resolution of synthetic intermediates(s)). Stereochemically, pure starting materials may require additional evaluation, compared to achiral starting materials to ensure appropriate configuration and purity. The significance of these requirements is exemplified in a series of publications concerned with the enantiomeric purity of over 270 chiral synthons, catalysts, and auxiliaries, ca. 20% and 5% of these reagents had enantiomeric impurities between 1% and 10% and in excess of 10%, respectively (28–30).

It is also noted in the policy that the unrequired enantiomer may be considered as an impurity (even in racemates) and as such require appropriate control during both manufacture and in the final drug substance. The guidance specifically addresses the issue of key intermediates, those compounds in which the essential molecular characteristics necessary for the desired pharmacological activity are first introduced into the structure. Obviously, for chiral compounds, key intermediates will be those where the chiral center of the required configuration is introduced, and as such they should be subjected to quantitative analysis to limit the content of undesired isomers. The control of drug substances is discussed in the policy but specifications and tests are now addressed in the ICH guidelines (see section "ICH Guidelines on Specifications and Tests"). The requirement for stereochemical characterization of reference materials used during analytical procedures is also noted.

The FDA's *Reviewer guidance on the validation of chromatographic methods* issued in November 1994 also refers to stereospecific methods. The guidance incorporates the ICH analytical terms (see section "ICH Guidelines on Analytical Validation"). It is noted that separation of enantiomers can be achieved by HPLC using chiral stationary phases, or with achiral stationary phases by the use of chiral mobile phase additives, or following derivatization with a unichiral reagent to yield diastereoisomeric derivatives. This latter point, considering the issues associated with the indirect approach to enantiomeric analysis (e.g., the requirement for a suitable derivatizable functional group in the analyte, rapid reproducible derivatization, stereochemically pure unichiral reagent, possible partial racemization of the unichiral reagent and/or the analyte, kinetic resolution, differential detector response), and the requirement for additional validation steps, seems somewhat surprising now. In this context, it is noteworthy that the latest editions (2011) of the United States (USP), British (BP), and European (EP) Pharmacopoeia monographs for determination of the enantiomeric impurity of the carbonic anhydrase inhibitor dorzolamide, the 4S,6S-stereoisomer (Fig. 13.3), involves derivatization with the unichiral derivatizing agent (−)-(S)-α-methylbenzyl isocyanate followed by chromatography using an achiral stationary phase. The enantiomeric impurity, the 4R,6R-stereoisomer, is limited to 0.5% and determined by comparison of chromatographic peak areas. Similarly, all three pharmacopoeias use (+)-(R)-α-methylbenzyl isocyanate followed by achiral chromatography for determination of the stereochemical purity of the antitubercular drug ethambutol [(2S,2′S)-2,2′-(ethylenediimino)dibutan-1-ol]. As a result of its structure, ethambutol exists as a pair of enantiomers and a *meso* form (Fig. 13.4), the activity and associated toxicity of which differ. The BP/EP monograph includes the *meso* form as a specific impurity with a limit of 1.0% by peak area whereas the enantiomer (2R,2′R)-ethambutol is an unspecified

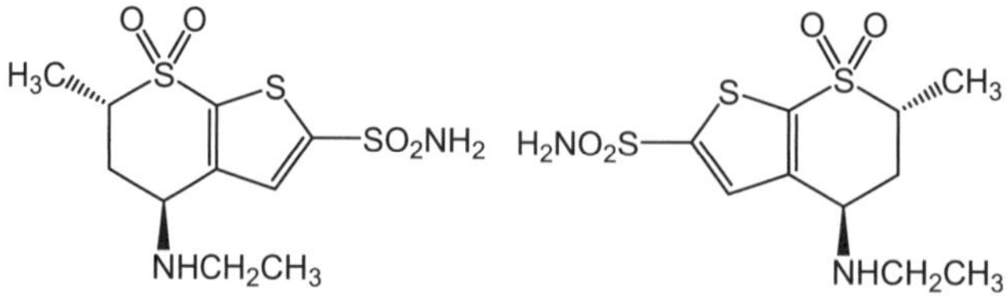

FIGURE 13.3 Structures of (4S,6S)-dorzolamide (*left*) and its 4R,6R-enantiomeric impurity.

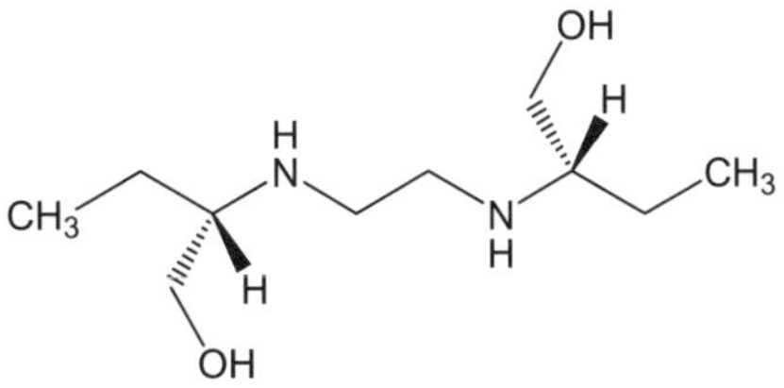

REGULATORY PERSPECTIVE

FIGURE 13.4 Structures of (2S,2'S)-ethambutol and its stereoisomeric impurities.

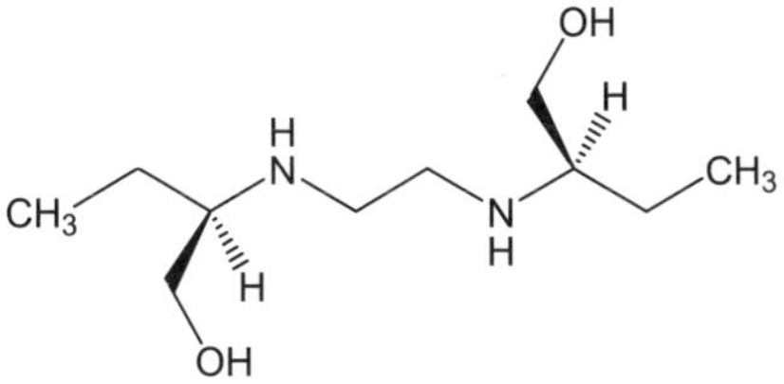

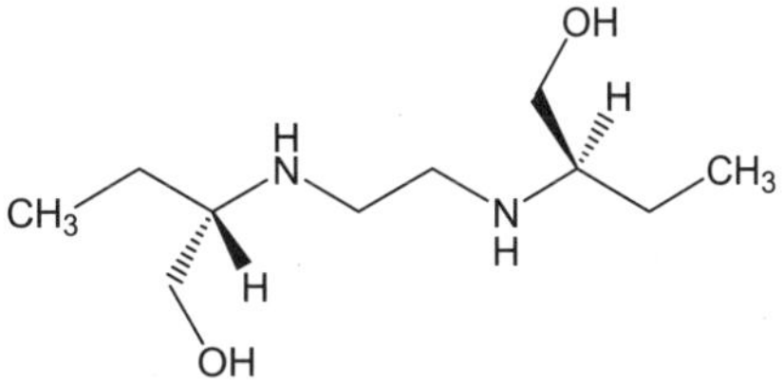

impurity covered by a general limit statement of 0.1% in the related substances test, the total impurity limit for related substances in this test being 1.0% including the *meso* form. In contrast, the USP monograph requires the individual calculation of both stereoisomeric impurities with a total limit of not more than 4.0% but without a specified limit on either individual stereoisomeric impurity. The formation of diastereoisomeric derivatives is uncommon in the USP and both the BP and EP where chiral stationary phases for both HPLC and gas chromatographic analysis are most frequently employed for the determination of enantiomeric purity as a result of the issues outlined above. When a chromatographic method is used as a limit test to determine stereochemical purity, the sensitivity of the methodology is enhanced if the enantiomeric impurity elutes before the drug (to avoid the tail of the main peak). This latter point can obviously be problematic using a CSP particularly those based on natural products, for example, the polysaccharide derivative, cyclodextrin, protein, and macrocyclic antibiotic-based phases.

REQUIREMENTS IN JAPAN
Introduction
The Japanese regulatory authority is the Ministry of Health, Labour and Welfare (MHLW; koseirodosho in Japanese), which was established in January 2001 as a result of a reorganization of government ministries and the merger of the Ministry of Health and Welfare (MHW), originally established in 1938, and the Ministry of Labour. The Pharmaceutical and Food Safety Bureau (PFSB) is one of 11 bureaus of the MHLW and is concerned with policies to assure the efficacy and safety of drugs, quasi-drugs, cosmetics, and medical devices. Additionally the PFSB is involved in policies related to general health including those relating to narcotic and stimulant drugs and blood products and their supply. Essentially the PFSB is concerned with clinical investigations, approval

reviews, licensing, and postmarketing safety. In contrast, the Health Policy Bureau is concerned with the promotion of research and development, production, distribution, and drug pricing policies (31).

The PFSB is divided into five divisions including the Evaluation and Licensing Division, the function of which is to provide guidance and supervision associated with the production of drugs; manufacturing licences and approval to manufacture and market; reexamination and reevaluation of drugs; and related issues. Other divisions include the Safety Division, concerned with policies associated with the assurance and safety of drugs, and the Compliance and Narcotics Division, which is concerned with the control of drug quality, testing and certification of drugs, and associated responsibilities.

An additional agency, the Pharmaceutical and Medical Devices Agency (PMDA; SOGO-KIKO), was established in 2004 by the integration and merger of the Pharmaceutical and Medical Devices Evaluation Center (PMDEC), the National Institutes of Health Sciences, the Organization for Pharmaceutical Safety and Research (OPSR), and part of the Medical Devices Center. The PMDEC was established in 1997 in order to "strengthen approval reviews," whereas the OPSR carried out reviews on application data and consultation services on clinical trial protocols. The formation of the PMDA resulted in the formation of an independent administrative organization, the function of which is to review the safety, efficacy, and quality of drugs for which regulatory approval applications have been submitted based on current scientific and technological standards. The PDMA also undertakes reexamination and reevaluation of drugs and devices and reviews of cell- and tissue-based products prior to first-in-man studies. The PMDA also interacts with the MHLW with respect to improvements in the safety of marketed products. Thus, the PMDA interacts with the three divisions of the PFSB, indicated above in areas associated with their specific functions, for example, the Evaluation and Licensing Division with respect to approval, licence to manufacture and market, the Safety Division with respect to drug safety, and the Compliance and Narcotics Division with regard to manufacturing site inspections (31).

Administration of pharmaceuticals in Japan is governed by numerous laws and regulations the most significant of which is the Pharmaceutical Affairs Law (PAL), the object of which is to improve public health by regulations to assure the quality and efficacy of drugs, quasi-drugs, cosmetics, and medical devices. PAL is also concerned with the promotion of research and development of drugs and medical devices that are particularly essential for health care (31). Modern pharmaceutical legislation in Japan dates from 1889 with the Regulations of Handling and Sales of Medicines; the PAL, of 1943, has been revised a considerable number of times since the 1940s with the revision of 2002 being concerned with safety considerations associated with the development of biotechnology products, genomics, postmarketing surveillance, and revision of the approval and licensing system (31).

Information concerning the regulation of pharmaceuticals and new drug development in Japan is presented in the document Pharmaceutical Administration and Regulations in Japan, which is updated annually by the English Regulatory Information Task Force, International Affairs Committee of the Japanese Pharmaceutical Manufacturers Association (JPMA) (31). This publication contains descriptions of the background to drug regulation, including the structure, functions, and responsibilities of the various government departments and agencies, together with current practice of the regulatory agencies and industry.

An overview of the regulations governing aspects of drug development in Japan has been published by Labbé (32). The article presents an outline of the organization of the Japanese health authorities, including the Pharmaceutical Affairs Bureau and Central Pharmaceutical Affairs Council (CPAC) together with the PAL, and an overview of the regulations governing drug development and associated procedures for new drug approval and postapproval requirements are also addressed.

Development of Chiral Drugs

The MHLW has not issued specific guidance on the development of chiral drugs but has nonetheless responded to the "enantiomer-versus-racemate" scientific debate. The attitude of the MHLW, formerly the MHW, has been discussed in two articles by Shindo and Caldwell published in the early and mid-1990s (33,34). The second of these articles also includes the results of a survey on stereochemical issues from the Japanese pharmaceutical industry. In a more recent publication, Shimazawa et al. (35) reported an analysis of the stereochemistry of drugs introduced in Japan between 1988 and 2002. These authors also carried out a detailed examination of the stereochemical information provided in NDAs submitted between January 2001 and July 2003 using data obtained from a review of the summaries provided in Module 2 of the CTDs. The information and analysis presented not only provide an indication of the change in attitude with respect to the development of chiral drugs over the time period examined but also, as two of the four authors are affiliated to the PMDA, a useful insight into the regulatory expectations.

The first specific reference to stereochemical issues in the Japanese Requirements for Drug Manufacturing Approval was in 1985 with the addition of a statement in the section on *Test data concerning absorption, distribution, metabolism and excretion* to the effect that when the drug concerned is a racemate, it is recommended to investigate the ADME of both enantiomers (33). Such a requirement obviously necessitates enantiomeric resolution and characterization of the properties of the individual enantiomers (36) or as a minimum the development of stereospecific bioanalytical methodologies. The above statement was subsequently extended in the following year with the additional requirements to include reference to the possibility of in vivo chiral inversion and particularly for mixtures of diastereoisomers the requirement to investigate the metabolism and disposition of each isomer and how each stereoisomer contributes to efficacy (33).

A second reference to chiral drugs was an amendment, published in 1989, in the section on *Data concerning physicochemical properties and standards and test methods*, which stated that, for mixtures of optical isomers, it is recommended that chromatographic tests are performed in addition to optical rotation, indicating the response of the MHLW to the developments with respect to chiral stationary phases in HPLC (33).

In their second article concerning the development of chiral drugs in Japan, Shindo and Caldwell (34) make reference to the *Japanese Guidelines for Nonclinical Pharmacokinetic Studies* and the associated documentation of 1991, which essentially restates the above points to the effect that when the drug is a racemate the recommendation is to investigate the ADME of each enantiomer, which may be done by their individual administration or by stereospecific analysis following administration of the racemate; that it is desirable to investigate isomeric interconversion; and that all studies are to be carried out at the current academic research level (34). Further references to chiral drugs have been found in the Japanese guidelines on establishing specifications for new active substances issued in 1994. These indicate that consideration should be given to the solvent used in a test for optical rotation and its effect on the result explained, and second, that where the active ingredient is a single stereoisomer, a method of discriminating between the enantiomers should be investigated and the enantiomeric ratio determined. The ICH Topic Q6A specifications and tests (see section "ICH Guidelines on Specifications and Tests") was officially implemented in Japan in 2001 (35).

Shindo and Caldwell noted in their 1995 article (34) that the Japanese authorities have stated, unlike other regulatory bodies, that they will not publish formal guidelines on the development of chiral drugs and this continues to be the position as noted in the recent publication of Shimazawa et al. (35). However, while the brief official statements provide limited guidance on the investigation of chiral drugs, there is considerable correspondence on individual cases either with individual applicants or pharmaceutical industry associations.

The Pharmaceutical Affairs and Food Sanitation Council (PAFSC), an advisory body to the MHLW, reviews and addresses matters relating to pharmaceuticals and was established by the merger of the CPAC and the Food Sanitation Investigation Council. Questions and responses to the PAFSC, and formally the CPAC, are published and provide insights into detailed requirements for drug approval and offer interpretations of the official guidelines. Shindo and Caldwell (34) provide examples of some responses from CPAC that indicate that the approval of a racemate is not precluded but that the selection of a single stereoisomer for marketing should be based on a consideration of the efficacy and toxicity of each enantiomer. Investigations should be performed on enantiomeric composition, pharmacological effect, metabolism, toxicity, in vivo interconversion, etc. following administration of the racemate and individual enantiomers.

The recent analysis of chiral drug introductions reported by Shimazawa et al. (35), on both single stereoisomers and racemates, using data from the common CTDs includes information concerning manufacturing, quality, pharmacology, toxicology, and pharmacokinetics. During the period under examination, January 2001 to July 2003, 76 new active ingredients were approved, of these 29 were achiral, 37 single stereoisomers, including 14 compounds with a single and 23 with multiple stereogenic centers as part of their structures, and 10 were racemates. The stereochemical composition of all ten racemates was confirmed by optical rotation, nine by stereospecific HPLC and four by X-ray analysis. The principle pharmacodynamic activity of the individual enantiomers was investigated for all ten agents with only "minor" differences in activity being reported for five, whereas four exhibited different "potencies," the relative magnitude of which was not reported but in three of the four cases the increase in toxicity mirrored the increase in pharmacodynamic activity. In one instance the toxicology of the single stereoisomer was similar to that of the racemate and for the remaining compound the stereochemical differences depended on the test system used but in vivo the differences were reported to be small.

Single-dose toxicity studies were carried out on the individual enantiomers of seven of the racemates, of these three resulted in differences that were related to the pharmacodynamic effects and the remaining four exhibited no differences. Similarly pharmacokinetic investigations were carried out in animals and/or man for nine compounds, data on both enantiomers being reported, and the possibility of chiral inversion was investigated in six cases but not observed. No stereochemical pharmacokinetic data was reported for one compound, which was not named, and the lack of information was rationalized by the authors as being due to the fact that the drug had been in common use throughout the world for a number of years (35).

Of the 37 single stereoisomers examined, 29 were produced using optically pure starting materials, 3 by asymmetric synthesis, and the remaining 5 by resolution, 1 via chromatographic methods, and the remainder by crystallization. The stereochemistry of 22 compounds was determined by X-ray crystallography, 4 by comparison with authentic samples, and no data were provided on the remaining 11 compounds. However, this latter result does not necessarily mean that the authorities were not provided with any appropriate data but that the information was not described/addressed in the summaries examined. Specification of the stereochemistry was determined by optical rotation (11 compounds), chromatography (2 compounds) or a combination of both methods (11 compounds), and in 3 instances no specification method was reported. Pharmacokinetic studies were obviously carried out for all compounds examined but only in 12 instances an evaluation relating to chirality was included, possible chiral inversion was examined in ten instances, only one of which indicated inversion taking place in a single test species, and one other compound indicated enantiomer-specific metabolism. Pharmacokinetic data relating to stereochemistry were not reported for the remaining 25 compounds (35).

As pointed out above, ICH-Q6A was officially implemented in Japan in July 2001, however, there was a transitional period from the previous guidelines until July 2003. Thus the information on the compounds examined in the article by Shimazawa et al. (35) is not subject to Q6A and the situation may have varied slightly. However, applicants should have been prepared to follow Q6A requirements as step 4 was attained in 1999. From the above data analysis it is obvious that the development of racemates is not precluded in Japan and it would appear that in the majority of instances for which data was available the regulatory requirements had not changed significantly from those outlined by Shindo and Caldwell (34). It is however noteworthy that in the two and a half-year period for which data were obtained, 37 single stereoisomers were developed compared to 10 racemates and thus the trend in Japan is increasingly toward single stereoisomers (see also section "Impact of Regulatory Guidelines on Chiral Drug Development").

While the Japanese guidelines do not present selection strategies for the development of single stereoisomers or racemates, similarly to other regulatory regions, sponsors are expected to provide scientific justification for the development of a racemate. Obviously given the expenditure involved, the development of pharmaceuticals is, of necessity, a global activity and the development of chiral drugs in Japan is influenced by ICH and United States and EU guidelines. Thus, although there is a lack of formal guidance in Japan, it is apparent that there is a considerable degree of concordance with the regulatory principles established elsewhere.

IMPACT OF REGULATORY GUIDELINES ON CHIRAL DRUG DEVELOPMENT

The global market for pharmaceuticals was estimated to be worth around US$550 billion in the top 10 markets in the year June 2007 to June 2008 [United States ~ $290, Europe (France, Germany, Italy, United Kingdom, Spain) ~ $150, Japan ~ $64, and others, Canada, China, and Brazil ~ $48; figures expressed as billions of U.S. dollars], which amounts to 78% of global sales. The contribution of chiral drugs to these global sales is significant with 5 of the top 10 marketed drugs being chiral, 4 as single stereoisomers, and 1 preparation being a combination of a single isomer and a racemate (Table 13.2); the total sales of these agents accounting for 12% in the top 10 markets. However, the cost of research and development in the industry has increased from ~$8.4 billion in 1990 to ~$48 billion resulting in an average cost of bringing a drug to market of $1.3 billion (37). As a result of such costs there is obviously a need to optimize the drug development process and there is considerable interest in the investigation of druggability and drug likeness—the likelihood of a compound having the characteristics required for successful development (38)—drug property profiles, and structural molecular complexity (39–41). The presence of stereogenic or chiral centers in a molecule is known to contribute to their selectivity of action and the removal of such features results in the introduction of additional flexibility and frequently leads to a reduction in specificity and activity (39). A recent extensive analysis of the progression of compounds from discovery, through development phases 1, 2, and 3 to drugs has indicated the significance of stereogenic centers in drug structure as molecules progress, the fraction of compounds with a center of chirality increasing from ~50% in discovery and phase 1, through ~60% in phase 2, to 64% in phase 3 and as drugs (41). Thus, from the above financial and scientific considerations it is apparent that the significance of drug chirality cannot be over emphasized. In addition, advances in the chemical technologies associated with stereoselective synthesis and stereospecific analysis have facilitated the pharmacological evaluation of the enantiomers of chiral drugs and resulted in an increasing awareness of the potential significance of their biological properties. This increased

TABLE 13.2   Stereochemistry of the Top Ten Marketed Drugs Worldwide (2007–2008)

Drug	Preparation	Nature	Stereochemistry	Sales (US$ billions)
Atorvastatin	Lipitor	Chiral	Single isomer	13.8
Clopidogrel	Plavix	Chiral	Single isomer	8.3
Esomeprazole	Nexium	Chiral	Single isomer	7.7
Fluticasone & Salmeterol	Seretide[a]	Chiral	Single isomer	7.5
		Chiral	Racemate	
Etanercept	Enbrel	Biologic	–	5.6
Quetiapine	Seroquel	Achiral	–	5.1
Olanzapine	Zyprexa	Achiral	–	5.1
Risperidone	Risperdal	Achiral	–	5.0
Infliximab	Remicade	Biologic	–	4.7
Montelukast	Singulair	Chiral	Single isomer	4.6

[a]Marketed as a combination product of the single isomer of fluticasone and racemic salmeterol formulated as either a dry powder for inhalation or an aerosol.

awareness resulted in the regulatory guidelines and attitudes addressed previously and it is of interest to examine the impact of these guidelines in the development of chiral drugs as either racemates or single stereoisomers.

Ariëns et al. (42) carried out an analysis of the stereochemistry of 1675 drugs from a compilation published in 1982. Of these agents, 1200 (72%) were classified as synthetic, the remainder being designated as natural products or semisynthetic (28%). Of the synthetic agents, 480 (29%) were chiral with 58 (~3.5%) and 422 (~25%) being marketed as single stereoisomers and racemates, respectively, that is, 88% of the synthetic chiral agents were marketed as racemates. From these figures it is obvious that drug chirality is not restricted to particular therapeutic groups of drugs but is an across-the-board problem. This extensive survey was the most comprehensive published at that time and, in the present context, may be used as an indication of how the situation with respect to chiral drugs has changed over the past two decades. Since the publication of the Ariëns et al. survey, a number of authors have examined the trends in chiral drug approvals by various regulatory authorities (33,34,43) and used alternative reference sources for the analysis of chiral drugs (44). The most recent of these surveys being the report by Shimazawa et al. (35) on an analysis of the stereochemistry of drugs undergoing approval in Japan between 1988 and 2002 (Table 13.3).

Ariëns et al. (42) extended their original analysis by an examination of the agents introduced over the period 1983 to 1985 using the "To market to market" information in the series *Annual Reports in Medicinal Chemistry* published initially by Academic Press and more recently by Elsevier. During this three-year period a total of 130 compounds were introduced, of which 91 were classified as synthetic, 38 of which were chiral, just 2 (~5%) of which were marketed as single isomers, the remaining 36 (~95%) being racemates. More recently Miller and Ullrich (45) used the same information source to examine new drug introductions in 2003 and 2004. Over this two-year period only one drug, gemifloxacin, was marketed in the United States as a racemate in 2004, the enantiomers being equipotent.

The most recent comprehensive survey of new drug introductions is that carried out by Murakami (46) using mainly the "To market to market" information over the twenty-year period 1985 to 2004, the data being analyzed in four-year periods. During this period 754 new chemical and biological entities were introduced of which 550 were classified as synthetic, of these 313 were chiral with 137 (~44%) racemates and 167 (~53%) as single stereoisomers. A

TABLE 13.3 Chirality of New Drug Introductions in Japan Between 1988 and 2002

Time period	Total	Achiral (%)	Racemates (%)	Single stereoisomers (%)
1988–1990	81	27 (33)	26 (32)	28 (35)
1991–1993	76	23 (30)	26 (34)	27 (36)
1994–1996	66	25 (38)	19 (29)	22 (33)
1997–1999	69	23 (33)	12 (17)	34 (49)
2000–2002	63	21 (33)	7 (11)	35 (55)
1988–2002	355	119 (33.5)	90 (25.4)	146 (41.1)

Source: Adapted from information presented in Ref. 35.

TABLE 13.4 Chirality of New Drug Introductions Between 1985 and 2004

Time period	Total	Synthetic	Chiral	Racemates (%)[a]	Single stereoisomers (%)[a]
1985–1988	204	146	76	50 (65.8)	24 (31.6)
1989–1992	141	104	58	32 (55.2)	21 (36.2)
1993–1996	160	116	67	30 (44.8)	37 (55.2)
1997–2000	136	109	63	22 (34.9)	41 (65.0)
2001–2004	113	75	49	3 (6.1)	44 (89.8)
1985–2004	754	550	313	137 (43.8)	167 (53.4)

[a]The figures presented under Racemates and Single stereoisomers as a percentage of the total number of chiral drugs do not add up to 100% (nor do the total numbers add up to those presented as Chiral) as in the original analysis an additional classification was given as Diastereoisomers. However, from the data presented, it is unclear if the compounds are single stereoisomers or diastereoisomeric mixtures and as such these agents have not been included here. The total number of compounds classified as such over the 20-year period was 9 with a maximum number of 5 introduced between 1989 and 1992 with none between 1993 and 2000.

Source: Adapted from information presented in Ref. 46.

detailed analysis, divided into four-year periods, is presented in Table 13.4. Interestingly, with the exception of the period 1993 to 1996 the total number of introductions decreased, as did the number of synthetic agents, in each four-year period. However, between 1989 and 2000 the total number of chiral compounds as a fraction of the synthetic agents remained relatively constant, between 56% and 58%, and increased to 65%, with only 3 racemates, compared to 44 single stereoisomers being introduced between 2001 and 2004. This situation is very similar to that in Japan where between 1988 and 2002, a total of 355 compounds were approved, 41% of which were as single stereoisomers and 25% racemates; the fraction of racemates decreasing from 32% to 11% whereas the number of single stereoisomers increased, particularly from 1996, to 55% between 2000 and 2002 (35). Again, with the exception of the period 1997 to 1999, the number of new introductions decreased (Table 13.3).

This decrease in new drug approvals between 1985 and 2004, in the United States at least, has continued, in 2008 the U.S. FDA CDER approved 21 new molecular entities (NMEs) and 3 biologic licence applications, which was a modest increase over the previous three-year period (47). Similarly in 2009, 19 and 6 NMEs and biologic applications were approved (48). Examination of the stereochemical nature of the 21 NMEs introduced in 2008 indicates that the trends indicated by the analysis of Murakami (46), together with the observations of Miller and Ullrich (45), have continued. Of the 21 agents approved by the FDA CDER, 7 were achiral and 14 chiral, and of these only 2, desvenlafaxine and tetrabenazine, were developed as racemates, the remaining 12 being single stereoisomers. The corresponding figures for the 19 NMEs approved in 2009 are achiral and racemates 5 of each, 7 single stereoisomers, an equal parts epimeric mixture of the dihydrofolate reductase inhibitor pralatrexate, and coartem, a mixture of the single isomer artemether and racemic lumefantrine (48). Similarly in 2010, four achiral compounds, one racemate and nine single stereoisomers were approved together with a combination oral contraceptive product consisting of a mixture of the single stereoisomers of estradiol valerate and dienogest (49) (Table 13.5).

REGULATORY PERSPECTIVE

TABLE 13.5 Stereochemical Nature of FDA Approved New Molecular Entities between 2008 and 2010

Year (number of NME)	Achiral	Racemate	Single stereoisomer	Others
2008 (21)	Etravirine Bendamustine Iobeguane Rufinamide Eltrombopag Fospropofol Plerixafor	Desvenlafaxine Tetrabenazine	Regadenoson Methylnaltrexone Alvimopan Difluprednate Gadoxetate disodium Clevidipine Silodosin Lacosamide Fesoterodine Tapentadol Gadofosveset Degarelix	
2009 (19)	Febuxostat Benzyl alcohol Iloperidone Dronedarone Pazopanib	Milnacipran Tolvaptan Prasugrel Asenapine (*RS,SR*) Vigabatrin	Everolimus Besifloxacin Bepotastine besilate Telavancin Romidepsin Saxagliptin Pitavastatin	Artemether- lumefantrine (Coartem) Pralatrexate
2010 (15)	Dalfampridine Polidocanol Fingolimed Dabigatran	Alcaftadine	Liraglutide Valaglucerase alfa Carglumic acid Cabazitaxel Ulipristal Lurasidone Ceftaroline fosamil Tesamorelin Eribulin	Estradiol valerate— dienogest (Natazia)

Source: Data from Refs. 47–49.

An informal analysis of new chemical entities reviewed by the U.K. Commission of Human Medicines shows similar trends (Table 13.6): there has been a decrease in overall numbers in the last 10- to 12-year period. Of 14 synthetic products assessed by MHRA in a two-year period 2008–2010, there were 7 achiral and 7 chiral molecules. Three of the latter were racemates and four were single enantiomers. Racemic mixtures were generally justified by equivalent pharmacodynamic and pharmacokinetic properties. The proportions are similar to the period 1996–1998 (43), but lower than in 1998–2000 when there was a much higher proportion of single stereoisomers. It should be noted that these figures do not represent all drugs finally marketed in the United Kingdom as they may not have reached final approval or may have been assessed only by other member states.

From the figures presented above, and the data presented in Tables 13.3 and 13.4, it is obvious that not only is the number of drug introductions decreasing but also that the number of single stereoisomers compared to racemic mixtures is increasing. When the figures presented in Tables 13.3 and 13.4 are considered in relation to the current timescales for drug development, it is also apparent that the pharmaceutical industry was taking the developing

TABLE 13.6 NCEs Assessed in United Kingdom

	NCEs assessed by MCA 1996–1998	NCEs assessed by MCA 1998–2000	NCEs assessed by MHRA 2008–2010
Natural/Semisynthetic	14	6	4
Racemate	1	0	0
Single enantiomer	12	5	4
Achiral	1	1	0
Synthetic	48	27	14
Racemate	11	3	3
	(41%)	(19%)	(43%)
Single enantiomer	16	13	4
	(59%)	(81%)	(57%)
Achiral	21	11	7
Total synthetic and natural/semisynthetic	62	33	18

stereochemical technologies into account and anticipating the regulatory guidelines of the early 1990s. The latter point is obvious as a result of the number of conferences and articles in both the serious (50–55) and popular scientific press in the late 1980s and early 1990s (56–58). This situation is readily illustrated by an examination of the timescales associated with the development of single stereoisomers of two drugs, one marketed originally as a single stereoisomer and another that has undergone the chiral switch process.

The initial synthesis of the racemic proton pump inhibitor omeprazole was achieved in 1979 with the launch of the racemate in 1988 and 1990 in Europe and the United States, respectively (59). However, the first analytical scale resolution of omeprazole was achieved in 1984 and the project, which led to the development of the single stereoisomer esomeprazole, and ultimately resulted in the marketing of the drug in 2000, was started in 1987 (59,60).

The selective serotonin reuptake inhibitor sertraline was originally marketed as a single stereoisomer. Sertraline contains two stereogenic centers in its structure and examination of the in vitro pharmacological activity of the stereoisomers indicated that the *trans*-(+)-1*R*,4*S*-stereoisomer was the most potent compound with respect to inhibition of biogenic amine uptake, but that the *cis*-(+)-1*S*,4*S*-isomer offers the greatest selectivity and the latter compound was selected for development and ultimately marketed as sertraline (Table 13.7) (61,62).

TABLE 13.7 Selectivity of the Stereoisomers of Sertraline for the Inhibition of Biogenic Amine Uptake

Stereoisomer	Inhibitory concentration (IC_{50}, μM)		
	Serotonin	Dopamine	Noradrenaline
trans-(+)-1*R*,4*S*	0.033	0.033	0.011
trans-(−)-1*S*,4*R*	0.45	0.23	0.050
cis-(+)-1*S*,4*S*[a]	0.06	1.1	1.2
cis-(−)-1*R*,4*R*	0.46	0.29	0.38

[a]Sertraline is the *cis*-(+)-1*S*,4*S*-stereoisomer.
Source: Data from Ref. 61.

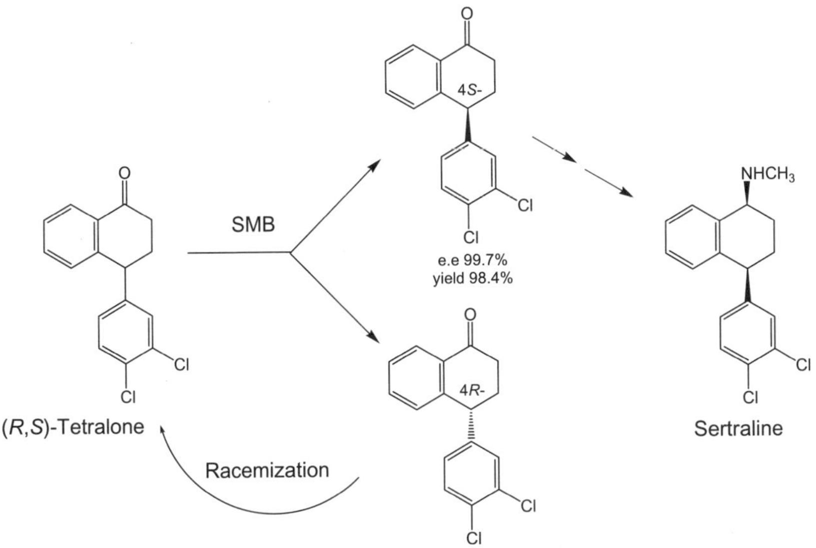

FIGURE 13.5 Combination of simulated moving bed (SMB) technology and synthetic approaches for the synthesis of sertraline.

The original methodology involved the synthesis of the racemate, which was subsequently resolved using D-mandelic acid. This approach was used to produce sufficient material for the initial toxicological evaluation and clinical investigation, and was patented in 1985 (62,63). However, in order to produce commercial-scale quantities of the required stereoisomer, a number of possible approaches were evaluated and ultimately the application of SMB technology for the resolution of a chiral synthetic intermediate followed by additional synthetic chemistry was adopted (Fig. 13.5). The patent for the preparation of the intermediate was published in 2002 (61).

The development of single stereoisomers from previously marketed racemates, the chiral switch, both in terms of commercial advantage and possible retention of market share of high-earning blockbuster drugs near to patent expiration was also apparent relatively early following the expression of regulatory interest. Such developments have been the focus of some criticism directed toward a number of compounds (2,3). However, the development and reevaluation of single isomers from previously market racemates is not entirely without risk, for example, the hepatotoxicity of the β-adrenoceptor dilevalol, the R,R-stereoisomer of the combined α,β-adrenoceptor antagonist labetalol (64,65); the disappointing results of the SWORD trial (Survival With ORal D-sotalol) with (+)-sotalol (66); and the adverse cardiac effects observed on the evaluation of (R)-fluoxetine (67,68). Such "failures" are not without financial risk; for example, it has been reported that the research and development costs of dilevalol were $100 million (64). Similarly, the termination of the development of (R)-fluoxetine resulted in a considerable reduction in the stock value of Sepracor (68). However, in comparison to the costs outlined above associated with the development of new drugs such financial risks are obviously worth serious consideration.

CONCLUDING COMMENT

In this chapter an attempt has been made to provide an overview of the background and current position with respect to the regulations associated with the development of chiral drugs. Since the renewed interest in drug stereochemistry in the mid-1980s, the development of stereochemical technologies and the greater appreciation of the significance of stereochemical considerations in clinical pharmacology, the number of single stereoisomeric drugs has increased considerably. The initial discussions in the "enantiomer-versus-racemate" argument resulted in considerable debate which, in some instances, became somewhat "heated." There are now similar discussions concerned with the relative merits, in therapeutic terms, of a number of agents that have undergone the chiral switch process. However, as a result of the developments in technology the argument has moved to a different level.

The issues, considerations, and requirements associated with the regulation of chiral drugs are now well established. Additionally, the synthetic and analytical methodologies involved in the production and characterization of single isomer chiral compounds are now well understood and advancing at such a pace that the technical challenges of the 1980s are not so daunting. In the future the development of single stereoisomers will increase and racemates will continue to require justification, older racemates will continue to be reevaluated, in some instances for purely academic interest, and possibly be reintroduced as single stereoisomers with "cleaner" pharmacological profiles, and in some instances new indications ultimately resulting in therapeutic benefit.

REFERENCES

1. Gal J. Chiral drugs from an historical point of view. In: Francotte E, Lindner W, eds. Chirality in Drug Research. Weinheim: Wiley-VCH, 2006:3–26.
2. Somogyi A, Bochner F, Foster D. Inside the isomers: the tale of chiral switches. Australian Prescriber 2004; 27:47–49.
3. Mansfield P, Henry D, Tonkin A. Single-enantiomer drugs, elegant science, disappointing effects. Clin Pharmacokin 2004; 43:287–290.
4. Anon. Do single stereoisomer drugs provide value? Therapeutics Letter, June-September 2002, British Columbia Ministry of Health. Available at: http://www.ti.ubc.ca/PDF/45.pdf.
5. Anon. MeReC Bulletin 2004; 14(5), The National Prescribing Centre, UK.
6. Malagueño de Santana FJ, Jabor VAP, Bonato PS. Chiral determination of antidepressant drugs and their metabolites in biological samples. Bioanalysis 2009; 1: 221–237.
7. Rules Governing Medicinal Products in the European Union, European Commission, DG Health & Consumers. Available at: http://ec.europa.eu/health/documents/eudralex/index_en.htm.
8. Scientific Guidelines on Medicinal Products for Human Use; Volume 3 Rules Governing Medicinal Products in the European Union, European Commission, DG Health & Consumers. Available at: http://www.emea.europa.eu/htms/human/humanguidelines/background.htm.
9. Investigation of Chiral Active Substances (originally CPMP/III/3501/91, current reference 3CC29A). Rules Governing Medicinal Products in the European Union, European Commission, DG Health & Consumers. Available at: http://www.ema.europa.eu/docs/en_GB/document_library/Scientific_guideline/2009/09/WC500002816.pdf.
10. Chemistry of Active Substances (Eudra/Q/87/011, current reference 3AQ5a) published in Volume 3 of the Rules, European Commission, DG Health & Consumers.

Available at: http://www.ema.europa.eu/docs/en_GB/document_library/Scientific_guideline/2009/09/WC500002817.pdf.
11. Chemistry of New Active Substances (CPMP/QWP/130/96 Rev 1), European Commission, DG Health & Consumers. Avaialble at: http://www.ema.europa.eu/docs/en_GB/document_library/Scientific_guideline/2009/09/WC500002815.pdf.
12. Summary of Requirements for Active Substances (CHMP/QWP/297/97 Rev 1), European Commission, DG Health & Consumers. Available at: http://www.ema.europa.eu/docs/en_GB/document_library/Scientific_guideline/2009/09/WC500002813.pdf.
13. European Pharmacopoeia. Technical Guide for the Elaboration of Monographs, 5th ed. 2010. Available at: http://www.edqm.eu/medias/fichiers/NEW_Technical_Guide_for_the_Elaboration_of_Monogra.pdf.
14. International Conference on Harmonisation. Available at: http://www.ich.org.
15. EMA. Available at: http://www.emea.europa.eu/htms/human/ich/background.htm.
16. FDA Centre for Drug Evaluation and Research. Available at: http://www.fda.gov/Drugs/GuidanceComplianceRegulatoryInformation/Guidances/ucm121568.htm.
17. Pharmaceutical and Medical Devices Agency (PMDA). Available at: http://www.pmda.go.jp/ich/ich_index.html.
18. ICH Topic Q6A Specifications: Test procedures and acceptance criteria for new drug substances and new drug products: chemical substances. Available at: http://www.ich.org/products/guidelines/quality/article/quality-guidelines.html.
19. ICH Guidelines: Topic Q3A Impurities in new drug substances; Topic Q3B Impurities in new medicinal products. Available at: http://www.ich.org/products/guidelines/quality/article/quality-guidelines.html.
20. ICH Topic Q2 (CPMP/ICH/381/95); ICH Topic Q2A Validation of analytical procedures: definitions and terminology; ICH Topic Q2B Validation of analytical procedures: methodology. Available at: http://www.ich.org/products/guidelines/quality/article/quality-guidelines.html.
21. FDA website: http://www.fda.gov/Drugs/GuidanceComplianceRegulatoryInformation/Guidances/default.htm.
22. FDA (1987). Guideline for submitting documentation in drug applications for the manufacture of drug substance. Rockville, MD.
23. Gross M. Enantioselective analysis and the regulation of chiral drugs. In: Aboul-Enein HY, Wainer IW, eds. The Impact of Stereochemistry on Drug Development and Use. New York: John Wiley, 1997:565–572.
24. Laganière, S. Current regulatory guidelines of stereoisomeric drugs: North American, European and Japanese point of view. In: Aboul-Enein HY, Wainer IW, eds. The Impact of Stereochemistry on Drug Development and Use. New York: John Wiley, 1997:545–564.
25. FDA. Policy statement for the development of new stereoisomeric drugs. Chirality 1992; 4:338–340.
26. Ariëns EJ. Stereochemistry, a basis for sophisticated nonsense in pharmacokinetics and clinical pharmacology. Eur J Clin Pharmacol 1984; 26:663–668.
27. Evans AM, Nation RL, Sansom LN, et al. Stereoselective drug disposition: potential for misinterpretation of drug disposition data. Br J Clin Pharmacol 1988; 26:771–780.
28. Armstrong DW, Lee JT, Chang LW. Enantiomeric impurities in chiral catalysts, auxiliaries and synthons used in enantioselective synthesis. Tetrahedron Asymmetry 1998; 9:2043–2064.
29. Armstrong DW, He L, Yu T, et al. Enantiomeric impurities in chiral catalysts, auxiliaries, synthons and resolving agents. Part 2. Tetrahedron Asymmetry 1999; 10:37–60.
30. Huang K, Breitbach ZS, Armstrong DW. Enantiomeric impurities in chiral synthons, catalysts and auxiliaries. Part 3. Tetrahedron Asymmetry 2006; 17:2821–2832.
31. Pharmaceutical Administration and Regulations in Japan 2010. English Regulatory Information Task Force, Japan Pharmaceutical manufacturers Association. Available at: http://www.jpma.or.jp/about/issue/gratis/index2.html (Japanese); http://www.jpma.or.jp/english/parj/1003.html (English).

32. Labbé E. Japanese regulations. In: Flectcher AJ, Edwards LD, Fox AW, et al., eds. Principles and Practice of Pharmaceutical Medicine. Chichester: John Wiley, 2002:307–324.
33. Shindo H, Caldwell J. Regulatory aspects of the development of chiral drugs in Japan: a status report. Chirality 1991; 3:91–93.
34. Shindo H, Caldwell J. Development of chiral drugs in Japan: an update on regulatory and industrial opinion. Chirality 1995; 7:349–352.
35. Shimazawa R, Nagai N, Toyoshima S, et al. Present state of new chiral drug development and review in Japan. J Health Sci 2008; 54:23–29.
36. Agranat I, Wainschtein SR. The strategy of enantiomer patents of drugs. Drug Discovery Today 2010; 15:163–170.
37. Alex AA, Storer RI. Drugs and their structural motifs. In: Smith DA, ed. Metabolism, Pharmacokinetics and Toxicity of Functional Groups: Impact of Chemical Building Blocks on ADMET. London: Royal Society of Chemistry, 2010:1–60.
38. Vistoli G, Pedretti A, Testa B. Assessing drug-likeness—what are we missing? Drug Discovery Today 2008; 13:285–294.
39. Henkel T, Brunne RM, Muller H, et al. Statistical investigation of structural complementarity of natural products and synthetic compounds. Angew Chem Int Ed 1999; 38:643–647.
40. Feher M, Schmidt JM. Property distributions: differences between drugs, natural products and molecules from combinatorial chemistry. J Chem Inf Comput Sci 2003; 43:218–227.
41. Lovering F, Bikker J, Humblet C. Escape from flatland: increasing saturation as an approach to improving clinical success. J Med Chem 2009; 52:6752–6756.
42. Ariëns EJ, Wuis EW, Veringa EJ. Stereoselectivity of bioactive xenobiotics. A pre-Pasteur attitude in medicinal chemistry, pharmacokinetics and clinical pharmacology. Biochem Pharmacol 1988; 37:9–18.
43. Shah RR, Branch SK. Regulatory requirements for the development of chirally active drugs. In: Eichelbaum M, Testa B, Somogyi A, eds. Stereochemical Aspects of Drug Action and Disposition. Berlin: Springer-Verlag, 2003:379–399.
44. Millership JS, Fitzpatrick A. Commonly used chiral drugs: a survey. Chirality 1993; 5:573–576.
45. Miller CP, Ullrich JW. A consideration of the patentability of enantiomers in the pharmaceutical industry in the United States. Chirality 2008; 20:762–770.
46. Murakami H. From racemates to single enantiomers—chiral synthetic drugs over the last 20 years. In: Sakai K, Hirayama N, Tamura R, eds. Novel Optical Resolution Technologies, Topics in Current Chemistry. Berlin: Springer-Verlag, 2007; 269:273–299.
47. Hughes B. 2008 FDA drug approvals. Nat Rev Drug Discovery 2009; 8:93–96.
48. Hughes B. 2009 FDA drug approvals. Nat Rev Drug Discovery 2010; 9:89–92.
49. Mullard A. 2010 FDA drug approvals. Nat Rev Drug Discovery 2011; 10:82–85.
50. De Camp WH. The FDA perspective on the development of stereoisomers. Chirality 1989; 1(1):2–6.
51. Cayen MH. Racemic mixtures and single stereoisomers: industrial concerns and issues in drug development. Chirality 1991; 3:94–98.
52. Nation RL. Chirality in new drug development. Clinical pharmacokinetic considerations. Clin Pharmacokin 1994; 27:249–255.
53. Caldwell J. St Mary's discussion forum: racemates and enantiomers: scientific and regulatory aspects. Chirality 1989; 1:249–250.
54. Hutt AJ. Drug chirality: impact on pharmaceutical regulation. Chirality 1991; 3:161–164.
55. Gross M, Cartwright A, Campbell B, et al. Regulatory requirements for chiral drugs. Drug Info J 1993; 193:453–457.
56. Mason S. The left hand of nature. New Sci 1984; 101:10–14.
57. Matteson D. Through the chemical looking glass. New Sci 1991; 132:35–39.
58. Amato I. Looking glass chemistry. Science 1992; 256:964–966.

59. Olbe L, Carlson E, Lindberg P. A proton-pump inhibitor expedition: the case histories of omeprazole and esomeprazole. Nat Rev Drug Discovery 2003; 2:132–139.
60. Federsal H-J. Facing chirality in the 21st century: approaching the challenges in the pharmaceutical industry. Chirality 2003; 15:S128–S142.
61. Koe BK, Weissman A, Welch WM, et al. Sertraline, 1S,4S-N-methyl-4-(3,4-dichlorophenyl)-1,2,3,4-tetrahydro-1-naphthylamine, a new uptake inhibitor with selectivity for serotonin. J Pharmacol Expt Ther 1983; 226:686–700.
62. Welch WM, Kraska AR, Sarges R, et al. Nontricyclic antidepressant agents derived from *cis* and *trans*-1-amino-4-aryltetralins. J Med Chem 1984; 27:1508–1515.
63. Quallich GJ. Development of the commercial process for Zoloft®/Sertraline. Chirality 2005; 17:S120–S126.
64. Anon. Schering-Plough withdraws dilevalol. Scrip Number 1990; 1540:24.
65. Shah RR, Midgley JM, Branch SK. Stereochemical origin of some clinically significant drug safety concerns: lessons for future drug development. Adverse Drug React Toxicol Rev 1998; 17:145–190.
66. Waldo AL, Camm AJ, de Ruyter H, et al. Effect of d-sotalol on mortality in patients with left ventricular dysfunction after recent and remote myocardial infarction. Lancet 1996; 348:7–12.
67. Anon. Side effects kill "new Prozac." Chem Br 2000; 36:11.
68. Thayer A. Eli Lilly pulls the plug on prozac isomer drug. Chem Eng News 2000; October 30:8.

14 Molecular analysis of agonist stereoisomers at β_2-adrenoceptors

Roland Seifert and Stefan Dove

β_xAR-MEDIATED SIGNAL TRANSDUCTION

β_x-adrenergic receptors (β_xARs) are prototypical G-protein-coupled receptors (GPCRs), that is, receptors that possess seven transmembrane (TM) domains, three extracellular and three intracellular loops with the N-terminus being localized extracellularly and the C-terminus being localized intracellularly (1–3). GPCRs interact with heterotrimeric G-proteins that act as signal transducers to regulate the activity of cellular effector systems (4,5). There are three β_xAR subtypes, referred to as β_1AR, β_2AR, and β_3AR (1,6). The β_1AR is predominantly localized in the heart to mediate positive inotropic, chronotropic, and dromotropic effects, while the β_2AR is broadly expressed (1,6). Activation of β_2ARs with selective agonists is of relevance for the treatment of bronchial asthma and preterm labor and potentially for the treatment of heart failure (7–10). The β_3AR is mainly expressed in fat cells and mediates lipolysis.

The G-protein cycle exemplified for the β_2AR (11) is shown in Figure 14.1. Binding of an agonist to the β_2AR induces a conformational change in the receptor. The agonist-bound β_2AR then interacts with the GDP-liganded G_s-protein (stimulatory G-protein of adenylyl cyclase). The G_s-protein consists of three subunits, that is, the name-giving $G_{s\alpha}$ subunit as well as a β- and γ-subunit that form a tight complex under physiological conditions (4,5). The agonist-bound β_2AR promotes dissociation of GDP from $G_{s\alpha}$, that is, the receptor acts as guanine nucleotide exchange factor (GEF). Subsequently, a ternary complex consisting of agonist, β_2AR and nucleotide-free G-protein forms (11). This complex possesses high agonist affinity and facilitates the binding of GTP to $G_{s\alpha}$. GTP-binding to $G_{s\alpha}$ causes a substantial conformational change in this G-protein subunit, resulting in the disruption of the ternary complex and dissociation of the G_s-protein heterotrimer into $G_{s\alpha}$ and the $\beta\gamma$-complex. GTP-bound $G_{s\alpha}$ activates the effector protein adenylyl cyclase (AC). AC catalyzes the conversion of ATP into the second messenger cAMP that subsequently activates downstream proteins such as cAMP-dependent protein kinase and cAMP-regulated cation channels. Termination of G_s-protein activation is achieved by the GTPase activity of $G_{s\alpha}$, cleaving GTP into GDP and inorganic phosphate (P_i). GDP-bound $G_{s\alpha}$ and the $\beta\gamma$-complex reassociate, closing the G-protein cycle. All β_xARs couple to G_s-proteins. Additionally, the β_2AR can also couple to inhibitory G-protein of adenylyl cyclase (G_i-proteins), resulting in activation of the mitogen-activated protein kinase pathway (12–14).

Methods have been developed to assess various steps of the G-protein cycle. The corresponding experimental protocols are described in detail elsewhere (11) and are, therefore, only briefly mentioned in this chapter. In intact cells, cAMP formation or the analysis of distal cAMP-dependent cell functions including gene expression can be assessed. In membrane preparations, allowing for the effective manipulation of the experimental conditions, the direct analysis of GDP dissociation is not feasible in most systems. In the absence of GTP, the

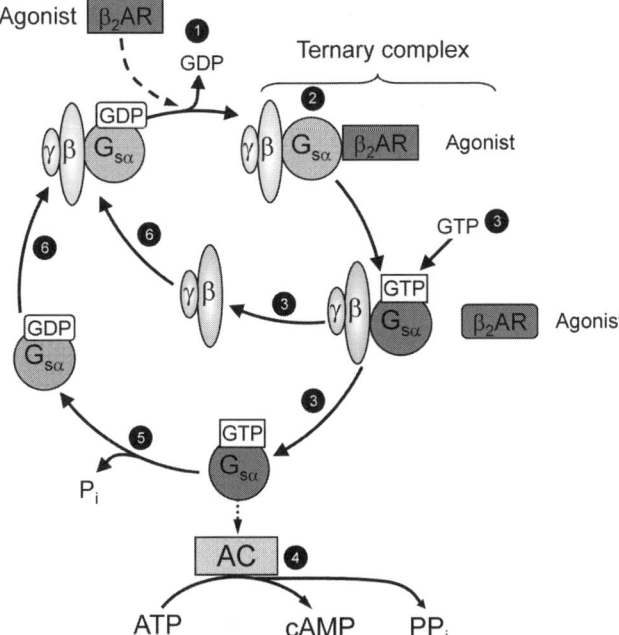

FIGURE 14.1 β_2AR-G_s-protein coupling. The G-protein activation/deactivation cycle is depicted. The numbers designate the individual steps of the cycle. 1, Agonist-catalyzed GDP dissociation. 2, Ternary complex formation (high-affinity agonist binding). 3, GTP binding with subsequent disruption of the ternary complex and dissociation of the G-protein subunits into $G_{s\alpha}$-GTP and the $\beta\gamma$-complex. 4, AC activation. 5, GTP hydrolysis. 6, Re-association of G-protein subunits.

ternary complex can be assessed in radioligand binding studies by determining high-affinity agonist binding. A prerequisite for such studies is that the system studied possesses a sufficiently large G-protein concentration or close GPCR/G-protein proximity. In general, high-affinity agonist binding is GTP sensitive, that is, GTP abrogates high-affinity binding. The binding of GTP to $G_{s\alpha}$ can be monitored by using the high-affinity ligand [^{35}S]GTPγS (guanosine 5'-[γ-thio] triphosphate) that is resistant to cleavage by the GTPase. The assessment of AC activity is a classic readout for determining β_xAR-G_s interaction and uses [α-^{32}P]ATP as substrate with subsequent solid phase separation of [α-^{32}P]ATP and the formed [^{32}P]cAMP. In systems with sufficiently tight receptor/G-protein coupling, the β_xAR-stimulated steady-state high-affinity GTPase activity can be determined. The advantage of this assay is that it monitors ligand/receptor interactions at a very proximal level, avoiding bias in data interpretation caused by the limited availability of AC molecules (15).

MODELS OF RECEPTOR ACTIVATION

Over the past decades, several models of GPCR activation have been developed and discussed elsewhere in substantial detail (16–21). For the purposes of this chapter, the two-state model versus the multistate model will be briefly

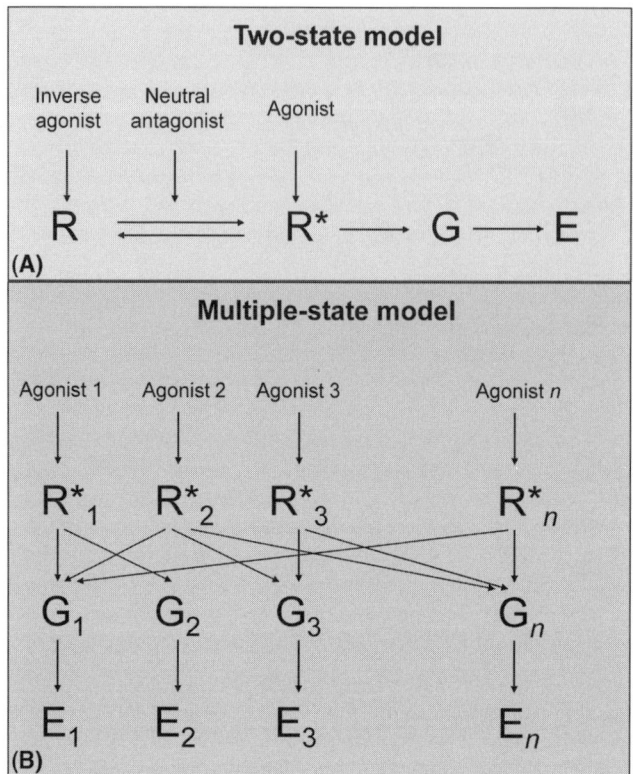

FIGURE 14.2 Models of receptor activation. (**A**) Two-state model of receptor activation assuming isomerization of the receptor from an inactive (R) state to an active (R*) state. (**B**) Multiple-state model assuming the existence of multiple (1-*n*) agonist-selective conformations that differ from each other in their efficacy at activating various downstream G-proteins and effectors. *Abbreviations*: G, G-protein; E, effector.

discussed. In the two-state model, it is assumed that receptors exist in two different states, that is, an inactive (R) state and an active (R*) state (Fig. 14.2A). Agonists stabilize the R* state and, thereby, promote G-protein and effector activation. The two-state model is particularly useful for explaining the agonist independent, that is, constitutive receptor activation (21). Specifically, in the two-state model it is assumed that receptors undergo R/R* isomerization not only in the presence of an agonist but also in the absence of an agonist. This constitutive R/R* isomerization gives rise to a basal G-protein and effector activity and has been observed for many wild-type GPCRs including the β_2AR (21). Inverse agonists are counterparts of agonists and stabilize the R state of GPCRs, thereby reducing basal G-protein and effector activity. Partial agonists are less effective than agonists at stabilizing the R* state, whereas partial inverse agonists are less effective than full inverse agonists at stabilizing the R state. Neutral antagonists do not alter the equilibrium between R and R* and, accordingly, do not change basal G-protein and effector activity.

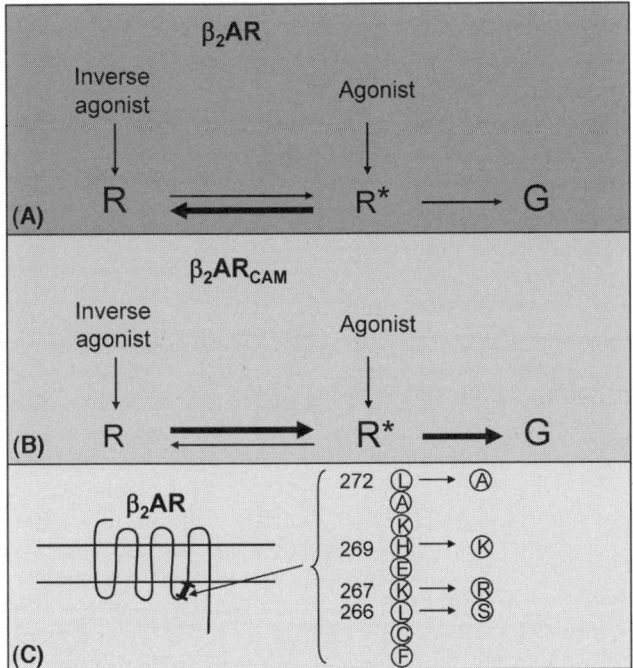

FIGURE 14.3 Constitutive activity of wild-type β_2AR and a constitutively active β_2AR mutant. (**A**) The β_2AR exhibits significant constitutive activity. (**B**) The β_2AR_{CAM} mutant possesses an increased constitutive activity compared to the wild-type β_2AR, that is, the equilibrium between R and R* is shifted toward R*. This change in equilibrium results in higher basal G-protein activity, increased inverse agonist efficacy, increased partial agonist efficacy, and increased agonist-affinity and agonist potency in functional assays. (**C**) Location of the four amino acid exchanges in β_2AR_{CAM} in the C-terminal end of the third intracellular loop.

Although constitutive activity is a property of several wild-type GPCRs, this physiological constitutive activity can be further enhanced by mutations (21). This concept for the β_2AR is illustrated in Figure 14.3. The β_2AR is a prototypical wild-type GPCR exhibiting constitutive activity (Fig. 14.3A). In a constitutively active mutant of the β_2AR, referred to as β_2AR_{CAM} (15), this constitutive activity is increased, resulting in a shift of the R/R* equilibrium toward R* (Fig. 14.3B). An important manifestation of this change in the R/R* equilibrium is an increased basal G-protein activity and increased inverse agonist efficacy at the β_2AR_{CAM} mutant (15). The structural basis for the increased constitutive activity in the β_2AR_{CAM} mutant is an exchange of four amino acids in the C-terminal portion of the third intracellular loop of the β_2AR (Fig. 14.3C).

While the two-state model is very useful at explaining basic concepts of receptor activation and constitutive activity, this model cannot explain all experimental observations. Specifically, over the past 15 years, pharmacological, biochemical, and biophysical studies revealed that numerous agonists at several receptors actually exhibit multiple efficacies that depend on the particular

G-protein and effector examined (16–21). Accordingly, a multiple-state model has been postulated (Fig. 14.2B). This model assumes that many ligands stabilize unique and ligand-specific receptor conformations that exhibit different efficacies at activating downstream G-proteins and effectors. This concept is also referred to as "functional selectivity" and offers numerous opportunities for drug development, that is, ligands that exhibit unique profiles at stimulating a subset of possible cellular responses.

OVERVIEW ON THE INTERACTION OF AGONIST STEREOISOMERS WITH β_XARs

The endogenous ligands for β_xARs are the neurotransmitter norepinephrine and the hormone epinephrine. Chemically, norepinephrine and epinephrine are catecholamines that bear an asymmetric carbon atom, that is, they are chiral compounds. Because of stability problems, light sensitivity and very limited receptor subtype selectivity among α-adrenoceptors and βARs, in most experimental studies, the epinephrine derivative isoproterenol is used. Isoproterenol possesses selectivity for β_xARs relative to α-adrenoceptors. Like epinephrine and norepinephrine, isoproterenol bears an asymmetric carbon atom (the β-carbon atom), and accordingly, two isoproterenol stereoisomers, that is, (−)-(R)-isoproterenol and (+)-(S)-isoproterenol, exist (Fig. 14.4). The nomenclature of stereoisomers is being dealt with in chapter 2 of this book. Isoproterenol isomers should be referred to as (−)-(R)-isoproterenol and (+)-(S)-isoproterenol, but this recommendation is actually not generally followed in the literature. Specifically, (−)-(R)-isoproterenol is also termed (−)-isoproterenol, l-isoproterenol, (L)-isoproterenol or (R)-isoproterenol. Conversely, (+)-(S)-isoproterenol is also referred to as (+)-isoproterenol, d-isoproterenol, (D)-isoproterenol, or (S)-isoproterenol. Evidently, the very inconsistent use of the isoproterenol stereoisomer nomenclature in the literature (15,22–24) is confusing for the nonexpert and has also lead to apparent mix-up of stereoisomers in publications (25).

It has been known for a century that (−)-(R)-stereoisomers of epinephrine and norepinephrine possess a higher potency in biological systems than the corresponding (+)-(S)-stereoisomers (26). The (−)-(R)-stereoisomers of catecholamines are, according to a suggestion by E. J. Ariëns, referred to as the "pharmacologically active" eutomers, and the (+)-(S)-stereoisomers as "pharmacologically inactive" distomers. However, while these definitions of the terms "eutomer" and "distomer" are commonly used in the literature (22,27),

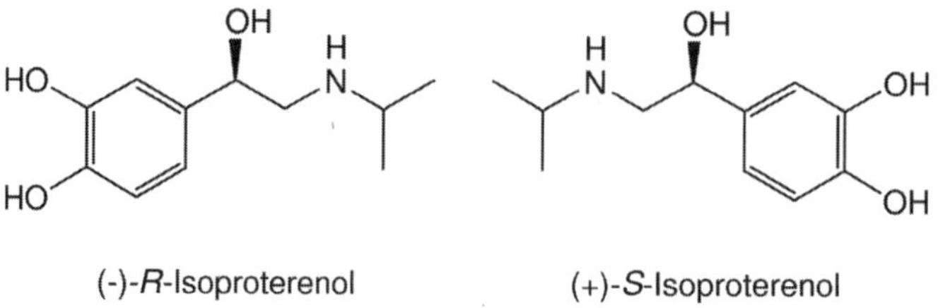

(-)-*R*-Isoproterenol   (+)-*S*-Isoproterenol

FIGURE 14.4 Structures of isoproterenol stereoisomers. Shown are the structures of the pharmacologically more potent "eutomer" (−)-(*R*)-isoproterenol and the pharmacologically less potent "distomer" (+)-(*S*)-isoproterenol. The β-carbon atom bearing a hydroxyl group critically involved in hydrogen bonding with the β_2AR (see Fig. 14.9) constitutes the chirality center.

the terms are actually obscuring rather than clarifying issues. Specifically, in several biological systems, the distomers exhibit important biological effects (14,15,28,29).

Patil et al. (30) have presented an excellent overview on chiral adrenoceptor ligands in general. Here, we will focus on stereoisomer agonists at β_xARs, specifically the β_2AR. Experiments with intact rat heart atrium, expressing β_1AR, and rat uterus, expressing β_2AR, revealed that $(-)$-(R)-isoproterenol is about 1000-fold more potent at eliciting functional responses than $(+)$-(S)-isoproterenol (31). Both isoproterenol stereoisomers are similarly efficacious agonists in both these systems. Thus, this example illustrates that the original definition by E. J. Ariëns of the term eutomer as pharmacologically active compound and the distomer as pharmacologically inactive compound is misleading. It is more correct to designate the eutomer as the "more potent" stereoisomer, and the distomer as the "less potent" stereoisomer. Similarly, at the turkey βAR in erythrocyte membranes, $(-)$-(R)-isoproterenol is about 1000-fold more potent than $(+)$-(S)-isoproterenol, but also a full agonist with regard to AC activation (32). At the purified turkey erythrocyte βAR reconstituted with G_s in phospholipid vesicles, $(-)$-(R)-isoproterenol is about 100-fold more potent than $(+)$-(S)-isoproterenol (33). Since in the latter study, the concentration-response curve for $(+)$-(S)-isoproterenol was not extended to saturation, a statement about efficacy differences between the stereoisomers cannot be made. However, from the studies with the turkey βAR (32,33), it is clear that the affinity of βARs for ligand stereoisomers is not an invariable property but a property that depends on the specific experimental conditions.

At the recombinant human β_2AR expressed in human embryonic HEK293 cells, $(-)$-(R)-isoproterenol possesses about 40-fold higher affinity than $(+)$-(S)-isoproterenol (34). The potency difference between the two stereoisomers with respect to AC activation is about 140-fold. Most intriguingly and in contrast to observations made for the turkey βAR (31,32), $(+)$-(S)-isoproterenol is clearly only a partial agonist at the human β_2AR expressed in Chinese hamster ovary (CHO) cells (34). These findings indicate that the two isoproterenol stereoisomers do not only differ from each other in potency and affinity but that they actually stabilize distinct β_2AR conformations. It is noteworthy that for the $(-)$-(R)- and $(+)$-(S)-stereoisomers of epinephrine and norepinephrine, similar differences in efficacy were observed as for isoproterenol stereoisomers (30).

So far, little is known about the molecular basis for the differential interaction of isoproterenol stereoisomers with β_xARs. At the human β_2AR, the exchange of Asn-293 against Leu-293 reduces the isoproterenol stereoselectivity of the receptor from 38-fold to 6-fold in terms of affinity and to 13-fold in terms of potency for AC activation (34). However, the reasonable hypothesis that the mutation abrogates a hydrogen bond between Asn-293 and the β-hydroxyl group of agonistic phenethanolamines (34) has not been verified by the recently resolved structures of β_xARs (see below).

Isoproterenol is a nonselective β_xAR agonist. Hence, its clinical use is very limited. In marked contrast, selective β_2AR agonists are of substantial clinical importance for short-term and long-term treatment of asthma and prevention of premature labor (8,9). Most interestingly, recent data indicate that selective β_2AR agonists may be beneficial drugs in the treatment of heart failure, complementing the well-established usefulness of β_1AR antagonists (10). Intriguingly, clinically used β_2AR agonists such as clenbuterol, albuterol, fenoterol, and formoterol are also stereoisomeric compounds (28,29,35–37). The clinically used drugs are actually racemic mixtures of the stereoisomers and there is an ongoing and still unresolved discussion whether or not the pure eutomer possesses any advantages compared to the racemic mixture (8,9).

Importantly, the differences between stereoisomers of β_2AR agonists go well beyond differences in potency and moderate differences in efficacy. For example, the pharmacologically active eutomer (R)-albuterol acts, expectedly, as β_2AR agonist and exhibits beneficial anti-inflammatory effects in asthma (28,29). In contrast, the distomer (S)-albuterol is not simply pharmacologically inactive but, rather, the compound exhibits undesired proinflammatory effects. The molecular basis for the adverse effects of (S)-albuterol is as yet unknown but it has been hypothesized that (S)-albuterol may either act as inverse β_2AR agonist (Fig. 14.3) or muscarinic receptor agonist (28,29). However, evidence for or against either hypothesis is not yet available since studies with (S)-albuterol and recombinant β_2AR or muscarinic receptors are still missing. This is an important issue since in native cell systems, it cannot be excluded that differences in metabolism account for the differential pharmacological effects of the ligand stereoisomers. Specifically, it is well known that various catecholamine-metabolizing enzymes exhibit marked selectivity for catecholamine stereoisomers (22,36).

The pharmacological effects of fenoterol stereoisomers are particularly intriguing. Fenoterol possesses two asymmetric carbon atoms, that is, there are four stereoisomers, referred to as (R,R)-, (R,S)-, (S,R)-, and (S,S)-fenoterol (37). The effects of fenoterol have been particularly well studied in cardiomyocytes. In isolated rat cardiomyocytes, the β_2AR activates both G_s- and G_i-proteins as assessed by the enhancing effects of pertussis toxin, uncoupling GPCRs from G_i-proteins (4,5), on positive inotropic responses induced by β_2AR agonists such as albuterol and zinterol (12,13). Noteworthy, the responses to fenoterol were not enhanced by pertussis toxin, indicating that this ligand may stabilize a specific β_2AR conformation (13). However, those early studies had been performed with racemic fenoterol but not with purified fenoterol stereoisomers. A seminal follow-up study then revealed that fenoterol stereoisomers actually differentially activate G_s- and G_i-proteins (14). Specifically, the more potent eutomer (R,R)-fenoterol activates only G_s-proteins, whereas the less potent distomer (S,R)-fenoterol is less efficient than (R,R)-fenoterol at activating G_s-proteins but quite efficient at stimulating G_i-proteins. This is a very striking example how misleading the terms eutomer and distomer are, since in the cardiomyocyte system, the so-called distomer exhibits actually more interesting pharmacological effects than the eutomer. These data indicate that (R,R)-fenoterol and (S,R)-fenoterol stabilize distinct active β_2AR conformations with different efficacies at activating G_s- and G_i-proteins and, thereby, show functional selectivity. However, relatively few studies have been performed with defined agonist stereoisomers using recombinant β_xARs. Such studies would be most important in order to exclude contributions of other receptors than β_xARs and differences in metabolism to the different pharmacological effects of ligand stereoisomers.

Therefore, a suitable recombinant expression system was characterized that would allow for the systematic analysis of agonist (and inverse agonist) stereoisomers at β_xARs. The system described below can be readily implemented in any modern pharmacological laboratory and does not require highly specialized expertise such as receptor purification skills or sophisticated instruments such as ultrasensitive fluorescence spectroscopy equipment for monitoring conformational changes.

$β_XAR$-$G_{Sα}$ FUSION PROTEINS AS MODEL SYSTEMS FOR THE ANALYSIS OF LIGAND STEREOISOMERS

The GPCR-Gα fusion protein technique was originally introduced by Bertin et al. (38) and subsequently developed by several laboratories including the group of Milligan (39) and the Seifert (40). In GPCR-Gα fusion proteins, through recombinant DNA technology (41), the open reading frame of a GPCR is directly connected to the open reading frame of the Gα subunit. The fusion proteins are efficiently translocated to the plasma membrane, folded properly and functionally active.

In the laboratory of the Seifert group, all fusion proteins are N-terminally tagged with a FLAG epitope, and the proteins also contain a 6His tag between the GPCR C-terminus and the Gα N-terminus. Fusion proteins of the human $β_1AR$ and $β_2AR$ were constructed with both the short and long splice variants of $G_{sα}$ (41). The two-dimensional structure of $β_2AR$-$G_{sα}$ fusion proteins is shown schematically in Figure 14.5. Moreover, a fusion protein of the $β_2AR_{CAM}$ mutant with the long splice variant of $G_{sα}$ was prepared (15). The epitope tags are very useful for sensitive immunological detection of fusion proteins, enhance stability of the proteins, and could also be used for protein purification. There is no evidence that the epitope tags interfere with the pharmacological properties of the receptors.

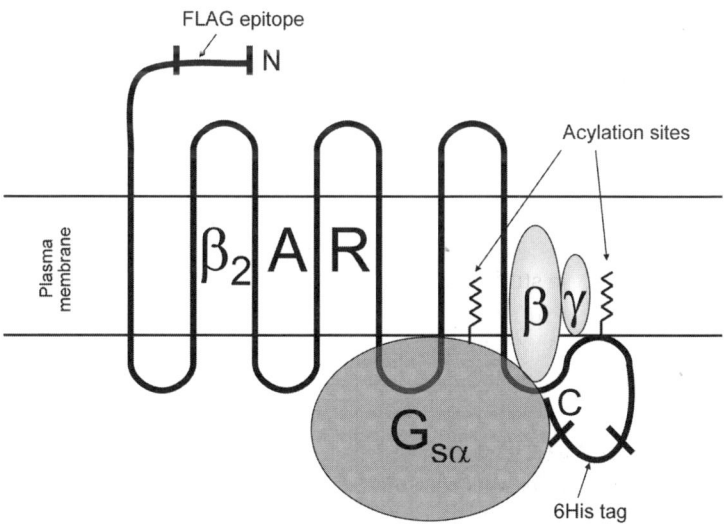

FIGURE 14.5 Structure of the $β_2AR$-$G_{sα}$ fusion protein used for studies with isoproterenol stereoisomers. The N-terminus of the $β_2AR$ is tagged with the FLAG epitope. The C-terminus of the $β_2AR$ is tagged with a 6His tag and is linked to the N-terminus of $G_{sα}$. $G_{sα}$ and the $β_2AR$ are palmitoylated. Acylation is important for membrane attachment of proteins. The covalent link between $β_2AR$ and $G_{sα}$ ensures close proximity between the signaling partners and prevents $G_{sα}$ dissociation from the membrane. The fusion protein can interact with the βγ-complex.

As expression system, the Sf9 insect cell/baculovirus system (11), which has several advantages, was used. First, the insect cells can be readily grown in large quantities using suspension cultures, and accordingly, large protein quantities can be obtained. Second, the expression levels of fusion proteins achievable in Sf9 cells are high; that is, they reach 5 to 10 pmol/mg of membrane protein. Third, Sf9 cells are essentially devoid of endogenous GPCRs, providing an excellent signal-to-noise ratio for the analysis of mammalian receptor/G-protein coupling. Fourth, an endogenous insect cell AC coupling well to mammalian $G_{s\alpha}$ can be used as distal readout. Fifth, unlike in bacterial expression systems, GPCRs are posttranslationally modified in Sf9 cells, which process is important for proper receptor function. Sixth, frozen Sf9 insect cell membranes can be stored for extended periods of time (several years) at $-80°C$ without substantial loss of biological activity. Seventh, the fusion protein technique allows the generation of various receptor/$G\alpha$ couples in a defined 1:1 stoichiometry, allowing for the comparison of the coupling efficiency of a given receptor to various G-proteins under defined experimental conditions (40). Last but not least, the fusion protein technique ensures close proximity of receptor and $G\alpha$, enhancing coupling efficiency and, thereby, signal intensity. This is particularly true for GPCR-$G_{s\alpha}$ fusion proteins since $G_{s\alpha}$, unlike $G_{i\alpha}$, dissociates from the plasma membrane following activation, with the fusion preventing this process (40). In aggregate, all these properties facilitate the analysis of ligand-specific GPCR conformations.

The pharmacological properties of several $\beta_x AR$-$G_{s\alpha}$ fusion proteins have been analyzed using ternary complex formation, the steady-state high-affinity GTPase assay, the GTPγS binding assay, and the AC assay as readout (15,42,43). To this end, multiple pairs of ligand stereoisomers in these systems have not yet been systematically examined although $\beta_x AR$-$G_{s\alpha}$ fusion proteins are certainly very well suited for such studies. The reason for this incomplete analysis is twofold. First, many stereoisomer pairs are not readily available to the scientific community, that is, many companies offer only racemic mixtures of ligands or the eutomer in their catalogs. Second, until very recently (7,14), it was not apparent how important the systematic analysis of multiple pairs of ligand stereoisomers is in terms of enhancing knowledge on mechanisms of receptor activation and identifying ligands that stabilize unique receptor conformations and induce distinct G-protein and effector activation patterns. Thus, in the following, it is pertinent to review studies on (−)-(R)-isoproterenol and (+)-(S)-isoproterenol. Although these studies, for the reasons stated above, are far from exhaustive, they are nonetheless very useful at illustrating the sensitivity of the Sf9 insect cell system and providing a paradigm for future studies addressing the stereoisomer issue.

ANALYSIS OF THE INTERACTION OF (−)-(R)-ISOPROTERENOL AND (+)-(S)-ISOPROTERENOL WITH $\beta_2 AR$ AND $\beta_2 AR_{CAM}$

Radioligand competition binding studies (step 2 in Fig. 14.1), high-affinity GTPase experiments (step 5 in Fig. 14.1), and AC experiments were conducted in the presence of various nucleoside 5′-triphosphates, that is, guanosine 5′-triphosphate (GTP), inosine 5′-triphosphate (ITP), and xanthosine 5′-triphosphate (XTP). The competition binding isotherms for (−)-(R)-isoproterenol and (+)-(S)-isoproterenol at $\beta_2 AR$-$G_{s\alpha}$ and $\beta_2 AR_{CAM}$-$G_{s\alpha}$ is shown in Figure 14.6, and the corresponding results of the nonlinear regression analysis are shown in Table 14.1. At

MOLECULAR ANALYSIS OF AGONIST STEREOISOMERS

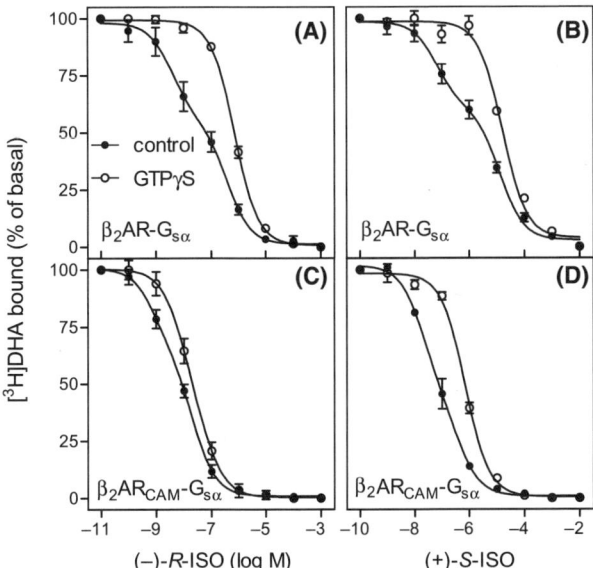

FIGURE 14.6 Competition binding isotherms for (−)-(R)-isoproterenol and (+)-(S)-isoproterenol at $\beta_2AR\text{-}G_{s\alpha}$ and $\beta_2AR_{CAM}\text{-}G_{s\alpha}$ expressed in Sf9 insect cell membranes. The antagonist radioligand [^{3}H]dihydroalprenolol (1 nM) was bound to $\beta_2AR\text{-}G_{s\alpha}$ and $\beta_2AR_{CAM}\text{-}G_{s\alpha}$ fusion proteins expressed in Sf9 insect cells as described. (−)-(R)-Isoproterenol and (+)-(S)-isoproterenol were added at increasing concentrations to compete with the radioligand. Assays were conducted either in the absence of a guanine nucleotide (control, ●) or in the presence of 10 µM GTPγS (○). Data are the means ± SD of four to seven independent experiments performed in triplicate and taken from Ref. 15. Data were analyzed by nonlinear regression for best fit to one-site or two-site binding. *Abbreviation*: ISO, Isoproterenol.

TABLE 14.1 Binding Properties of (−)-(R)-Isoproterenol and (+)-(S)-Isoproterenol at $\beta_2AR\text{-}G_{s\alpha}$ and $\beta_2AR_{CAM}\text{-}G_{s\alpha}$

Ligand	K_h (nM)	K_l (nM)	R_h (%)	$K_{hGTP\gamma S}$ (nM)	$K_{lGTP\gamma S}$ (nM)	$R_{hGTP\gamma S}$ (%)
$\beta_2AR\text{-}G_{s\alpha}$						
(−)-(R)-ISO	1.0 ± 0.6	100 ± 38	44.3 ± 3.8	−	190 ± 20	−
(+)-(S)-ISO	17 ± 4.0	3,400 ± 310	38.9 ± 4.4	−	4,300 ± 500	−
$\beta_2AR_{CAM}\text{-}G_{s\alpha}$						
(−)-(R)-ISO	0.2 ± 0.1	3.8 ± 1.3	35.4 ± 4.0	2.0 ± 1.0	13 ± 2.2	65.5 ± 8.9
(+)-(S)-ISO	3.8 ± 1.0	75 ± 13	56.1 ± 5.3	−	135 ± 21	−

The antagonist radioligand [^{3}H]dihydroalprenolol (1 nM) was bound to $\beta_2AR\text{-}G_{s\alpha}$ and $\beta_2AR_{CAM}\text{-}G_{s\alpha}$ fusion proteins expressed in Sf9 insect cells as described. (−)-(R)-isoproterenol and (+)-(S)-isoproterenol were added at increasing concentrations to compete with the radioligand. Assays were conducted either in the absence of a guanine nucleotide or in the presence of 10 µM GTPγS. Data shown in Figure 14.6 were analyzed by nonlinear regression for best fit to one-site or two-site binding and were taken from Ref. 15. Data are the means ±SD of four to seven independent experiments performed in triplicate. K_h and K_l designate the dissociation constants for high- and low-affinity agonist binding, respectively. %R_h represents the percentage of receptors displaying high agonist affinity. The corresponding values in the presence of GTPγS are referred to as $K_{hGTP\gamma S}$, $K_{lGTP\gamma S}$, and %$R_{hGTP\gamma S}$.

β_2AR-$G_{s\alpha}$, (−)-(R)-isoproterenol and (+)-(S)-isoproterenol inhibited binding of the antagonist [^{3}H]-dihydroalprenolol according to biphasic isotherms. The fraction of high-affinity binding sites (ternary complexes) for both ligands amounted to 40% to 45%, and (+)-(S)-isoproterenol exhibited 17-fold lower affinity for the β_2AR than (−)-(R)-isoproterenol. The addition of the hydrolysis-resistant GTP analog GTPγS disrupted the ternary complexes with both ligands at β_2AR and shifted the inhibition isotherm to the right. Moreover, the competition isotherm became monophasic instead of biphasic. Under the GTPγS conditions, (−)-(R)-isoproterenol showed a 23-fold higher affinity for the β_2AR than (+)-(S)-isoproterenol.

A consequence of the shift of the equilibrium between R and R* toward R* in β_2AR$_{CAM}$ is an increase in agonist-affinity/potency and partial agonist efficacy (15). Accordingly, in β_2AR$_{CAM}$-$G_{s\alpha}$, the K_i values for high-affinity binding of (−)-(R)-isoproterenol and (+)-(S)-isoproterenol at β_2AR$_{CAM}$ were decreased four- to fivefold relative to β_2AR (Fig. 14.6 and Table 14.1). Noteworthy, the distomer (+)-(S)-isoproterenol was more efficacious at stabilizing ternary complex (56% of the β_2AR$_{CAM}$ population) than the eutomer (−)-(R)-isoproterenol (35% of the β_2AR$_{CAM}$ population). Even more intriguingly, GTPγS was much less effective at disrupting ternary complex formation at β_2AR$_{CAM}$ than at β_2AR. Specifically, at β_2AR$_{CAM}$, GTPγS induced only smaller right-shifts of the competition binding isotherms than at β_2AR, and the isotherm for (−)-(R)-isoproterenol was still biphasic, that is, in the presence of GTPγS, 65% of the (−)-(R)-isoproterenol binding sites were in a high-affinity state. The small GTPγS shift for (+)-(S)-isoproterenol at β_2AR$_{CAM}$ indicates partially preserved high-affinity binding that is, however, not evident by a biphasic competition isotherm.

In the next step, how these differences in ternary complex formation translate into guanine nucleotide exchange as assessed by steady-state GTPase activity were examined (Fig. 14.7). At β_2AR, (−)-(R)-isoproterenol activated GTP hydrolysis with 10-fold higher potency than (+)-(S)-isoproterenol and both stereoisomers were full agonists. At β_2AR$_{CAM}$, the potencies of

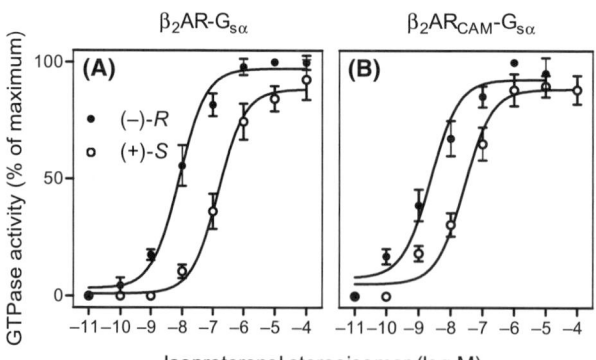

FIGURE 14.7 Concentration-response curves for the stimulatory effects of (−)-(R)-isoproterenol and (+)-(S)-isoproterenol on GTPase activity in Sf9 insect cell membranes expressing β_2AR-$G_{s\alpha}$ and β_2AR$_{CAM}$-$G_{s\alpha}$. The steady-state high-affinity GTPase activity in Sf9 insect cell membranes expressing β_2AR-$G_{s\alpha}$ and β_2AR$_{CAM}$-$G_{s\alpha}$ was determined in the presence of isoproterenol stereoisomers at various concentrations. Data are the means ±SD of three to six independent experiments performed in triplicate and taken from Ref. 15. Data were analyzed by nonlinear regression and were best fitted to sigmoidal concentration-response curves.

(−)-(R)-isoproterenol and (+)-(S)-isoproterenol were increased about fivefold relative to β₂AR, and again, both ligands were full agonists. The similar efficacies of isoproterenol stereoisomers at β_2AR_{CAM} in the GTPase assay are in marked contrast to their different effects on ternary complex formation. An explanation for this discrepancy is that at β_2AR_{CAM}, (−)-(R)-isoproterenol stabilizes nonsignaling or "frozen" ternary complexes that are inefficient at promoting downstream steps of the G-protein cycle (15). Such frozen ternary complexes have been observed for other GPCRs as well (18) and are a clear indication for the notion that at β_2AR_{CAM}, (−)-(R)-isoproterenol and (+)-(S)-isoproterenol stabilize distinct active receptor conformations. Although the differences in the effects of the two isoproterenol stereoisomers at wt-β₂AR on ternary complex formation were only small (Fig. 14.6), we suggest that also in this case each stereoisomer stabilizes a unique receptor conformation (Fig. 14.8).

Previous studies with nonfused β₂AR had revealed differences in potency between the isoproterenol stereoisomers with respect to AC activation (34). Therefore, AC experiments were conducted as well, although, so far, only with β₂AR and not β_2AR_{CAM}. In those studies advantage was taken of the fact that

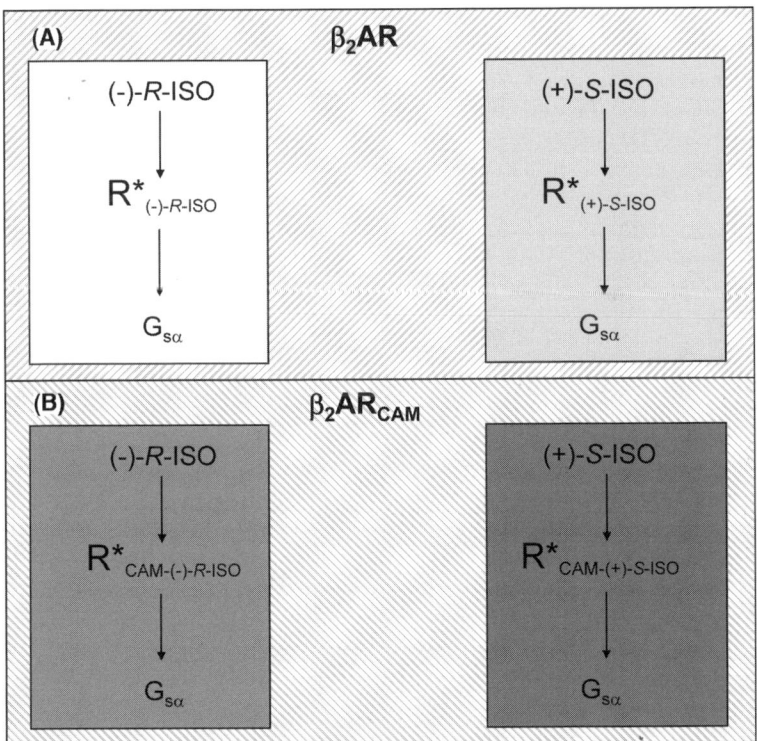

FIGURE 14.8 Stereoisomer-specific active conformations of β₂AR and β_2AR_{CAM}. (**A**) β₂AR conformations stabilized by isoproterenol stereoisomers. (**B**) β_2AR_{CAM} conformations stabilized by isoproterenol stereoisomers. The distinct effects of (−)-(R)-isoproterenol and (+)-(S)-isoproterenol in radioligand competition binding experiments versus GTPase experiments indicate that each stereoisomer stabilizes a unique conformation in β₂AR and β_2AR_{CAM}. The figure also illustrates that the terms "eutomer" and "distomer" are inappropriate to account for the pharmacological effects of the stereoisomers. *Abbreviation*: ISO, isoproterenol.

TABLE 14.2 Effects of (−)-(R)-Isoproterenol and (+)-(S)-Isoproterenol on AC Activity in Sf9 Insect Cell Membranes Expressing β_2AR-$G_{s\alpha}$

Parameter	(−)-(R)-Isoproterenol	(+)-(S)-isoproterenol
EC_{50} with GTP (nM)	18 ± 8	115 ± 7
EC_{50} with ITP (nM)	233 ± 34	1600 ± 750
EC_{50} with XTP (nM)	416 ± 44	5700 ± 2100
Efficacy with GTP	1.00	0.98 ± 0.06
Efficacy with ITP	1.00	0.64 ± 0.01
Efficacy with XTP	1.00	0.78 ± 0.09

AC activity was determined in Sf9 cell membranes expressing β_2AR-$G_{s\alpha}$ (long splice variant of $G_{s\alpha}$) in the presence of 1 μM GTP, 10 μM ITP, or 100 μM XTP. Reaction mixtures contained isoproterenol stereoisomers at increasing concentrations, and ligand stereoisomer potencies and efficacies were calculated by nonlinear regression. Data were taken from Ref. 42 and are the means ±SD of three to seven experiments performed in duplicate or triplicate.

G-proteins do not only bind GTP, but also, with lower affinity, the purine nucleotides ITP and XTP (Table 14.2) (42). In the presence of ITP and XTP, (−)-(R)-isoproterenol was 13-fold and 23-fold, respectively, less potent than in the presence of GTP at activating AC. This decrease in potency of (−)-(R)-isoproterenol in the presence of ITP and XTP is explained by a model in which higher occupancy rates of the β_2AR are required for promoting binding of the low-affinity nucleotides ITP and XTP to the $G_{s\alpha}$ protein compared to the high-affinity nucleotide GTP (42). Compared to (−)-(R)-isoproterenol, (+)-(S)-isoproterenol was 6- to 14-fold less potent at activating AC in the presence of GTP, ITP, and XTP. Under all experimental conditions, (−)-(R)-isoproterenol was a full agonist, that is, it exhibited an efficacy of 1.00. In contrast, (+)-(S)-isoproterenol was only a full agonist for AC activation in the presence of GTP but only a moderately strong partial agonist in the presence of ITP and XTP. These data show that the β_2AR conformation stabilized by (+)-(S)-isoproterenol is also less efficacious at promoting ITP and XTP binding to, and hence AC activation by, $G_{s\alpha}$. These data show that ITP and XTP are valuable experimental tools at dissecting functional differences between ligand stereoisomers at the β_2AR and support the model of stereoisomer-specific receptor conformations (Fig. 14.8).

MODELS OF THE INTERACTION OF (−)-(R)-ISOPROTERENOL AND (+)-(S)-ISOPROTERENOL WITH β_2AR

The experiments described above provide functional evidence for the existence of ligand stereoisomer-specific β_2AR conformations. Evidently, these data raise the question what the structural basis for these differences may be. Fortunately, a recently resolved structure of (−)-(R)-isoproterenol in complex with the turkey β_1AR (44) provides detailed information about the binding mode. The 13 amino acids of the binding site are identical and closely aligned (rms deviation of backbone atoms: 0.65 Å) with those of the β_2AR in a recent crystal structure of the active state, stabilized by a nanobody in place of $G_{s\alpha}$ (45). Therefore (−)-(R)-isoproterenol may be easily docked to the active β_2AR in the same pose as present in complex with the β_1AR. Figure 14.9A, C, E shows the putative interactions with the β_2AR according to a model of the minimized ligand-receptor complex. The protonated amine is involved in a salt bridge with Asp-113$^{3.32}$ (Ballesteros/Weinstein numbering scheme (46) as superscript) and

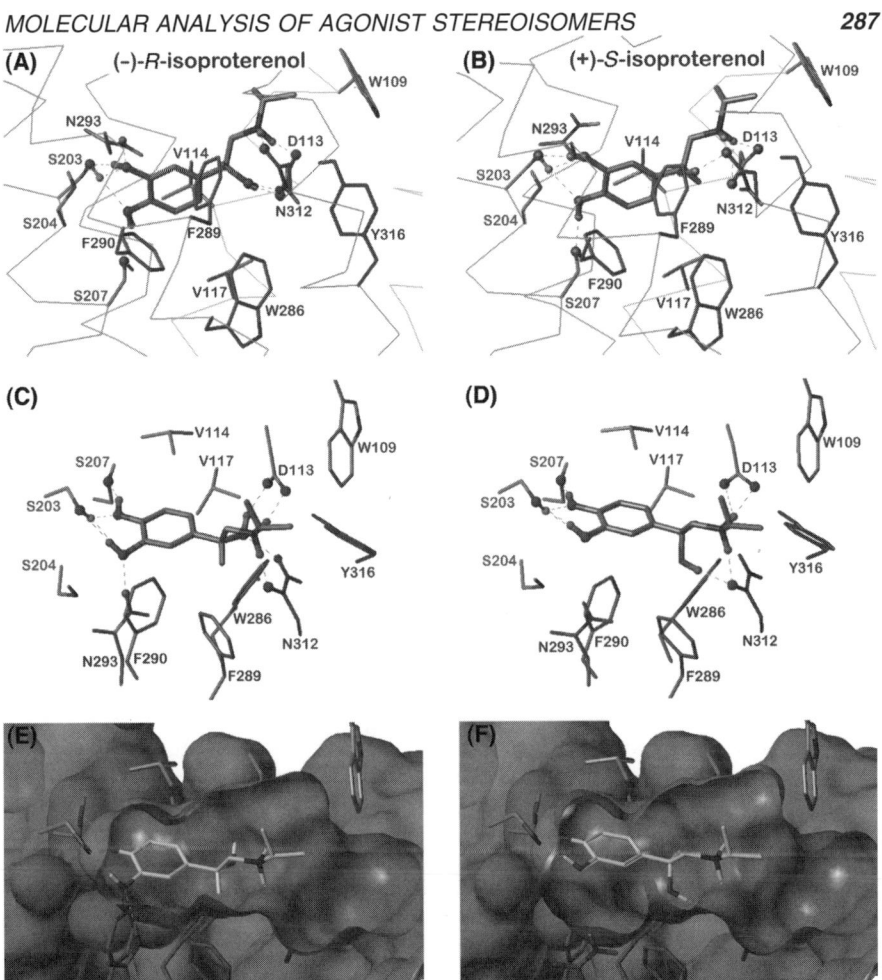

FIGURE 14.9 Model of the interaction of (−)-(R)-isoproterenol and (+)-(S)-isoproterenol with the β_2AR. The isomers were manually docked to the structure of the active β_2AR state (PDB 3P0G) according to the binding mode in the crystal structures of the β_1AR in complex with (−)-(R)-isoproterenol (PDB 2Y03). The complexes were then minimized with the Amber-FF99 force field (47) (β_2AR with fixed ligand) and the Tripos force field (48) (ligands and a "hot" receptor region of amino acids up to 6 Å distant from the ligand). Modeling was performed with SYBYL 7.3 (Tripos, L.P., St. Louis, Missouri, U.S.). Backbone traces, carbon atoms, and some essential hydrogen atoms of important amino acids are individually drawn in spectral colors: TM2-*orange*, TM3-*yellow*, TM4-*green*, TM5-*greenblue*, TM6-*blue*, TM7-*purple*. All nitrogen atoms, *blue*; oxygen atoms, *red*; carbon and some essential hydrogen atoms of the isoproterenol isomers, *grey*. (**A–D**) Detailed illustration of the docking modes of (−)-(R)-isoproterenol (**A, C**) and (+)-(S)-isoproterenol (**B, D**); shown are the ligands as well as the side chains and Cα atoms of 13 amino acids of the binding site as sticks; hydrogen bonds are drawn as dashed lines and the H bonding atoms as balls. (**A, B**) Side views, including Cα trace of the TM regions as lines. (**C, D**) Extracellular views. (**E, F**) Lipophilic potential (hydrophobic-*brown*, polar-*greenblue*) of the binding site [generated with MOLCAD, Darmstadt, Germany, using the protein variant with the new Crippen parameter table (49, 50)] with docked (−)-(R)- (**E**) and (+)-(S)-isoproterenol (**F**); the extracellular view corresponds to panels **C** and **D**, respectively.

You can view this figure in color online at http://www.informahealthcare.com/9781420092394/figure14.9

an amine hydrogen additionally interacts with an amide oxygen of Asn-312$^{7.39}$. Intriguingly, the β-hydroxyl group forms two hydrogen bonds as donor and acceptor, respectively, with the same residues Asp-113$^{3.32}$ and Asn-312$^{7.39}$. The ligand binds in a favorable conformation with an extended ethanolamine side chain nearly perpendicular to the catechol moiety. Each catechol hydroxyl group is involved in two hydrogen bonds: *m*-OH with the OH oxygen of Ser-203$^{5.42}$ and the amide NH of Asn-293$^{6.55}$, *p*-OH with the OH hydrogen of Ser-203$^{5.42}$ and the OH oxygen of Ser-207$^{5.46}$. The ethanolamine moiety and parts of the phenyl ring present an apolar edge that interacts with a hydrophobic "wall" mainly formed by the side chains of Trp-286$^{6.48}$, Phe-289$^{6.51}$, and Phe-290$^{6.52}$. Val-114$^{3.33}$ and Val-117$^{3.36}$ enable further hydrophobic contacts. The isopropyl group is directed to the indole plane of Trp-109$^{3.28}$.

(+)-(S)-Isoproterenol may be docked in a pose closely corresponding to that of the (−)-(R)-isomer (Fig. 14.9B, D, F). The interactions of the protonated amine with Asp-113$^{3.32}$ and Asn-312$^{7.39}$ are unchanged. However, the β-hydroxyl group, that is, the critical substituent at the chirality center of the ligands (Fig. 14.4), is now in a position where the two hydrogen bonds of the (−)-(R)-isomer with Asp-113$^{3.32}$ and Asn-312$^{7.39}$ are impossible. Instead the β-hydroxyl serves as donor in a new hydrogen bond with the amide oxygen of Asn-312$^{7.39}$. The interactions of the catechol OH groups with Ser-203$^{5.42}$ and Ser-207$^{5.46}$ are retained, but the hydrogen bond with Asn-293$^{6.55}$ is lacking. The polar β-hydroxyl contacts the hydrophobic wall of aromatic amino acids in TM6, leading to weakened hydrophobic interactions and to a slightly altered fit that enlarges the distance between the *m*-OH group and Asn-293$^{6.55}$.

These binding modes generally confirm the experimental findings that (+)-(S)-isoproterenol is a full or partial β$_2$AR agonist, depending on the specific experimental conditions (Table 14.2) (34), with lower affinity than the (−)-(R)-isomer (Figs. 14.6 and 14.7 and Tables 14.1 and 14.2). Specifically, the loss of affinity results from a deficit of two hydrogen bonds (β-OH with Asp-113$^{3.32}$, *m*-OH with Asn-293$^{6.55}$) and from weaker hydrophobic interactions. According to Warne et al. (44), the stabilization of the contracted catecholamine binding pocket is the main function of agonists on receptor activation. This stabilizing effect seems to directly depend on the number and strength of hydrogen bonds. Therefore, the loss of the two hydrogen bonds may also account for partial agonism of (−)-(S)-isoproterenol. The lower, but partially retained, stereospecificity of a β$_2$AR Asn-293-Leu mutant for (−)-(R)-isoproterenol versus (+)-(S)-isoproterenol (34) probably results from similar interactions of both isomers with the leucine side chain, whereas the hydrogen bonds of the β-hydroxyl group remain different like in the wild type receptor.

CONCLUSIONS AND FUTURE STUDIES

The studies discussed above show that GPCR-Gα fusion proteins constitute a very sensitive system for the systematic analysis of β$_2$AR ligand stereoisomers at the molecular level and that the assessment of various parameters of the G-protein cycle is very useful at revealing ligand-specific receptor conformations. The functional studies are solidly complemented by molecular modeling studies based on recently resolved GPCR crystal structures. It may be anticipated that systematic studies with stereoisomer pairs using the above-described design will corroborate the concept of ligand stereoisomer-specific G-protein activation and provide important information for the improvement of pharmacological treatments of bronchial asthma, preterm labor, and heart failure with β_2AR agonists. Studies with stereoisomers of albuterol, clenbuterol, fenoterol, and salmeterol will be particularly important.

Analysis of β_2AR-$G\alpha$ fusion proteins and molecular modeling are only the first steps in the molecular analysis of ligand stereoisomers. Particularly, fluorescence studies with purified β_2AR and ligand stereoisomers will be particularly informative (51). The generation of β_2AR crystal structures bound to agonist stereoisomers will be a formidable task since such structures, according to published data (Figs. 14.6 and 14.7 and Tables 14.1 and 14.2) (15,42), represent active receptor states. However, it cannot be excluded that certain ligand stereoisomers such as (S)-albuterol stabilize inactive β_2AR conformations (28), potentially facilitating crystallographic studies. It will also be very important not only to focus on the molecular aspects of ligand stereoisomer-receptor interactions but also on the cellular aspects. Specifically, it will be crucial to examine the effects of ligand stereoisomers on receptor desensitization and receptor trafficking (51). So far, this issue has only marginally been studied in context with ligand stereoisomers. Finally, experiments with intact organs and whole animals will be necessary.

In this chapter, the focus has been on the β_2AR and β_2AR_{CAM} as prototypical GPCRs exhibiting different degrees of constitutive activity. However, the studies reviewed herein are not limited to these GPCRs. β_1AR-$G_{s\alpha}$ fusion proteins also ensure efficient coupling (41) and the studies can easily be extended to other G_s-coupled biogenic amine GPCRs such as the histamine H_2-receptor (52). Moreover, the Sf9 insect cell expression system is not only useful for the analysis of G_s-coupled GPCRs, but also G_i-coupled and stimulatory G-protein of phospholipase C (G_q)-coupled GPCRs (53,54). Over the past decade, sensitive analysis systems for G_s-, G_i-, and G_q-coupled GPCRs (Table 14.3) have been established. Certainly, the analysis of ligand stereoisomers can be performed in greatest detail for G_s-coupled receptors since four parameters (high-affinity agonist binding, GTPγS binding, steady-state GTPase activity, and AC activation) are readily assessed. For G_i-coupled receptors, GPCR/G-protein coexpression or GPCR-$G\alpha$ fusion protein systems may be studied with similar sensitivity measuring ternary complex formation, GTPγS binding and GTPase activity, but for those GPCRs, inhibition of AC activity cannot be assessed in Sf9 cell membranes (54). For G_q-coupled GPCRs, steady-state GTPase activity (but not GTPγS binding) can be assessed when the GPCRs are coexpressed with the GTPase-activating regulator of G-protein signaling (RGS) proteins (53). Additionally, for G_q-protein-coupled GPCRs, only regular monophasic radioligand competition binding studies without the option to detect ternary complexes are possible (54). Despite limitations with respect to some parameters of the G-protein cycle, the measurement of steady-state GTPase activity provides a sensitive and proximal readout of receptor/G-protein coupling that allows for the analysis of ligand stereoisomers for G_s-, G_i-, and G_q-coupled GPCRs at a molecular level and with high sensitivity. The rigorous use of sensitive proximal readouts of ligand/receptor interaction such as GTPase activity in recombinant expression systems should facilitate future characterization of ligand stereoisomers since complicating factors such as differential metabolism of stereoisomers (22), presence of other interfering receptors (29), and problems arising from the limited availability of effector systems (40) are avoided.

TABLE 14.3 Analysis of Receptor/G-Protein Coupling in Sf9 Insect Cell Membranes

Parameter	G_s-coupled GPCRs	G_i-coupled GPCRs	G_q-coupled GPCRs
Most sensitive system(s)	GPCR-$G\alpha$ fusion protein	GPCR-$G\alpha$ fusion protein or coexpression	GPCR coexpressed with RGS-protein (insect cell G_q-protein as coupling partner)
High-affinity agonist binding	Yes	Yes	No
GTPγS binding	Yes	Yes	No
Steady-state GTPase	*Yes*	*Yes*	*Yes*
Effector regulation	AC↑	No	No
Representative GPCRs for which systems were explored	β_xARs, histamine H_2-receptor	Histamine H_3- and H_4-receptor, chemoattractant receptors	Histamine H_1-receptor

The G-protein cycle shown in Figure 14.1 applies to all GPCRs, regardless to which G-protein(s) they are coupled. Nonetheless, not all steps can be assessed experimentally for all classes of GPCRs, limiting the ability to dissect stereoisomer-specific GPCR conformations. The most comprehensive analysis is possible for G_s-coupled GPCRs (11, 40), followed by G_i-coupled GPCRs (54). The analysis of G_q-coupled GPCRs is limited to steady-state GTP hydrolysis in the presence of RGS-proteins (53). Nonetheless, the steady-state high-affinity GTPase activity represents a universally applicable proximal readout of receptor/G-protein coupling to assess the effects of ligand stereoisomers at G_s-, G_i-, and G_q-coupled GPCRs. *Abbreviations*: GPCR, G-protein-coupled receptor; GTP, guanosine 5'-triphosphate.

Finally, it cannot be reiterated enough that the uncritical use of the terms eutomer and distomer as originally defined by Ariëns causes confusion and is misleading. Thus, it may be suggested that the two terms, although historically useful, should be avoided or only be used with the clear definition that the eutomer possesses a higher potency/affinity than the distomer without any implications about pharmacological effects in terms of efficacy such as partial agonism or inverse agonism, differential modulation of various steps of the G-protein cycle, differential activation of G-protein subtypes or even activation of various receptors. It is to be hoped that this chapter makes a contribution toward the clarification of important nomenclature issues.

ACKNOWLEDGMENTS

R.S. would like to thank Dr Brian Kobilka (Stanford University, California, U.S.) for many stimulating discussions on mechanisms of β_2AR activation and Dr Katharina Wenzel-Seifert for fruitful experimental collaboration on the β_2AR. The authors' research was supported by the Deutsche Forschungsgemeinschaft (Graduiertenkolleg 760 "Medicinal Chemistry: Molecular Recognition—Ligand-Receptor Interactions"). While working at the University of Kansas, Lawrence, Kansas, U.S., the β_2AR work of R.S. was supported by Grants 005140Z and 0450120Z from the Heartland Affiliate of the American Heart Association.

REFERENCES

1. Brodde OE. β-1 and β-2 Adrenoceptor polymorphisms: functional importance, impact on cardiovascular diseases and drug responses. Pharmacol Ther 2008; 117:1–29.

2. Audet M, Bouvier M. Insights into signaling from the β₂-adrenergic receptor structure. Nat Chem Biol 2008; 4:397–403.
3. Gether U, Kobilka BK. G protein-coupled receptors. II. Mechanism of activation. J Biol Chem 1998; 273:17979–17982.
4. Gilman AG. G proteins: transducers of receptor-generated signals. Annu Rev Biochem 1987; 56:615–649.
5. Birnbaumer L. Expansion of signal transduction by G proteins. The second 15 years or so: from 3 to 16 α subunits plus βγ dimers. Biochim Biophys Acta 2007; 1768:772–793.
6. Rohrer DK, Kobilka BK. Insights from in vivo modification of adrenergic receptor gene expression. Annu Rev Pharmacol Toxicol 1998; 38:351–373.
7. Seifert R, Dove S. Functional selectivity of GPCR ligand stereoisomers: new pharmacological opportunities. Mol Pharmacol 2009; 75:13–18.
8. Boulton DW, Fawcell JP. β₂-Agonist eutomers: a rational option for the treatment of asthma? Am J Respir Med 2002; 1:305–311.
9. Broadley KJ. β-Adrenoceptor responses of the airways: for better or worse? Eur J Pharmacol 2006; 533:15–27.
10. Ahmet I, Krawczyk M, Zhu W, Cardioprotective and survival benefits of long-term combined therapy with β₂ adrenoceptor (AR) agonist and β₁ AR blocker in dilated cardiomyopathy postmyocardial infarction. J Pharmacol Exp Ther 2008; 325:491–499.
11. Wieland T, Seifert R. Methodological approaches. In: Seifert R, Wieland T, eds. G-Protein-Coupled Receptors as Drug Targets. Vol. 24. Weinheim: Wiley-Verlag Chemie, 2005:81–120.
12. Xiao RP, Ji X, Lakatta EG. Functional coupling of the β₂-adrenoceptor to a pertussis toxin-sensitive G protein in cardiac myocytes. Mol Pharmacol 1995; 47:322–329.
13. Xiao RP, Zhang SJ, Chakir K, Enhanced G_i signaling selectively negates β₂-adrenergic receptor (AR)- but not β₁-AR-mediated positive inotropic effect in myocytes from failing rat hearts. Circulation 2003; 108:1633–1639.
14. Woo AYH, Wang TB, Zeng X, Stereochemistry of an agonist determines coupling preference of β₂-adrenoceptor to different G proteins in cardiomyocytes. Mol Pharmacol 2009; 75:158–165.
15. Seifert R, Wenzel-Seifert K, Gether U, Functional differences between full and partial agonists: evidence for ligand-specific receptor conformations. J Pharmacol Exp Ther 2001; 297:1218–1226.
16. Perez DM, Karnik SS. Multiple signaling states of G-protein-coupled receptors. Pharmacol Rev 2005; 57:147–161.
17. Galandrin S, Oligny-Longpré G, Bouvier M. The evasive nature of drug efficacy: implications for drug discovery. Trends Pharmacol Sci 2007; 28:423–430.
18. Kenakin T. Functional selectivity through protean and biased agonism. Who steers the ship? Mol Pharmacol 2007; 72:1393–1401.
19. Kobilka BK, Deupi X. Conformational complexity of G-protein-coupled receptors. Trends Pharmacol Sci 2007; 28:397–406.
20. Urban JD, Clarke WP, von Zastrow M, Functional selectivity and classical concepts of quantitative pharmacology. J Pharmacol Exp Ther 2007; 320:1–13.
21. Seifert R, Wenzel-Seifert K. Constitutive activity of G-protein-coupled receptors: cause of disease and common property of wild-type receptors. Naunyn-Schmiedeberg's Arch Pharmacol 2002; 366:381–416.
22. Hartman AP, Wilson AA, Wilson HM, Enantioselective sulfation of β₂-receptor agonists by the human intestine and the recombinant M-form phenolsulfotransferase. Chirality 1998; 10:800–803.
23. Rouveix B, Badenoch-Jones P, Larno S, Lymphokine-induced macrophage aggregation: the possible role of cyclic nucleotides. Immunopharmacology 1980; 2:319–326.
24. Potter DE, Nicholson HT, Rowland JM. Ocular hypertensive response to β-adrenoceptor agonists. Curr Eye Res 1982–1983; 2:711–719.
25. Koike K, Hagiwara H, Takayanagi I. Comparison of interactions of R-(+)- and S-(-)-isomers of β-adrenergic partial agonists, befunolol and carteolol, with high affinity site of β-adrenoceptors in isolated rabbit ciliary body and guinea-pig taenia caeci. Can J Physiol Pharmacol 1991; 69:951–957.

26. Cushny AR. The action of optical isomers. III. Adrenalin. J Physiol (London) 1908; 37:130–138.
27. Causon RC, Desjardins R, Brown MJ, Determination of d-isoproterenol sulphate by high-performance liquid chromatography with amperometric detection. J Chromatogr 1984; 306:257–268.
28. Baramki D, Koester J, Anderson AJ, Modulation of T-cell function by (R)- and (S)-isomers of albuterol: anti-inflammatory influences of (R)-isomers are negated in the presence of the (S)-isomer. J Allergy Clin Immunol 2002; 109:449–454.
29. Mitra S, Ugur M, Ugur O, (S)-Albuterol increases free intracellular calcium by muscarinic activation and a phospholipase C-dependent mechanism in airway smooth muscle. Mol Pharmacol 1998; 53:347–354.
30. Patil PN, Li C, Kumari V, Analysis of efficacy of chiral adrenergic agonists. Chirality 2008; 20:529–543.
31. Birnbaum JE, Abel PW, Amidon GL, Changes in mechanical events and adenosine 3′,5′-monophosphate levels induced by enantiomers of isoproterenol in isolated rat atria and uteri. J Pharmacol Exp Ther 1975; 194:396–409.
32. Pike LJ, Lefkowitz RJ. Agonist-specific alterations in receptor binding affinity associated with solubilization of turkey erythrocyte membrane Beta adrenergic receptors. Mol Pharmacol 1978; 14:370–375.
33. May DC, Ross EM, Gilman AG, Reconstitution of catecholamine-stimulated adenylate cyclase activity using three purified proteins. J Biol Chem 1985; 260:15829–15833.
34. Wieland K, Zuurmond HM, Krasel C, Involvement of Asn-293 in stereospecific agonist recognition and in activation of the β_2-adrenergic receptor. Proc Natl Acad Sci U S A 1996; 93:9276–9281.
35. Culmsee C, Junker V, Thal S, Enantio-selective effects of clenbuterol in cultured neurons and astrocytes, and in a mouse model of cerebral ischemia. Eur J Pharmacol 2007; 575:57–65.
36. Zhang M, Fawcett JP, Kennedy JM, Stereoselective glucuronidation of formoterol by human liver microsomes. Br J Clin Pharmacol 2000; 49:152–157.
37. Jozwiak K, Khalid C, Tanga MJ, Comparative molecular field analysis of the binding of the stereoisomers of fenoterol and fenoterol derivatives to the β_2 adrenergic receptor. J Med Chem 2007; 50:2903–2915.
38. Bertin B, Freissmuth M, Jockers R, Cellular signaling by an agonist-activated receptor/$G_{s\alpha}$ fusion protein. Proc Natl Acad Sci USA 1994; 91:8827–8831.
39. Milligan G, Parenty G, Stoddart LA, Novel pharmacological applications of G-protein-coupled receptor-G protein fusions. Curr Opin Pharmacol 2007; 7:521–526.
40. Seifert R, Wenzel-Seifert K, Kobilka BK. GPCR-Gα fusion proteins: molecular analysis of receptor-G-protein coupling. Trends Pharmacol Sci 1999; 20:383–389.
41. Weitl N, Seifert R. Distinct interactions of human β_1- and β_2-adrenoceptors with isoproterenol, epinephrine, norepinephrine and dopamine. J Pharmacol Exp Ther 2008; 327:760–769.
42. Seifert R, Gether U, Wenzel-Seifert K, Effects of guanine, inosine, and xanthine nucleotides on β_2-adrenergic receptor/G_s interactions: evidence for multiple receptor conformations. Mol Pharmacol 1999; 56:348–358.
43. Wenzel-Seifert K, Seifert R. Molecular analysis of β_2-adrenoceptor coupling to G_s-, G_i-, and G_q-proteins. Mol Pharmacol 2000; 58:954–966.
44. Warne T, Moukhametzianov R, Baker JG, The structural basis for agonist and partial agonist action on a β_1-adrenergic receptor. Nature 2011; 469:241–244.
45. Rasmussen SGF, Choi HJ, Fung JJ, Structure of a nanobody-stabilized active state of the β_2 adrenoceptor. Nature 2011; 469:175–180.
46. Ballesteros JA, Weinstein H. Integrated methods for the construction of three-dimensional models and computational probing of structure-function relations in G protein-coupled receptors. Methods Neurosci 1995; 25:366–428.
47. Wang J, Wolf RM, Caldwell JW, Development and testing of a general amber force field. J Comput Chem 2004; 25:1157–1174.
48. Clark M, Cramer RDI, Van Opdenbosch N. Validation of the general purpose tripos 5.2 force field. J Comp Chem 1989; 10:982–1012.

49. Heiden W, Moeckel G, Brickmann J. A new approach to analysis and display of local lipophilicity/hydrophilicity mapped on molecular surfaces. J Comput Aided Mol Des 1993; 7:503–514.
50. Ghose AK, Viswanadhan VN, Wendoloski JJ. Prediction of hydrophobic (lipophilic) properties of small organic molecules using fragmental methods: An analysis of ALOGP and CLOGP methods. J Phys Chem 1998; 102:3762–3772.
51. Swaminath G, Xiang Y, Lee TW, Sequential binding of agonists to the β_2 adrenoceptor. Kinetic evidence for intermediate conformational states. J Biol Chem 2004; 279:686–691.
52. Preuss H, Ghorai P, Kraus A, Constitutive activity and ligand selectivity of human, guinea pig, rat, and canine histamine H_2 receptors. J Pharmacol Exp Ther 2007; 321:983–995.
53. Strasser A, Striegl B, Wittmann HJ, Pharmacological profile of histaprodifens at four recombinant histamine H_1 receptor species isoforms. J Pharmacol Exp Ther 2008; 324:60–71.
54. Wenzel-Seifert K, Arthur JM, Liu HY, Quantitative analysis of formyl peptide receptor coupling to $G_i\alpha_1$, $G_i\alpha_2$, and $G_i\alpha_3$. J Biol Chem 1999; 274:33259–33266.

15 Development of chiral drugs from a U.S. legal patentability perspective: Enantiomers and racemates

Svetlana M. Ivanova

INTRODUCTION

By the very close of the past century, the chiral drug industry soared so high that the annual sales in this industry sector topped $100 billion for the first time (1). Chiral drugs boldly entered into this century representing a stunning about one-third of all drug sales worldwide (1). From a patentability perspective, this has spurred an ongoing debate on what are the proper means for patent protection of chiral drugs. Namely, companies have needed to make decisions whether to develop and seek protection of either the racemic mixture or the single enantiomers, or whether to pursue a chiral switching route instead by seeking to first patent the racemic drug, and then years down the road seeking to extend a drug product's life cycle by subsequently patenting the single enantiomer form of the drug.

The industry stakes in addressing these patentability questions in accord with a well orchestrated business strategy are enormous. An example with one of the best selling drugs of all times, Prozac, or fluoxetine, is used to illustrate the lucrative and complex market that these patentability and development questions tap into. Nowadays Eli Lilly, Indianapolis, Indiana, U.S.A.'s blockbuster antidepressant Prozac stands as a famous and unfortunate example of an aborted and unsuccessful chiral switch. Once it had achieved enormous success with its blockbuster antidepressant Prozac, Eli Lilly had hoped it could continue the drug's life cycle by switching patients to the single enantiomer (R)-fluoxetine by December 2003 prior to the expiration of a key patent on fluoxetine (1). Eli Lilly itself did not actively pursue a chiral switch from the racemic drug to the enantiomer (R)-fluoxetine on its own. Nor apparently could it, as it needed instead to enter into a deal with Sepracor, Marlborough, Massachusetts, U.S.A. which had already obtained patents for (S)- and (R)-fluoxetine for treatment of migraine and depression (1). In 1998, the two companies struck a deal to codevelop (R)-fluoxetine as a side effect–free version of Prozac. By 2000, however, the deal was called off following clinical studies, which showed side effects at higher concentrations that were not observed with the racemate. In August 2000, a district court ruled the key patent on fluoxetine at issue invalid, and by May 2001 an appellate court upheld the ruling. As a result, Eli Lilly, Indianapolis, Indiana, U.S.A. saw its sales plunge precipitously from $2.6 billion in 2000 to $734 million in 2002 instead (1).

On the other end of the spectrum of success, just at the time of submission of this book chapter, the blockbuster dollar enantiomer drug Plavix emerged from a reexamination at the United States Patent and Trademark Office (USPTO) unscathed (2), scoring a major victory for Sanofi-Aventis and Bristol Myers Squibb, Plainsboro, New Jersey, U.S.A.—the companies that comarket the drug. Sanofi-Aventis first patented the racemic mixture claiming the active ingredient clopidogrel bisulfate before subsequently patenting the dextro isomer substantially separated from its levo isomer (U.S. patent 4,529,596—hereafter "the '596 patent") several years later, thus obtaining eight additional years of exclusivity over the initial patent (3). The request for reexamination was initiated by Apotex Corp. on allegations that the prior art rendered the claims of the '596 obvious, and thus invalid (3). The Reexamination Certificate acknowledged that the dextro isomer would have been obvious in view of the racemate or the pharmaceutical salt thereof, the recommendations of the regulatory authorities to separate and test the individual enantiomers, and the general knowledge in the art of resolution methods and of the likelihood of differences in the properties of the individual enantiomers (4). It nonetheless went on to conclude that the evidence of record supported the unpredictability of success in the resolution of racemic thienopyridine compounds, as well as the "unpredicted and unusual" therapeutic properties of the claimed dextro isomer (4).

The above and many other cases clearly raise the questions: Are enantiomers worth pursuing over their racemic counterparts? Are patents of single enantiomers valid in view of prior art disclosure of their respective racemates? And finally, if so, what is the best strategy for enantiomer drug development from a patentability perspective? Here, the author will briefly explore some of the fascinating factual and legal issues that have made this area of intellectual property continually pervade through decades of our times.

FOCUS ON PATENTABILITY: ENANTIOMERS AND/OR RACEMATES?

A glimpse at the legal framework will show some of the factors that are necessary ingredients for going from failure to success in pursuing and upholding patent protection of chiral drugs. This legal framework is very complex, and a single short chapter can do no service on the point. Instead, the author will attempt to illustrate key legal precedents where the rubber hits the road in the areas of anticipation and obviousness of enantiomers.

Anticipation

The first line of legal analysis in a patentability/validity determination is whether a prior art claim to, or disclosure of the racemic mixture of a patented drug product, and/or specific disclosure of the individual enantiomers anticipates a later claim to an individual enantiomer of the compound. In general, a patent is invalid for anticipation if a single prior art reference discloses each and every limitation of the claimed invention (5). Anticipation requires identity, namely, that "every element of the claimed invention must be *identically* shown in a single reference" (6). However, disclosure of every property or attribute of a composition of matter is not necessary for a reference to be valid anticipation (7). Moreover, a prior art reference may anticipate without disclosing a feature of the claimed invention if that missing characteristic is necessarily present, or inherent, in the single anticipating reference (8). There are two general legal standards that are applied by the courts: (*i*) anticipation of a species by a prior art genus, and (*ii*) anticipation of an enantiomer over a prior art disclosure of the racemate, the alternative enantiomer, or the same enantiomer.

The Law on Anticipation of Species by a Prior Art Genus

There is a split in the case law on the question of whether a prior art genus anticipates a later claimed species (corresponding by analogy to a racemic mixture and its individual enantiomers). The general rule is that earlier disclosure of a chemical genus is not an anticipation of a later-claimed species (9). The case law, however, recognizes an exception in the context of the "limited" subgenus (species) anticipation. The distinction is primarily based on considerations as to the size of the prior art genus and the sufficiency of its description, and may further include an analysis of properties and other relevant factors.

A number of cases have held that a genus may be so small that, when considered in light of the totality of the circumstances, it would anticipate the claimed species or subgenus (10). Determination of anticipation of species in view of prior art disclosure of the genus involves an evaluation of the totality of the circumstances, with primary considerations on the size of the genus and the sufficiency of its description. "[I]t is not the mere number of compounds in this limited class which is significant here but, rather, the *total circumstances* involved, including such factors as limited variations for R, only two alternatives for Y and Z, no alternatives for the ring position, and a large unchanging parent structural nucleus" (11). Additionally, while clearly important for an obviousness analysis, evaluation of properties is not always implicated (12).

However, a number of cases have held that a claimed species may not be anticipated by a prior art genus, where hindsight anticipations would be created by improper on the facts mechanistic application of *Petering* (13). As illustrated by one court, the level of guidance necessary for a finding of anticipation is akin to pointing out to blaze marks on particular trees as necessary for finding a trail through the woods (14).

Anticipation of Enantiomers by a Prior Art Disclosure of the Racemate

This line of cases clearly supports the proposition that an enantiomer is not anticipated by a prior art disclosure of the racemate (15). None of these cases, however, addresses the issue of whether there is anticipation in the context of a prior art disclosure of *both* the racemic mixture and its optical isomers, as was subsequently the case in *Forest Labs., Inc. v. Ivax Pharms., Inc* (16). In that case a prior article by Donald F. Smith ("Smith reference") specifically disclosed that citalopram was a racemic compound with two different enantiomers. The Federal Circuit held, however, that the Smith reference did not anticipate claim 1 of the patent at issue, U.S. Patent No. RE 34,712 patent ("the '712 patent") because it did not disclose "substantially pure" escitalopram as claimed in independent claim 1 and it did not enable a person having ordinary skill in the art to obtain the enantiomer compound escitalopram. This analysis needs to be approached with caution, however, as it was premised on the rare and particular factual finding made by the lower Delaware district court, which the Federal Circuit merely affirmed on appeal as not "clearly erroneous" (17). Moreover, since this decision, at least one other court, the Federal Patent Court in Germany, has found on a different factual record that the prior art would have enabled a person of skill in the art to obtain the invention of the foreign counterpart of the '712 patent and has struck the patent as invalid.

Obviousness

The next line of legal analysis in a patentability/validity determination is whether a prior art claim to, or disclosure of the racemic mixture of a patented drug product, and/or specific disclosure of the individual enantiomers renders obvious a later claim to an individual enantiomer of the compound. The statutory requirement of nonobviousness is set forth in Title 35, section 103(a) of the United States Code:

> A patent may not be obtained though the invention is not identically disclosed or described as set forth in section 102 of this title, if the differences between the subject matter sought to be patented and the prior art are such that the subject matter as a whole would have been obvious at the time the invention was made to a person having ordinary skill in the art to which said subject matter pertains. Patentability shall not be negatived by the manner in which the invention was made.

Whether or not a claim is obvious is a legal determination based on the underlying facts, which must be established by clear and convincing evidence (18). This legal determination includes four factual inquiries commonly referred to as the "*Graham* factors": (*i*) the scope and content of the prior art, (*ii*) the differences between the prior art and the claims at issue, (*iii*) the level of ordinary skill in the art, and (*iv*) objective indicia of nonobviousness (19). The initial three Graham factors are of primary importance to the prima facie case of obviousness, which the defendants bear the initial burden of establishing (20). Courts may also consider other relevant factual considerations such as commercial success, long-felt but unsolved need, and the failure of others (21). Courts have generally applied the following three legal standards with respect to enantiomer cases: (*i*) obviousness of a species by a prior art genus, (*ii*) obviousness of an enantiomer over a prior art disclosure of the racemate, and (*iii*) obviousness of a chemical compound in view of structurally similar prior art chemical compounds.

The Law on Obviousness of a Species in View of a Prior Art Disclosure of the Genus

The Federal Circuit has held that earlier disclosure of a chemical genus does not as a general rule render obvious a later-claimed species, as each case must be determined on its own facts (22). This is particularly the case where that disclosure indicates preference leading away from the claimed compounds (23). It is not the mere number of compounds that is significant, but rather the total circumstances involved (23).

Prima Facie Obviousness of Enantiomers in View of Prior Art Disclosure of the Racemate

It is a well-established law from the time of the predecessor court to the Federal Circuit that enantiomers "discovered" after discovery of the racemic compound are prima facie obvious over either the isomer of opposite rotation or the racemic compound itself in the absence of unexpected or unobvious beneficial properties (24). "[U]nder existing law a stereoisomer is not patentable over its known racemic mixture unless it possesses unexpected properties not possessed by the racemic mixture" (25). Moreover, when the purported invention is "an active ingredient" of "a mixture that existed in the prior art[,]" the active ingredient "is *prima facie obvious* over the mixture *even without an explicit teaching* that the ingredient should be concentrated or purified" (26). As the Federal Circuit explained, "[o]rdinarily, one expects a concentrated or purified ingredient to retain the same properties it exhibited in a mixture, and for those properties to be amplified when the ingredient is concentrated or purified; *isolation of interesting compounds is a mainstay of the chemist's art*. If it is known how to perform such an isolation, doing so 'is likely the product not of innovation but of ordinary skill and common sense'" (27). It should be noted, however, that a small minority of panels or courts hold that prior art disclosure of the racemic mixture does not automatically create a prima facie case of obviousness for the constituent enantiomers (28).

By way of example, one of the earliest historic pronouncements on establishing of prima facie obviousness came from the U.S. Court of Customs and Patent Appeals (CCPA), which is the predecessor court to the Federal Circuit, in the 1960 decision in *Adamson* (29). Claim 1 of *Adamson* was directed to a levo isomer of a compound selected from the class consisting of 1-cyclohexyl-1-phenyl-3-piperidinopropan-1-ol and 1-cyclohexyl-1-phenyl-3-pyrrolidinopropan-1-ol and their acid addition salts and quaternary ammonium salts substantially separated from the dextro isomer. In seeking to establish patentability, appellants claimed that the levo isomers have substantially higher spasmolytic activity than either the dextro isomer or the racemic mixture and slightly higher toxicity than the same quantity of the racemate. The CCPA found the claimed compound unpatentable over a prior art reference, which disclosed the racemic compound (albeit without a statement that it was racemic, or that dextro and levo isomers existed), and a second organic chemistry text, which disclosed that optical activity is attributable to asymmetric molecular structure about a carbon, and that a racemic mixture could be resolved into its levorotatory and dextrorotatory components via numerous methods (30). In concluding that the record did not support appellants' argument "that the levo isomer exhibits surprisingly superior spasmolytic activity," the court weighed the facts that the claimed enantiomer exhibited *twice* the activity as the racemate, and that it was known that different enantiomers may have different properties (31).

Establishing a Prima Facie Case of Obviousness Within the Context of the Law on Chemical Inventions

Establishing a prima facie case of obviousness for enantiomers within the framework of the broader law of chemical inventions follows a two-prong analysis: (*i*) establishing structural similarity, and (*ii*) establishing motivation to make the structurally similar compounds where the prior art offers a reasonable expectation of success (32). The motivation to make the claimed compounds is with the expectation that compounds similar in structure will have similar properties (33). As the Federal Circuit stated in *Deuel*, a known compound may suggest its analogs or isomers, either geometric isomers (*cis* vs. *trans*) or position isomers (e.g., *ortho* vs. *para*) (34). This motivation need not be express and may flow from the prior art references, the knowledge of one of ordinary skill in the art, or the nature of the problem to be solved (35).

Whether the motivation to resolve racemates into enantiomers is found in the prior art, or not, is a contradictory area of the law. Similar to *Adamson*, some courts have found that the prior art provides *"ample* motivation to separate the optical isomers" (36). Other courts have held that the requisite motivation to separate the individual enantiomers of a racemate is not found in the prior art. Thus, in *Pfizer Inc. v. Ranbaxy Labs*, the district court reasoned that resolution of a racemate into its component enantiomers would be at best expected to yield a twofold difference in activity, the benefit of which could be offset by drawbacks such as the difficulty and complexity of the resolution process, such that the prior art motivated at the time to develop racemates and make structural changes to the compounds to increase their activity, not to resolve those racemates into individual isomers (37). Similarly, in *Sanofi-Synthelabo v. Apotex, Inc.*, the Federal Circuit stated that nothing existed in the prior art that would make separating the enantiomer of MATTPCA an obvious choice, particularly in light of the unpredictability of the pharmaceutical properties of the enantiomers, the potential for enantiomers to racemize in the body, the extensive time and money Sanofi spent developing the racemate before directing its efforts to the enantiomer, and the unpredictability of salt formation (38).

The motivation in a prima facie case of obviousness need only be founded on a reasonable expectation of success, and absolute predictability is not required in motivating and guiding the skilled artisan toward making the claimed compound at the time of the invention (39). Reasonable expectation of success is assessed from the perspective of the person of ordinary skill in the art (40). With respect to racemates, there must be a reasonable expectation not only that the single enantiomer will exhibit desirable properties, but that it will also be possible to resolve the racemate in the first place (41).

Historically, courts such as the above have used the so-called teaching, suggestion, or motivation (TSM) test, under which a patent claim is only proved obvious if "some motivation or suggestion to combine the prior art teachings" can be found in the prior art, the nature of the problem, or the knowledge of a person having ordinary skill in the art (42). However, the U.S. Supreme Court recently rejected the Federal Circuit's approach as rigid in *KSR* stating that "[t]he obviousness analysis cannot be confined by a formalistic conception of the words teaching, suggestion, and motivation, or by overemphasis on the importance of published articles and the explicit content of issued patents" (43). The Court instructed that obviousness may be proved by showing that the combination of elements was "obvious to try" (44). In view of *KSR*, one may accordingly expect newly decided cases will have a heightened slant in favor of finding of obviousness, including enantiomer patent cases.

Secondary Considerations That May Overcome a Prima Facie Case of Obviousness

Once a prima facie case is established, the burden of going forward shifts to the plaintiffs to rebut the prima facie case with evidence of the fourth *Graham* factor, often referred to as "secondary considerations" (45). Secondary considerations considered by courts to suggest nonobviousness include the following: unexpected results (46), commercial success (47), long-felt need (48), failure of others (49), copying by competitors (50), praise by others/industry recognition (51), industry acquiescence (52), skepticism by those in the art, and licensing by others (53). As a matter of law, unexpected results must be compared to the closest prior art (54). The most commonly shown factors are those of unexpected properties/results and commercial success.

Establishing unexpected results is a fact-driven and very hotly contested area of patent law. Unexpected results must be established by reliable, objective, factual evidence (55). Greater effectiveness in terms of the same property could be relied upon only if shown by "clear and convincing evidence" (56). Under Federal Circuit law, mere improvement in properties does not always suffice to show unexpected result (57). Unexpected results must be "truly unexpected," for example, "a difference in kind as opposed to degree" (58). For instance, in *May* the Court found unexpected results, which rebutted a prima facie claim of obviousness of the claimed analgesic compound over a structurally similar prior art isomer where the evidence demonstrated the presence of a new and distinct chemical property of the claimed compound—nonaddictiveness, and where the analgesic and addiction properties could not be reliably predicted on the basis of chemical structure (59). A number of court decisions have addressed the issue of what level of improved property/increased efficacy constitutes unexpected results. In *U.S. v. Ciba-Geigy Corp.*, the court held that 10 times greater potency, but otherwise the same effects, resulted in "a significant enough [difference] to be deserving of a patent" (60). In *Wichert*, the CCPA also upheld a similar range. "In the case at bar, we are impressed by the 7-fold improvements in activity and, in the absence of valid countervailing evidence, we find the claimed compounds to be unobvious" (61).

To establish commercial success, the patentee must merely show significant sales in a relevant market, and that the successful product is the invention disclosed and claimed in the patent (62). "If the patentee makes the requisite showing of nexus between commercial success and the patented invention, it is presumed that the commercial success is due to the patented invention, and the burden then shifts to the challenger to prove that the commercial success is instead due to other factors extraneous to the patented invention, such as advertising (63).

CONCLUSION

In view of their commercial significance, enantiomers of chiral drugs with purity limitations are frequently patented or claimed. Patents on enantiomers are also frequently challenged, particularly as the above cases outline on grounds of obviousness and lack of unexpected properties/results. A recent progeny of legal cases with a heightened slant in favor of finding obviousness after *KSR* will undoubtedly continually lead to an increase in invalidity challenges to existing chiral drug patents. Accordingly, scrutinized decisions on patentability and chiral switching of racemates and enantiomers will continue to be made in this fascinating area at the intersection of science and patent law, continually reshaping and redefining the landscape of successful chiral drug enterprises.

REFERENCES

1. Stinson SC. Chiral drugs. Science/Technology 2000; 78:43:55–78. Available at: http://pubs.acs.org/cen/coverstory/7843/7843scit1.html.
2. See USPTO Notice of Intent to Issue Ex Parte Reexamination Certificate ("Reexamination Certificate"), dated March 26, 2010. Available at: http://portal.uspto.gov/external/portal/pair.
3. See USPTO Grants Apotex's Bid to Review Plavix Patent. Available at: http://www.law360.com/articles/117241.
4. Reexamination Certificate at 3. (See fn. 2)
5. *Lewmar Marine Inc. v. Barient Inc.*, 827 F.2d 744, 747 (Fed Cir 1987).

6. *Diversitech v. Century Steps, Inc.*, 850 F.2d 675, 677 (Fed Cir 1988).
7. *In re Kalm*, 378 F.2d. 959 (C.C.P.A. 1967).
8. *Continental Can Co. v. Monsanto Co.*, 948 F.2d 1264, 1268 (Fed Cir 1991).
9. *Corning Glass Works v. Sumitomo Elec. U.S.A., Inc.*, 868 F.2d 1251, 1262–1263 (Fed Cir 1989).
10. See *In re Petering*, 301 F.2d 676, 681, 133 U.S.P.Q. 275, 280 (C.C.P.A. 1962) (holding that a prior art genus containing only 20 compounds and a limited number of variations in the generic chemical formula inherently anticipates a claimed species within the genus because "one skilled in [the] art would ... envisage each member" of the genus); see also *In re Schaumann*, 572 F.2d 312, 316, 197 U.S.P.Q. 5, 9 (C.C.P.A. 1978) (holding that a prior art genus encompassing claimed species, which disclosed preference for lower alkyl secondary amines and properties possessed by the claimed compound, constitutes description of the claimed compound for purposes of 35 U.S.C. § 102(b)); *In re Parameswar Sivaramakrishnan*, 673 F.2d 1383, 213 U.S.P.Q. 441 (C.C.P.A. 1982) (holding that a prior art genus claiming a combination of a resin with approximately 70 other salts anticipates a claim to a combination of the resin with 1 of the 70 salts where the prior art description of the specific salt would not lead one of skill in the art to speculate and judiciously choose possible combinations from the genus in order to obtain the later claimed species); *Bristol-Myers Squibb Co. v. Ben Venue Labs., Inc.*, 246 F.3d 1368, 1380 (Fed Cir 2001). ("The disclosure of a small genus may anticipate the species of that genus even if the species are not themselves recited.")
11. *Petering*, 301 F.2d at 681–682 (emphasis added).
12. On one end of the spectrum, in *Petering* the prior art disclosed a genus of compounds with vitamin activity, whereas appellant's claimed compound had antivitamin activity instead. The court stated that the prior art purpose for disclosing the vitamin activity was immaterial where the facts were clear that the prior art disclosed the limited subclass. *Petering*, 301 F.2d at 682. Per the court in *Schaumann*, however, the fact that the properties of the prior art and claimed compounds were the same, together with the disclosed preference in the prior art for "lower alkyl" compounds, were factors that established "a far stronger foundation on which to support a finding of anticipation than did the circumstances in *Petering*." *Schaumann*, 572 F.2d at 316. Conversely, in this vein, where the properties of the prior art disclosed compounds within the scope of appellant's claims possessed diametrically opposite properties, this warranted against a holding of anticipation. *Kalm*, 378 F.2d at 963.
13. See *In re Ruschig*, 343 F.2d 965, 974, 145 U.S.P.Q. 274, 282 (C.C.P.A. 1965) (holding that a rejection of a claimed compound in light of prior art genus based on *Petering* is not appropriate where the prior art does not disclose a small recognizable class of compounds with common properties); see also *In re Wiggins*, 488 F.2d 538, 543, 179 U.S.P.Q. 421, 425 (C.C.P.A. 1973) (holding that a prior art listing of specific compounds within the scope of the appealed claims "constituted nothing more than a speculation about their potential or theoretical existence," and hence, not a "description" of the compounds within the meaning of § 102(b)); *Ex parte Raymond*, 2000 WL 34227019 (BPAI 2000) [returning a patent application back to the examiner for reevaluation of an anticipation rejection of a stereoisomer (alone or in combination with its enantiomer) over prior art unresolved mixtures of enantiomers, in light of (i) the contrary to each other holdings on the issue in *Schaumann* and *May* (infra), and (ii) the question of whether the prior art provides an enabling disclosure of the individual stereoisomers "just as surely as if they were identified in the reference by name" to persons of ordinary skill in the art]; *Sanofi-Synthelabo v. Apotex*, No. 02-2255, 2007 U.S. Dist. LEXIS 44033, at *89–92 (S.D.N.Y. June 19, 2007) [holding that the analysis of *Petering* and *Schaumann* does not apply to a subgenus of nine possible combinations (PCR 4099, its levorotatory and dextrorotatory enantiomers, and three possible salts—the hydrobromide, the hydrochloride, and the bisulfate) where the general prior art formula covered millions of compounds with no guidance to this particular subset].
14. "It is an old custom in the woods to mark trails by making blaze marks on the trees. It is no help in finding a trail or in finding one's way through the woods where the

trails have disappeared—or have not yet been made, which is more like the case here—to be marked simply by a large number of unmarked trees. Appellants are pointing to trees. We are looking for blaze marks which single out particular trees. We see none." *In re Ruschig*, 379 F.2d 990, 994–995 (C.C.P.A. 1967) ("*Ruschig II*").
15. See *Ortho-McNeil Pharmaceutical, Inc. v. Mylan Labs., Inc.*, 267 F. Supp.2d 533, 545 (N.D. W. Va. 2003) (holding that the prior art disclosure of racemic ofloxacin did not anticipate its constituent enantiomer levofloxacin); see also, *In re May*, 574F.2d1082 (C.C.P.A. 1978) ("As recognized the novelty of an optical isomer is not negated by the prior art disclosure of its racemate."); *In re Williams*, 171 F.2d 319, 320 (C.C.P.A.1948) ("[t]he existence of a compound as an ingredient of another substance does not negative novelty in a claim to the pure compound, although it may, of course, render the claim unpatentable for lack of invention."); *Pfizer Inc. v. Ranbaxy Labs.*, 405 F. Supp.2d 495, 519 (D. Del. 2005) ("[C]ourts considering issues related to racemates and their individual isomers have concluded that a prior art disclosure of a racemate does not anticipate the individual isomers of the racemate.").
16. 2007 U.S.App. LEXIS 21165, 84 U.S.P.Q.2d (BNA) 1099 (Fed Cir 2007).
17. *Forest Labs., Inc. v. Ivax Pharms., Inc.*, 438 F. Supp.2d 479 (D. Del. 2006).
18. *Graham v. John Deere Co.*, 383 U.S. 1, 17–18 (1966); *Greenwood v. Hattori Seiko Co., Ltd.*, 900 F.2d 238, 241 (Fed Cir 1990).
19. *Graham*, 383 U.S. at 17–18.
20. *Ashland Oil, Inc. v. Delta Resins & Refractories, Inc.*, 776 F.2d 281, 291–292 (Fed Cir 1985).
21. *Graham*, 383 U.S. at 17–18.
22. *In re Baird*, 16 F.3d 380, 382 (Fed Cir 1994).
23. *Baird*, 16 F.3d at 383.
24. *In re Adamson*, 275 F.2d 952, 954 (C.C.P.A. 1960).
25. *Sterling Drug, Inc. v. Watson*, 135 F. Supp. 173, 176 (D.D.C. 1955) (citations omitted).
26. *Aventis Pharma v. Lupin*, 499 F.3d 1293, 2007 U.S.App. LEXIS 21753, at *22 (Fed Cir 2007).
27. *Aventis Pharma v. Lupin*, 499 F.3d 1293, 2007 U.S.App. LEXIS 21753, at *23 [citing *KSR International Co. v. Teleflex Inc.*, 127 S. Ct. 1727, 1742 167 L. Ed. 2d 705(2007)] (emphasis added).
28. See, for example, *Ex parte Bonfils*, 64 U.S.P.Q.2d (BNA) 1456, at *15–16 (B.P.A.I. 2002) (a nonbinding precedent holding that the disclosure of one enantiomer does not necessarily create a prima facie case of obviousness as to the other enantiomer where there is evidence of unpredictability and no evidence for one of skill in the art to conclude with a reasonable expectation of success that there are common biological or pharmaceutical properties of the two enantiomers); *Ortho-McNeil Pharm. v. Mylan Labs.*, 348 F. Supp. 2d 713, 749, n.19 (N.D.W.Va 2004) (declining to accept the generic manufacturer's assertion that enantiomers are prima facie obvious vis-à-vis the racemic mixture and reasoning that cases supporting this contention are inconsistent with the Federal Circuit's directive to make *Graham* findings in every case to establish a prima facie case of obviousness).
29. *Adamson*, 275 F.2d at 954.
30. *Adamson*, 275 F.2d at 953.
31. *Adamson*, 275 F.2d at 954–955.
32. *In re Dillon*, 919 F.2d 688, 692 (Fed Cir 1990).
33. *In re Geiger*, 815 F.2d 686, 688 (Fed Cir 1987); *In re Gyurik*, 596 F.2d 1012, 1018 (C.C.P.A. 1979); *May*, 574 F.2d at1094; *In re Soni*, 54 F.3d 746, 750, 34 U.S.P.Q.2d 1684, 1687 (Fed Cir 1995).
34. *In re Deuel*, 51 F.3d 1552, 1558, 34 U.S.P.Q.2d 1210, 1214 (Fed Cir 1995).
35. *Brown & Williamson Tobacco Corp. v. Philip Morris Inc.*, 229 F.3d 1120, 1125 (Fed Cir 2000).
36. *Ortho-McNeil*, 348 F. Supp. 2d at 752 (emphasis added); *Sanofi-Synthelabo*, 2007 U.S. Dist. LEXIS at *36 and *106 (S.D.N.Y. June 19, 2007) (finding that a person of ordinary skill in the art in the mid-1980s would have known that the enantiomers of a racemate could exhibit different biological activity).
37. 405 F. Supp. 2d 495, 517 (D. Del. 2005).
38. 470 F.3d 1368, 81 U.S.P.Q.2d 1097 (Fed Cir 2006).

39. *Yamanouchi Pharm. Co. v. Danbury Pharmacal, Inc.*, 231 F.3d 1339, 1343 (Fed Cir 2000).
40. *Life Techs., Inc. v. Clontech Labs., Inc.*, 224 F.3d 1320, 1326 (Fed Cir 2000).
41. *Ortho-McNeil*, 348 F. Supp. 2d at 752–753 (concluding that, as of 1984, the resolution of the particular enantiomers in question would have been a "logical extension of the prior art" though not "a routine matter").
42. *KSR*, 550 U.S. at 407 (citations omitted).
43. *KSR*, 550 U.S. at 1741.
44. Such a situation, according to the Supreme Court arises "[w]hen there is a design need or market pressure to solve a problem and there are a finite number of identified, *predictable solutions*, a person of ordinary skill has good reason to pursue the known options within his or her technical grasp. *If this leads to the anticipated success,* it is likely the product not of innovation but of ordinary skill and common sense. In that instance the fact that a combination was obvious to try might show that it was obvious under § 103." *KSR*, 550 U.S. at 1740 (emphasis added).
45. *Mayne*, 104 F.3d at 1343.
46. *In re Piasecki*, 745 F.2d 1468, 1473 (Fed Cir 1984).
47. *Sterling Drug*, 135 F. Supp. 173.
48. *May*, 574 F.2d at 1092.
49. *Sterling Drug*, 135 F. Supp. 173.
50. *In re GPAC*, 57 F.3d 1573, 1580 (Fed Cir 1995); *Hybritech Inc. v. Monoclonal Antibodies*, 802 F.2d 1367, 1380 (Fed Cir 1986).
51. *Brenner*, 247 F. Supp. 51 (D.D.C. 1965).
52. *Eli Lilly & Co. v. Generix*, 324 F. Supp. 715, 718 (S.D. Fla. 1971), *aff'd* 460 F.2d 1096 (5th Cir. 1972).
53. *Graham*, 383 U.S. at 17–18.
54. *In re Baxter Travenol*, 952 F.2d 383, 392 (Fed Cir 1991).
55. *Soni*, 54 F.3d at 750; *In re Johnson*, 747 F.2d 1456, 1460 (Fed Cir 1984).
56. *In re Lohr*, 317 F.2d 388, 137 U.S.P.Q. 548 (C.C.P.A. 1963). "When a new compound so closely related to a prior art compound as to be structurally obvious is sought to be patented based on the alleged greater effectiveness of the new compound for the same purpose as the old compound, clear and convincing evidence of substantially greater effectiveness is needed. Here there are no new properties, but merely an alleged improvement in the same property for use against the same pests." *Lohr*, 317 F.2d at 392, 137 U.S.P.Q. at 550–551.Cf. *Ex parte Gelles*, 22 U.S.P.Q.2d 1318, 1319 (Bd. Pat. App. & Int'f 1992) ("It should ... be established that the differences in results are in fact unexpected and unobvious and of both statistical and practical significance.").
57. *In re Geisler*, 116 F.3d 1465, 1471 (Fed Cir 1997); *Soni*, 54 F.3d at 751 (rejecting unexpected results argument because it showed 26% increase and was lacking in objective, factual support).
58. *In re Merck & Co., Inc.*, 800 F.2d 1091, 1099 (Fed Cir 1986); see also, *In re Huang*, 100 F.3d 135, 139 (Fed Cir 1996) ("even though applicant's modification results in great improvement and utility over the prior art, it may still not be patentable if the modification was within the capabilities of one skilled in the art, unless the claimed ranges "produce a new and unexpected result which is different in kind and not merely in degree from the results of the prior art.") [quoting *In re Aller*, 220 F.2d 454, 456 (C.C.P.A. 1955); *In re Woodruff*, 919 F.2d 1575, 1578 (Fed Cir 1990)].
59. 574 F.2d at 1094.
60. *U.S. v. Ciba-Geigy Corp.*, 508 F. Supp. 1157, 1169 (D. N.J. 1979).
61. *In re Wiechert*, 370 F.2d 927, 932 (C.C.P.A. 1967).
62. *Alza Corp. v. Mylan Labs. Inc.*, 388 F. Supp.2d 717, 740 (N.D. W. Va. 2005) [citing *J.T. Eaton & Co. Inc., v. Atlantic Paste & Glue Co.*, 106 F.3d 1563, 1571 (Fed Cir 1997)].
63. *J.T. Eaton*, 106 F.3d at 1571 (holding the claimed invention obvious in spite of a finding of "at least moderate commercial success").

Disclaimer: This article was accepted for publication prior to the author joining the USPTO, and represents the author's own views, and not those of any of her past, present, or future employers, affiliates, clients, or any party.

16 The importance of chiral separations in single enantiomer patent cases

Charlotte Weekes

INTRODUCTION
The interest in the current state of the art in chiral separations of analytical chemists and other scientists working in the pharmaceutical industry is well known. However, to the pharmaceutical companies themselves, the state of the art of chiral separations in years gone by may be just as critical to their business. The field of chiral separations has featured strongly in recent pharmaceutical patent cases and has had a bearing on the outcome regarding the "inventiveness" of chiral drug molecules when patent law is applied. One case found its way to the then highest U.K. court, the House of Lords (now the Supreme Court), and in doing so redefined the scope of patent law.

PATENTS ACT 1977–A BRIEF INTRODUCTION
Revocation of a patent is most commonly sought on the following grounds:

a. The invention lacks novelty (anticipation)
b. The invention lacks an inventive step (obviousness)
c. The invention is not capable of industrial application
d. The specification of the patent does not disclose the invention clearly and completely enough for it to be performed by a person skilled in the art (insufficiency).

For an invention to be **anticipated** there are two requirements:

1. DISCLOSURE—the prior art (1) relied on must disclose subject matter that, if performed, would necessarily result in an infringement of the patent; and
2. ENABLEMENT—the ordinary skilled person must have been able to perform the invention if the person attempted to do so using the disclosed matter and *common general knowledge* (CGK).

If an invention is obvious to the *person skilled in the art* (the notional addressee of a patent) it **lacks inventive step**. The skilled person is a legal fiction. He is deemed to possess the CGK in the field to which the invention relates (2) but to lack inventive capacity. He is deemed to have read all publicly available documents, in whatever jurisdiction or language, but is not deemed to think laterally or know everything that comprises "state of the art." The addressee may be a skilled team whose combined knowledge allows the instructions in the patent to be carried out (3).

To be CGK, the information must be generally known and accepted as a good basis for further action by the bulk of those engaged in that art (2).

The Cases
Atorvastatin, 2005 (4)
(a "statin" for inhibiting cholesterol synthesis)

Although chiral separation of atorvastatin (Fig. 16.1) was not the issue here, the Judge commented that where a drug is a chiral molecule, the skilled person at the priority date (5) (May 1986) would expect only one of the enantiomers or diastereomers to be responsible for its pharmaceutical activity. This did not mean that chiral drugs had to be administered as single enantiomers but there was undoubtedly a tendency to prefer single enantiomers where resolution of the racemate is practicable.

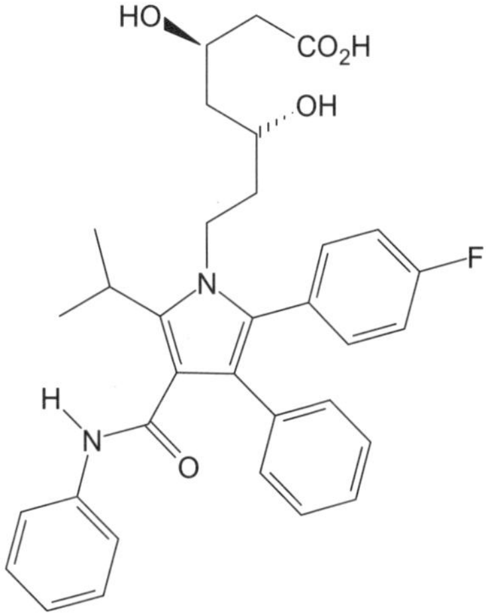

FIGURE 16.1 Atorvastatin, the active pharmaceutical ingredient of the Pfizer product Lipitor®.

Escitalopram, 2007–2009 (6)
(an antidepressant)

This notable case concerned an attempt by three generic manufacturers to revoke Lundbeck's patent. Escitalopram (Fig. 16.2A) was first marketed in 2002, the *priority date* of the patent (at which the state of knowledge of those skilled in the art had to be considered) was June 1988. The revocation arguments were based on the prior art racemate, citalopram, which was synthesized by Lundbeck in 1972 and launched as an antidepressant in 1989.

The argument of **lack of novelty** advanced by the generic manufacturers was that the claims in Lundbeck's patent did not exclude (+)-citalopram within the prior disclosed racemic mixture, therefore the racemate would infringe this later patent. The correct approach to a product claim in a patent is that it covers the product wherever found (7). The generic manufacturers did not suggest that disclosure of a racemate of itself amounted to disclosure of each of the enantiomers, the Judge agreed (previously considered by the European Patent Office (8)). The Judge said "if the claim properly construed, was simply to a product as such then the monopoly would indeed cover that product wherever it might be found." However, Lundbeck was clearly not laying claim to an

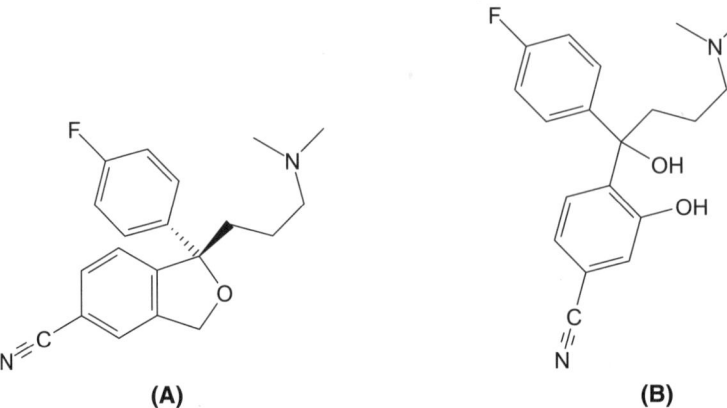

FIGURE 16.2 (A) Escitalopram (B) racemic diol intermediate in the synthetic route to citalopram.

unresolved moiety of the racemate—the patent title was "new enantiomers and their isolation" and the specification was clear that the racemate is not new—it related to the isolated product.

The **obviousness** attacks were based largely on what was said to have been CGK and, given that this invention had been made many years ago and the state of the art had advanced considerably since, the Judge emphasized the importance of avoiding hindsight when considering apparently obvious steps from something known.

The Judge found that a motive existed to resolve citalopram into its enantiomers and that testing their activity was trivial. Investigation of citalopram's enantiomers was an obvious goal for the ordinary skilled medicinal chemist in 1988. The activity of citalopram was well known; any medicinal chemist in 1988 would have appreciated that the enantiomers might have different activities; an inactive enantiomer was, at best, ballast but may be toxic; the regulators considered that an investigation of the enantiomers was desirable and this might in due course become mandatory.

Could the skilled person have achieved resolution of citalopram in June 1988?

- Citalopram could not easily be resolved by making diastereoisomers, the method the skilled person would naturally try.
- An earlier patent revealed a diol (Fig. 16.2B) as a chiral intermediate in the citalopram synthesis. A route from the diol to (+)-citalopram was argued as being obvious to the skilled person.
  - Attempting to form salts with the diol was routine albeit unpredictable. The salts crystallized differentially permitting resolution of the diol enantiomers. However, the Judge felt that the skilled person would not blindly go ahead with crystallization experiments without first satisfying himself that there was a real prospect of an S_N2 reaction working for the final step, and without the benefit of hindsight, the CGK was that the S_N1 reaction would be favored with the expectation of

losing stereochemistry. This was not a route the skilled person would attempt.
- The obviousness argument based on **chiral HPLC** included prior art showing the available chiral stationary phases (CSP) and expert evidence as to the identification of the appropriate CSP and parameters to conduct this separation. Post-priority publications were also relied on to show that separation could actually be obtained on CSP available at the relevant time.
- The Judge concluded:

> "this is one of those cases where each step seems very simple and logical with the benefit of hindsight. ... by the mid 1990s many techniques were routine which were still very much at an experimental stage in 1988 ... I do not believe it was obvious to resolve citalopram on a preparative scale using chiral HPLC in 1988 ... The ordinary skilled analytical chemist would have had no practical experience of preparative chiral HPLC and the ordinary skilled medicinal chemist would probably not have heard of it. The team would have been faced with a research program with an uncertain outcome."

Insufficiency was argued based on a House of Lords decision, *Biogen v. Medeva* (9). The Judge held that the claim to (+)-citalopram and the claim to a pharmaceutical composition containing (+)-citalopram cover all ways of making (+)-citalopram, which was an obviously desirable goal. The Judge's interpretation of *Biogen* was that the first person to find a way of achieving an obviously desirable goal is not permitted to monopolize every way of doing so. Lundbeck found a way of making (+)-citalopram, this was their technical contribution, and these claims extended beyond that.

The Judge therefore found the patent novel and nonobvious (i.e., inventive) but invalid for insufficiency

The Court of Appeal (10) overturned the Judge's insufficiency finding, stating that *Biogen* was limited to the form of claim in that case. When a product is new and inventive, the *technical contribution* is the product, and not the process by which it is made, even if that process is the only *inventive step*. This point was appealed to the House of Lords.

In the House of Lords (11), the Lordships agreed with the Court of Appeal that the way in which the Judge had applied *Biogen* went beyond that which was intended—*Biogen* only applied to the very unusual type of claim in that case and not to a straightforward product claim.

Lundbeck's patent for (+)-citalopram is valid.

The U.K. Intellectual Property Office granted a Supplementary Protection Certificate (SPC) for escitalopram, so despite the racemate being synthesized in 1972, no person other than Lundbeck can produce or market (+)-citalopram in the United Kingdom until May 2014. Clearly this raises policy issues. Should an obviously desirable product, resolved using known tools be entitled to such protection as to prevent others obtaining it in ways owing nothing to the "invention"?

Germany, April 2007 (12)*–2009* (13)

German case law states that a chiral compound is no longer novel in the form of one of its enantiomers if the skilled person's attention is concretely drawn in a prior publication to the enantiomer and if said person can prepare the compound by virtue of this concrete instruction and his general expert knowledge. The single enantiomer does not have to have been prepared before the priority date (14). At first instance, the Federal Patent Court held that obtaining the single enantiomer from an already known mixture of enantiomers is not questioned by the fact that its isolation involves a number of complicated ideas and analyses. If these are routine measures, standard for the skilled person, attributed to his average expert skill and do not require undue efforts of an inventive quality, the single enantiomer is regarded as readily accessible.

The German court considered that even if the common method of forming diastereomeric salts with a chiral acid is inadequate for resolution, the skilled person would employ the chiral chromatography method known to him and established pre-priority. *Wainer* (15) was referred to for commercially available CSP that the skilled person would use at the priority date. The ability to obtain (+)-citalopram was further evidenced by later published documents reporting chiral separation of citalopram (16,17).

In contrast to the United Kingdom, the German court held that analytical amounts were sufficient. The testing of commercially available CSP, including selecting the appropriate parameters, was not considered unreasonable in terms of the tests and analyses the skilled person would conduct, whereas the U.K. court considered this to extend the CGK too far.

Had the case ended there what impact might this divergence in application of patent law between territories have? If there was a European Court to consider the matter overall, which decision would have been applied?

On appeal, the Federal Court of Justice was unable to agree with the conclusion reached by the Patent Court, and in delivering its judgment cited various extracts from the U.K. decision. The court concluded that the person skilled in the art had reason to carry out attempts at producing or isolating the citalopram enantiomers but this did not result in an overwhelming need to provide the enantiomers. There was no obvious way for the person skilled in the art to obtain the citalopram enantiomers at the priority date and this included the use of chiral HPLC to achieve separation which was not an obvious method to resort to.

Levofloxacin, 2008 (18)

(the (−)-enantiomer of ofloxacin, a member of the quinolone class of antimicrobial agents)

The rejection of the insufficiency attack by the Court of Appeal in escitalopram was acknowledged, as were the findings of the Judge himself in that case on novelty. The patent expired in June 2006 but an SPC existed to protect levofloxacin (Fig. 16.3) until June 2011. The SPC's validity was challenged in two ways: (*i*) the patent was invalid and (*ii*) the SPC should not have been granted.

Preparative HPLC experiments were relied on to demonstrate that chiral separation could have been achieved but it was not possible to show that the equipment used was available, let alone CGK in 1985. Although the same outcome may well have been achieved on this equipment, proving it is problematic.

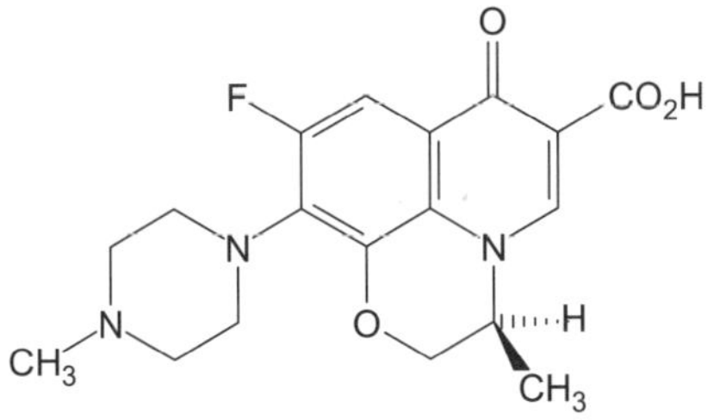

FIGURE 16.3 Levofloxacin (commonly prepared as its hemihydrate).

The Judge concluded:

"by 1985 the skilled person would have been aware of the particular promise of ofloxacin as a pharmaceutical and its chiral nature. It was possible one enantiomer might have more activity than the other and ... would retain the other beneficial qualities of the racemate ... [H]e would have considered it worthwhile exploring whether ofloxacin could be resolved, but only to a point."

The Judge did not believe it was a goal obvious to pursue relentlessly.

He was not persuaded that the skilled person would have achieved separation of the diastereomers by preparative HPLC in 1985. Chiral separation would have involved a research program with an uncertain outcome, and in the Judge's opinion, the person skilled in the art would have turned attention to development of new molecules if the enantiomers could not be separated relatively easily.

For an **SPC** to be valid, it must be for a product that has not already been subject to a certificate and be based on the first marketing authorization to place the product on the market as a medicinal product. The argument advanced was that the authorization to sell ofloxacin was the first to sell levofloxacin because it is an active component of ofloxacin. The Judge rejected this. It is to be noted that in recent years the U.K. courts have referred many questions of interpretation relating to the SPC Regulation (Council Regulation (EEC) No. 1768/92, now 469/2009) to the Court of Justice of the European Union and many remain to be answered, but in this instance, the court considered the question to be clear.

The Court of Appeal (19) noted that from all practical points of view and that of U.K. patent law, levofloxacin was a new product and the earlier authorization did not entitle Daiichi to market it in the U.K. as such. Once the ofloxacin patent expired anyone could market ofloxacin; levofloxacin could not be marketed due to the levofloxacin patent and prior to this no one could market it because invention was needed to make it.

The authorization for levofloxacin was the first for that active ingredient alone. One Judge remarked that the position *might* have been different if the other component of the racemic mixture had been inactive biologically, but it was a moot point.

CONCLUSION

Not only can a single enantiomer of a known racemate be entitled to 20 years patent protection, it can also be protected for a further five years by an SPC.

It can be seen that when patent law is strictly applied to the facts where chiral separations are concerned, many policy issues arise as to what is the appropriate scope of protection for "inventions" of this type. It is evident that, even though patent law is supposed to be the same throughout Europe, courts in different jurisdictions consider chiral separations differently, which can lead to conflicting outcomes.

A point of note is how important publications of chiral separations are to patent litigation. One problem that lawyers and expert witnesses face is where subject matter has been considered too obvious to be published! Without publications to support expert opinion on what was CGK at the relevant time, it can be difficult to prove that separation could "easily" be achieved in instances when it, in fact, was.

It is unlikely that the cases discussed above will be the last to be seen in relation to chiral separations. Under the U.K. law which has resulted from these cases, if a racemate could easily be resolved using known methods then arguably the enantiomers are obvious and not entitled to patent protection. Furthermore, if a patent claims an enantiomer but does not disclose the method by which to obtain it clearly and completely enough, arguably it is insufficient.

There may also be instances where the infringement of claims relating to enantiomers is unclear and the court's assistance is required to assess the proper interpretation of the claims. For example, a claim to the use of magnesium esomeprazole with an enantiomeric purity of $\geq 99.8\%$ enantiomeric excess for the manufacture of a medicament for the inhibition of gastric acid secretion was asserted against a product that did not contain magnesium esomeprazole of that enantiomeric purity (albeit that the process of manufacture did begin with magnesium esomeprazole of that purity to which a quantity of the omeprazole racemate was added). The product was held not to infringe (20).

REFERENCES

1. Prior art, or state of the art, constitutes all information that has been made available to the public in any form before the priority date.
2. *Beloit Technologies Inc v. Valmet Paper Machinery*, 1997, RPC 489.
3. Genentech Inc's patent, 1989, RPC 147.
4. *Ranbaxy UK Ltd. v. Warner-Lambert Co.*, 2006, FSR 14.
5. The priority date is the date at which a patent application is filed. It is the cutoff point for determining what is included in the "state of the art" against which the novelty or inventive step of the claimed patent is assessed.
6. *Generics (UK) Limited & Ors v. H. Lundbeck A/S*, 2007, RPC 32.
7. *Merrell Dow Inc v. H. N. Norton & Co. Ltd.*, 1996, RPC 76.
8. T1046/97&T0296/87 (*Technical Board of Appeal*).
9. *Biogen Inc. v. Medeva Plc*, 1997, RPC 1.
10. *H. Lundbeck A/S v. Generics (UK) Limited & Ors*, 2008, EWCA Civ 311.
11. *Generics (UK) Limited & Ors v. H. Lundbeck A/S*, 2009, EWHL 12.
12. *Neolab Ltd. & Ors v. H. Lundbeck A/S* (Ni 352 Federal Patent Court).
13. *H. Lundbeck A/S v. Neolab Ltd. & Ors* (Docket No. Xa ZR 130/07).
14. Cf. BGH GRUR 1978, 696, 698—Aminobenzylpenicillin; BPatG (Federal Patent Court) GRUR Int. 1996, 822—Herbicid wirksames Enantiomer.
15. Wainer IW. Classification of chiral stationary phases. Trends Anal Chem 1987; 6: 125–134.
16. Haupt D. Determination of citalopram enantiomers in human plasma by liquid chromatographic separation on. J Chromatogr B 1996; 685:299–305.
17. Rochat B, Amey M, Baumann P. Analysis of enantiomers of citalopram and its demethylated metabolites in plasma of depressive patients using chiral reverse-

phase liquid chromatography. Ther Drug Monit 1995; 17(3):273–279; Rochat B, Amey M, Van Gelderen H, et al. Determination of the enantiomers of citalopram, its demethylated and propionic acid metabolites in human plasma by chiral HPLC. Chirality 1995; 7(6):389–395.
18. *Generics (UK) Limited v. Daiichi Pharmaceutical Co. Ltd. & Daiichi Sankyo Co Ltd.*, 2008, EWHC 2413.
19. *Generics (UK) Limited v. Daiichi Pharmaceutical Co. Ltd. & Daiichi Sankyo Co Ltd.*, 2009, EWCA Civ 646.
20. *Ranbaxy (UK) Limited v. AstraZeneca AB*, 2011, EWHC 1831.

FOOTNOTE

Article adapted with permission from *The Importance of Chiral Separations in Single Enantiomer Patent Cases*, Weekes C, Volume 2 Issue 3, pp. 12–14, printed in *Chromatography Today* published on behalf of *The Chromatographic* Society.

Index

ABC. *see* ATP-binding cassette (ABC)
ABC transporters, 175–178
  domains of, 175
  MRP1, 177–178
  Pgp, 176–177
Abridged applications, 247
Absorption
  intestinal absorptive/secretory transporters, 207–208
  oral, 206–208
  passive intestinal, 206–207
Absorption, distribution, metabolism, excretion, and toxicity (ADME/Tox) parameters, 182
AC. *see* Adenylyl cyclase (AC)
ACE. *see* Affinity capillary electrophoresis (ACE)
Acetate, 150
Acetic acid, 127
Achiral, 20
Active substance
  chemical development, 246
  finished medicinal product, 246
  quality of, 245–246
  synthesis of, 244
Additives
  BGE, effect of organic modifiers as, 155
  and enantioselective chromatography screening, 126–127
  in SFC, 142–143
Adenylyl cyclase (AC), 274
ADME/Tox parameters. *see* Absorption, distribution, metabolism, excretion, and toxicity (ADME/Tox) parameters
Affinity capillary electrophoresis (ACE), 190, 193–195
  chiral selectors in, 194
  drug-protein interactions in, 193
  principle of mobility shift assay in, 194
Aging
  and plasma protein binding, 198
  and stereoisomer disposition, 217
AGP. *see* α_1-acid glycoprotein (AGP)
Albumin, binding strength of, 186, 208
Albuterol, 225
Alcohol-acetonitrile systems, 124
α–methoxy–α–(trifluoromethyl)-phenylacetyl chloride (MTPA-Cl), 82
α-methyl-α-phenylsuccinimide
  HPLC/SMB separations of, 126
  loading studies using ethyl acetate/chloroform, 125
α-methylbenzyl isocyanate (MBIC), 85

α_1-acid glycoprotein (AGP), 95, 96, 182, 186–189, 194
  glycosylation of, 195
  plasma concentration of, 189
Amides
  formation of, 82–85
Amines
  derivatization with isocyanates, 85
Amino acids (AA)
  determination of, 87–88
Amino group, derivatization of, 76–86
Amitriptyline, 189
Amlodipine, 221
Ammonium citrate, 150
Amphetamine, 84
Amphetamine-type stimulants (ATS), 84
Anion-exchange (AX) CSP, 106–107
  structure of chiral, 107
Anticipation, enantiomers, 295–296
  by prior art disclosure of racemate, 296
  of species by prior art genus, 296
Applied voltage
  effect on enantioseparation, 154–155
Aptamers, 159
Astec, 102–103
Asymmetric carbons, 48
Asymmetry
  molecular, and optical activity, 2–6, 15
Atorvastatin, 305
ATP-binding cassette (ABC), 171
ATS. *see* Amphetamine-type stimulants (ATS)
AX CSP. *see* Anion-exchange (AX) CSP

Background electrolyte (BGE) solution, 147
  composition effect on enantioseparation, 150–151
  effect of organic modifiers as additives to, 155
  ionic strength, and enantioseparation, 151–152
  ph of, and enantioseparation, 151
β-adrenoreceptor antagonists (β–blockers), 220–221
β_2AR agonists, 40, 277, 279, 280
  interaction with (–)-(R)-isoproterenol/(+)-(S)-isoproterenol, 282–288
β_2AR$_{CAM}$, 277
  interaction with (-)-(R)-isoproterenol/(+)-(S)-isoproterenol, 282–286
β_2AR-G$_{s\alpha}$ fusion protein, 281–282
BCRP. *see* Breast cancer resistance protein (BCRP)
β-cyclodextrin, 42
β–cyclodextrin, 168
Benzodiazepines, 42
BGE solution. *see* Background electrolyte (BGE) solution

313

Biliary clearance, racemic drugs, 216–217
Bimoclomol, 188
Bioavailability, 213
Blockbuster drugs, 241
Borate, 150
Breast cancer resistance protein (BCRP), 175, 207
BTCC. see N-(tert-butylthiocarbamoyl)-L-cysteine ethyl ester (BTCC)
Buffers, CE
   used for enantioseparations, 150–151
Bupivacaine, 223
β_x-adrenergic receptors (β_xARs), 274–275.
   see also G-protein-coupled receptors (GPCRs)
   interaction of agonist stereoisomers with, 278–281
   subtypes, 274
β_xARs. see β_x-adrenergic receptors (β_xARs)

Cahn-Ingold-Prelog (C.I.P.) convention, 24, 25–27, 28
   sequence rule, 26
Calcium channel antagonists, 221–222
Candesartan, 178
Capillary electrochromatography (CEC), 158–161
   HPFA, 190
   molecularly imprinted polymers, 160–161
   monolithic stationary phase, 159–160
   open-tubular, 158–159
   packed, 159
Capillary electrophoresis (CE), 42, 71, 99, 147
   buffers, 150–151
   chiral selectors used in, 149, 152–154
   enantioseparation by, 150–156
   instrumentation, schematic representation of, 148
Capillary gel electrophoresis (CGE), 148
Capillary isoelectric focusing (CIEF), 148
Capillary isotachophoresis (CITP), 148
Capillary zone electrophoresis (CZE), 147, 245
CAPS. see 3-cyclohexylamino-1-propanesulphonic acid (CAPS)
Carbamates, formation of, 80–82
Carbon atom
   asymmetric, 12
   tetrahedral model for, 12
Carbon dioxide, 141
Carboxylic groups, derivatization of, 87
Cardiotoxicity, 223
Carvedilol, 221
CCPA. see Court of Customs and Patent Appeals (CCPA)
CD. see Cyclodextrins (CD)
CDA. see Chiral derivatizing agent (CDA)
CDER. see Center for Drug Evaluation and Research (CDER)
CDR. see Chiral derivatizing reagent (CDR)
CE. see Capillary electrophoresis (CE)
CEC. see Capillary electrochromatography (CEC)
Cellcoat, 100
Cellular membrane affinity chromatography (CMAC), 171–172
CelluloZe, 101

Center for Drug Evaluation and Research (CDER), 250
Center of chirality, 20, 26
Cetirizine, 226
CGE. see Capillary gel electrophoresis (CGE)
CGK. see Common general knowledge (CGK)
Chemistry, manufacturing, and controls (CMC), 256
Chemistry of Active Substances, 243
Chemistry of New Active Substances, 243–244
CHES. see Tris, 2-(N-cyclohexylamino) ethanesulfonic acid (CHES)
Chinese hamster lung (CHL), 77
Chinese hamster ovary (CHO), 279
Chiral
   capillary electrochromatography (see Capillary electrochromatography (CEC))
   discrimination, 64
   discrimination in metabolism, 212
   drugs
      analysis of, 70, 167–169
      development of, 261–269
      U.S. patentability perspective and, 294–300
   free-energy differences, 41
   HPLC screening, 98–99
   inversion, and metabolism, 213–215
   plane, axis, 21
   separation, 98
   stereoisomers, 20
   surfactants, 156
Chiral Active Substances, 243
Chiral-AGP, 106
Chiral-CBH, 106
Chiralcel OD, 98, 100–101
Chiralcel OJ, 98
Chiral derivatizing agent (CDA), 69
Chiral derivatizing reagent (CDR), 69–70, 72–75, 78–79, 88
   chemical structure for derivatization, 73, 74–75
   derivatized enantiomers, 72
   groups, 71
   selection of, 73
Chiral-HSA, 106
Chiral inversion, 213–215, 246
Chirality, 43
   classification of crystal structures, 50
   of new drug introductions in Japan, 265–266
Chiralpak AD, 98
Chiralpak AS, 98
Chiralpak MA(+), 109
Chiralpak QD-AX, 106
Chiralpak QN-AX, 106
   method development on, 108
Chiralpak WH, 109
Chiral recognition, 30–45
   conformationally driven models, 36–37
   four-point models, 34–36
   models of, 30–38
   multiple-point model, 36
   thermodynamics and (see Thermodynamics)
   three-point models, 31–33

INDEX

Chiral selectors
 in ACE, 194
 in CE, 149, 152–154
Chiral stationary phases (CSP), 69, 113–114
 classes of, 100
 commercially available, 100–109
 cyclofructan, 104–105
 DACH-DNB, 102
 derivatized polysaccharide, 100–101
 HPLC, trends in development, 95–98
 ion-exchange, 106–107
 macrocyclic antibiotics, 102–103
 oligosaccharide, 103–105
 ovomucoid, 106
 Pirkle-type, 95, 96, 101–102
 polymeric, 109
 protein, 106
 research/future directions, 109
 Whelk O1, 101, 102
Chiral switching, 223–224, 226, 241, 248
 clinical examples of, 224
Chiral Technologies, 98, 99
Chirosil RCA(+), 108
Chirosil SCA(–), 108
CHL. see Chinese hamster lung (CHL)
Chlorpheniramine, 187
CHMP. see Committee for Human Medicinal
 Products (CHMP)
CHO. see Chinese hamster ovary (CHO)
Chromatography, 113–144
 analytical, 115
 capillary electrochromatography, 158–161
  (see also Capillary electrochromatography
  (CEC))
 column efficiency, 117, 118–121, 128
 column saturation capacity, 117
 enantioselective, 121–127 (see also
  Enantioselective chromatography)
 high-performance affinity (see High-performance
  affinity chromatography (HPAC))
 micellar electrokinetic, 147, 156–157
 microemulsion electrokinetic, 158
 optimization/scale-up, 127–128
 overlapping injections, 128–131
 parameters, 71
 principles of, 114–121
 of racemic mixtures, 75–76
 shave recycle, 131–132
 simulated moving bed, 133–140 (see also
  Simulated moving bed (SMB)
  chromatography)
 Steady-State Recycling, 133
 supercritical fluid (see Supercritical fluid
  chromatography (SFC))
CIEF. see Capillary isoelectric focusing (CIEF)
C.I.P. convention. see Cahn-Ingold-Prelog (C.I.P.)
 convention
Circular dichroism (CD), 167–168
 vibrational, 168
Cis-trans isomers, 22–24
CITP. see Capillary isotachophoresis (CITP)
Clomipramine, 189

CMAC. see Cellular membrane affinity
 chromatography (CMAC)
CMC. see Chemistry, manufacturing, and controls
 (CMC); Critical micellar concentration (CMC)
CoA thioester. see Coenzyme A (CoA) thioester
Coenzyme A (CoA) thioester, 213
Column efficiency, chromatography, 117,
 118–121, 128
CoMFA. see Comparative molecular fields analysis
 (CoMFA)
Committee for Human Medicinal Products
 (CHMP), 243, 250
Common general knowledge (CGK), 304
Common technical document (CTD), 250
Comparative molecular fields analysis (CoMFA),
 43–44
Compliance and Narcotics Division, PFSB, 260
Configuration, 28
Conformationally driven models, chiral recognition,
 36–37
Conglomerate systems, 53–59
 characterization of, approach for, 55
Court of Customs and Patent Appeals (CCPA), 298
Critical micellar concentration (CMC), 156
Crown ethers, 154
Crownpak® CR(+) CSP, 107
Crownpak® CR(–) CSP, 108
Crystallization, 36
Crystallography, of chiral compounds, 49–52
CSP. see Chiral stationary phase (CSP)
CTD. see Common technical document (CTD)
C-terminus, 274
Cyclobond I SN CSP, 104
Cyclobond™ I 2000 DNP, 104
Cyclodextrin, 103–104
Cyclodextrins (CD), 152–153
Cyclofructan CSP, 104–105
3-cyclohexylamino-1-propanesulphonic acid
 (CAPS), 150
CYP. see Cytochrome P450 (CYP)
Cytochrome P450 (CYP), 212, 219
CZE. see Capillary zone electrophoresis (CZE)

DACH-DNB CSP, 102
DBD-PyNCS. see 4-(3-isothiocyanatopyrrolidin-1-
 yl)-7-(N,N-dimethylaminosulfonyl)-2,1,3-
 benzoxadiazole (DBD-PyNCS)
DCC. see Dicyclohexylcarbodiimide (DCC)
DDITC. see (S,S)-N-3,5-dinitrobenzoyltrans-
 diaminocyclohexane-isothiocyanate
 (DDITC)
Debrisoquine, 215
Dehydrosilybin, 178
Delmopinol, 86
Derivatization, 71
 of amines with isocyanates, 85
 of amino group, 76–86
 of amphetamines, 82
 of carboxylic groups, 87
 chemical structure of CDR used for, 73, 74–75
 of hydroxyl groups, 86–87
 reaction of enantiomer, 72

Derwent World Drug Index, 176
Desipramine, 189
Desmethylclomipramine, 189
Deuel, 298
Dextran sulfate, 154
Dextromethorphan, 40
Dextromethorphan molecule, 26
Dextrorotary, 28
Diastereomers, 20, 22, 28, 49, 69, 175, 305
  chemical formation of, 7–8
  enantiomers *vs.*, 22
  examples of, 23
1,4-dichlorocyclohexane, 24
Dichloromethane, 124
Dicyclohexylcarbodiimide (DCC), 84
Diet, and stereoisomer disposition, 217–218
Dihydropyridines, 221
Dilevalol, 221
Dimethylsulfoxide (DMSO), 196
Directive 2001/83/EC, EU, 243, 248
Dissymmetry, 17–19
Distomer, 28
Distribution
  drugs, 208–211
  tissue, 210–211
DMSO. *see* Dimethylsulfoxide (DMSO)
Dobutamine, 219
3D QSAR models, 43–45
Drug-drug interactions, pharmacokinetics, 218–219
Drugs
  absorption, 206–208
  chiral recognition of, 195
  distribution, 208–211
  excretion, 216–217
  interaction with plasma proteins, 183
  metabolism, 211–215
  stereochemistry of, 264, 265
  transporter (*see* Transporter, drug)

Electromigration techniques, 147–162
  disadvantage of, 161
  instrumentation for, 147
  overview of, 147–150
  vs. HPLC, 161–162
Electrospray ionization–mass spectrometry (ESI-MS), 167
EMA. *see* European Medicines Agency (EMA)
Enantiomeric purity (EP), 52
Enantiomers, 6, 14, 19, 20–21, 28, 31, 69, 175, 209, 219–220, 294–300, 305
  anticipation, 295–296
  derivatization reaction of, 72
  development of, 223
  drug, techniques for determination/isolation of, 167–169
  3D structures of, 30
  examples of, 21
  of ibuprofen, 87
  naming, 24
  physical properties of, 7–8
  prima facie case of obviousness for, 298–299
  procedure for separation of, 62
  of propoxyphene, 220
  separation/resolution of, 48–66
  single, 246, 247
    patent cases, 304–309
  structures of, 21
  verapamil, fit of, 174
  vs. diastereomers, 22
Enantioselective chromatography, 121–127
  screening, 121–122
    additives, 126–127
    solvent mixtures for, 122–126
Enantioselectivity, 42
  QSAR (*see* Quantitative structure-activity relationships (QSAR))
Enantioseparation
  of 2-aryloxypropionic acids, 159–160
  by capillary electrophoresis, 150–156
    effect of applied voltage/temperature on, 154–155
    effect of BGE on
      composition, 150–151
      ionic strength, 151–152
      pH, 151
    optimization of, 155–156
Enthalpic factor, 39
Entropic factor, 39
EP. *see* Enantiomeric purity (EP)
Ephedrine, 81, 85
Epimers, 49
Epinephrine, 278
Eprosartan, 178
Equilibrium dialysis, 190
Erythrose, 25
Escitalopram, 305–307
ESI-MS. *see* Electrospray ionization-mass spectrometry (ESI-MS)
Ethambutol, 222–223
Ethyl acetate, 124
Ethyl diacetyltartrate
  melting point phase diagram for, 59, 61
EU. *see* European Union (EU)
Europe
  implementation of ICH in, 250
European Medicines Agency (EMA), 243, 250
European Union (EU)
  current legislation, 242–243
  Investigation of Chiral Active Substances, 243–248
  principle, 241–242
  regulatory aspects, 248
  regulatory requirements for pharmaceuticals in, 241–248
Eutomer, 28
Evaluation and Licensing Division, PFSB, 260
Evanescent wave, 196
EWG. *see* Expert Working Group (EWG)
Excretion
  biliary clearance, 216–217
  renal clearance, 216
Expert Working Group (EWG), 249

Famprofazone, 82
FDA. *see* Food and Drug Administration (FDA)

INDEX

Fenfluramine, 84
Fenoterol stereoisomers
  pharmacological effects of, 280
Fexofenadine, 178, 207
Fischer convention, 24–25
FLEC. see 1-(9-fluorenyl) ethyl chloroformate (FLEC)
1-(9-fluorenyl) ethyl chloroformate (FLEC), 80–82, 86
  and (R)-amphetamine, 80
Flurbiprofen, 199
Food and Drug Administration (FDA), 223, 250, 254–259
  approved new molecular entities, 267
  guidance, 257–259
  policy statement, 255–257
Forest Labs., Inc. v. Ivax Pharms., Inc, 296
Four-point models, chiral recognition, 34–36
Free-energy differences, chiral, 41
Frontal analysis, 191, 192
Functional selectivity, 278

Gas chromatography (GC), 69, 76, 99, 169
GC. see Gas chromatography (GC)
GC-MS (SIM) chromatograms, 82, 83
GEF. see Guanine nucleotide exchange factor (GEF)
Gender, and stereoisomer disposition, 217
Geometric isomers
  cis-trans, 22–24
GITC. see 2,3,4,6-tetra-O-acetyl-β-D-glucopyranosyl-isothiocyanate (GITC)
Glut1, 171
Glyceraldehyde, 24
Glycosylation, of AGP, 195
GPCR-Gα fusion proteins, 281–282
GPCRs. see G-protein-coupled receptors (GPCRs)
G-protein-coupled receptors (GPCRs), 274. see also β$_x$-adrenergic receptors (β$_x$ARs)
  models of activation, 275–278
  two-state model vs. multistate model, 275–278
G-proteins, 274, 286
  cycle, 275
Graham factors, 297
GTP. see Guanosine 5'-triphosphate (GTP)
Guanine nucleotide exchange factor (GEF), 274
Guanosine 5'-triphosphate (GTP), 282, 286

Hatch-Waxman Act, 223
HDL. see High-density lipoprotein (HDL)
HEPES. see N-2-hydroxyethylpiperazine-N-2-ethanesulfonic acid (HEPES)
HFBOPCl. see (2S,4R)-N-heptafluorobutyryl-4-heptafluoro-butyroxy-prolyl chloride (HFBOPCl)
High-density lipoprotein (HDL), 190, 195
Highest occupied molecular orbital (HOMO), 42
High-performance affinity chromatography (HPAC), 186, 190–193
High-performance frontal analysis (HPFA), 190
High-performance liquid chromatography (HPLC), 69, 76, 86, 113, 140–141, 245
  chiral screening, 98–99
  chiral stationary phases, trends in development, 95–98
  disadvantages of, 161
  electromigration techniques vs., 161–162
  high-performance affinity chromatography (see High-performance affinity chromatography (HPAC))
Hill function, 227
HNEA. see Trans-4-hydroxy-2-nonenoic acid (HNEA)
hOCT1, 172
HOMO. see Highest occupied molecular orbital (HOMO)
HPAC. see High-performance affinity chromatography (HPAC)
HPFA. see High-performance frontal analysis (HPFA)
HPLC. see High-performance liquid chromatography (HPLC)
HSA. see Human serum albumin (HSA)
Human serum albumin (HSA), 182, 184–186
Hydroxyl groups
  derivatization of, 86–87
4-hydroxymethyl-2-oxazolidinone, 53
5-hydroxymethyl-2-oxazolidinone, 53
4-hydroxypropranolol, 77
Hydroxypropyl-β-cyclodextrin, 42
Hypnotics, 222
Hypoalbuminemia, 199

IAM-PC. see Immobilized artificial membrane-phosphatidylcholine (IAM-PC)
Ibuprofen, 87, 192
ICH. see International Conference on Harmonisation (ICH)
ICH Steering Committee, 249
Imipramine, 189
Immobilized artificial membrane-phosphatidylcholine (IAM-PC), 171, 172
Impurities, ICH guidelines on, 252–253
Indole-benzodiazepine binding site, 186
Indolopyrimidine, 178
Injections, sample
  overlapping, and chromatography, 128–131
Inosine 5'-triphosphate (ITP), 282, 286
Interaction patterns, molecular, 30, 31
International Conference on Harmonisation (ICH), 241, 248–254
  common technical document, 250
  guidelines
    on analytical validation, 253–254
    development of, 249
    on impurities, 252–253
    on specifications and tests, 250–252
  implementation in Europe/United States/Japan, 250
  overview, 248–249
International Union of Pure and Applied Chemistry (IUPAC), 17
Investigation of Chiral Active Substances, 243–248
  chemistry/pharmacy aspects, 243–246
Ion-exchange CSP, 106–107

Isomers
    cis-trans, 22–24
    optical, 48
Isoproterenol, 278, 279–280
4-(3-isothiocyanatopyrrolidin-1-yl)-7-(N,N-dimethylaminosulfonyl)-2,1,3-benzoxadiazole (DBD-PyNCS), 79–80
ITP. see Inosine 5′-triphosphate (ITP)
IUPAC. see International Union of Pure and Applied Chemistry (IUPAC)

Japan
    chirality of new drug introductions in, 265
    implementation of ICH in, 250
    regulatory requirements for pharmaceuticals in, 259–263

Ketamine, 225
    concentration-effect relationship, 227, 228
Ketoprofen, 199

Labetalol, 221
Lactic acid, 13
Langmuir isotherm, 115, 138
L-carnitine, 86
LDL. see Low-density lipoprotein (LDL)
Levofloxacin, 308–309
Levomethorphan, 40
LFER. see Linear free energy relationships (LFER)
Lifestyle, and stereoisomer disposition, 217–218
Ligand-binding process, 39
Linear free energy relationships (LFER), 42
Lipophilicity, 42
Lipoproteins, 189–190
Lisdexamfetamine, 226
losartan, 178
Low-density lipoprotein (LDL), 190, 195
Lowest unoccupied molecular orbital (LUMO), 42
LUMO. see Lowest unoccupied molecular orbital (LUMO)
Lundbeck, 305–307
LY329146, MRP1 inhibitors, 178
LY402913, MRP1 inhibitors, 178

Macrocyclic antibiotics, 102–103, 153
    chemical structures of, 153
Macromolecule, 37
Major facilitator superfamily (MFS), 171, 172
Maprotiline, 189
MBIC. see α-methylbenzyl isocyanate (MBIC)
MDA. see 3,4-methylenedioxyamphetamine (MDA)
MDEA. see 3,4-methylenedioxyethylamphetamine (MDEA)
MDMA. see 3,4-methylenedioxymethamphetamine (MDMA)
MDR. see Multidrug resistance (MDR)
MEEKC. see Microemulsion electrokinetic chromatography (MEEKC)
Mefloquine, 85
Mefloquine (MQ), 177

MEKC. see Micellar electrokinetic chromatography (MEKC)
Mephobarbital (MPB), 198
Mepivacaine, 223
MES. see Morpholinoethanesulfonic acid (MES)
Meso-diastereomer, 22
Meso-tartaric acid, 22
Metabolism, drugs, 211–215
    chiral inversion, 213–215
    first-pass, 213
    genetic polymorphisms in, 215
    phase I/phase II, 212–213
Methadone, 189
Methamphetamine, 81, 84
Methaqualone, 21
1-(6-methoxy-2-naphthyl)ethyl isothiocyanate (NAP-IT), 79
2-methoxy-2-(1-naphthyl)propionic acid (M-α-NPA), 86
Methyl diacetyltartrate, 54, 55
Methyl dipropionyltartrate
    melting point phase diagram for, 59, 60
3,4-methylenedioxyamphetamine (MDA), 81
3,4-methylenedioxyethylamphetamine (MDEA), 82
3,4-methylenedioxymethamphetamine (MDMA), 81
Methyl t-butyl ether (MTBE), 117, 124
MFS. see Major facilitator superfamily (MFS)
MHLW. see Ministry of Health, Labour and Welfare (MHLW)
MHW. see Ministry of Health and Welfare (MHW)
Micellar electrokinetic chromatography (MEKC), 147, 156–157
Microemulsion electrokinetic chromatography (MEEKC), 158
Ministry of Health, Labour and Welfare (MHLW), 250
Ministry of Health and Welfare (MHW), 259
MIP. see Molecularly imprinted polymers (MIP)
Mirror images, 4, 17, 18, 19, 48
Mobile phase
    choice of, 123, 142
    effect on adsorption, 125
Molecular asymmetry
    and optical activity, 2–6, 15
Molecularly imprinted polymers (MIP), 160–161
Molecular micelles, 157
Monolithic stationary phase capillaries, 159–160
Morbidelli Triangle, 138
Morpholinoethanesulfonic acid (MES), 150
MPB. see Mephobarbital (MPB)
MQ. see Mefloquine (MQ)
MRP2. see Multidrug resistance-associated protein 2 (MRP2)
MRP1 transporter, 177–178
    inhibitors, 178
    Pgp vs., 178
MSTFA. see N-methyl-N-triethylsilyl trifluoroacetamide (MSTFA)
MTBE. see Methyl t-butyl ether (MTBE)
MTPA. see (R)-(+)-α-methoxy-α-(trifluoromethyl)phenylacetic acid (MTPA)

INDEX

MTPA-Cl. see α–methoxy–α–(trifluoromethyl)-phenylacetyl chloride (MTPA-Cl)
Multidrug resistance–associated protein 2 (MRP2), 207, 216
Multidrug resistance (MDR), 176

1-(1-naphthyl)ethylisocyanate (NEIC), 85
NAP-IT. see 1-(6-methoxy-2-naphthyl)ethyl isothiocyanate (NAP-IT)
Naproxen, 69–70
NBD. see Nucleotide binding domains (NBD)
NDA. see New Drug Application (NDA)
NEIC. see 1-(1-naphthyl)ethylisocyanate (NEIC)
New Drug Application (NDA), 254
New molecular entities (NMEs), 266
N-2-hydroxyethylpiperazine-N-2-ethanesulfonic acid (HEPES), 150
Nitrogen, 19
NMEs. see New molecular entities (NMEs)
N-methyl-N-triethylsilyl trifluoroacetamide (MSTFA), 82
N-methylquinidine, 176–177
N-methylquinine, 176–177
NMR spectroscopy. see Nuclear magnetic resonance (NMR) spectroscopy
Nonsteroidal anti-inflammatory drugs (NSAIDs), 186
Norepinephrine, 278
Nortriptyline, 189
NSAIDs. see Nonsteroidal anti-inflammatory drugs (NSAIDs)
N-(tert-butylthiocarbamoyl)-L-cysteine ethyl ester (BTCC), 88
Nuclear magnetic resonance (NMR) spectroscopy, 71, 168, 245
Nucleotide binding domains (NBD), 175

OATPs. see Organic anion-transporting polypeptides (OATPs)
OATs. see Organic anion transporters (OATs)
Oligosaccharide CSP, 103–105
Omeprazole, 177, 226
OP. see Optical purity (OP)
Open-tubular CEC (OT-CEC), 158–159
OPSR. see Organization for Pharmaceutical Safety and Research (OPSR)
Optical activity, 1, 20, 28
  molecular asymmetry and, 2–6
Optical purity (OP), 52
Organic anion transporters (OATs), 216
Organic anion-transporting polypeptides (OATPs), 207, 216
Organic cation transporters (OCTs), 216
Organisms
  usage of, 8–15
Organization for Pharmaceutical Safety and Research (OPSR), 260
Orosomucoid (ORM). see α_1-acid glycoprotein (AGP)
OT-CEC. see Open-tubular CEC (OT-CEC)
OVM CSP, 106
Ovomucoid CSP (OVM CSP), 106

Packed capillaries, 159
PAL. see Pharmaceutical Affairs Law (PAL)
Pantoprazole, 226
Paratartaric acid
  methods to resolve, 6–15
Patentability, enantiomers/racemates, 294–300
  anticipation, 295–296
  obviousness, 297–300
  overview, 294–295
Patents act 1977, 304–309
PEIC. see 1-phenylethyl isocyanate (PEIC)
Penicillamine, 222
PEPT1. see Peptide transporter (PEPT1)
Peptide transporter (PEPT1), 207
Pfizer Inc. v. Ranbaxy Labs, 299
PFSB. see Pharmaceutical and Food Safety Bureau (PFSB)
P-glycoprotein, 211, 216
Pgp transporter, 176–177
  MRP1 vs., 178
Pharmaceutical Affairs Law (PAL), 260
Pharmaceutical and Food Safety Bureau (PFSB), 259–260
  divisions of, 260
Pharmaceutical and Medical Devices Agency (PMDA), 250, 260
Pharmaceutical and Medical Devices Evaluation Center (PMDEC), 260
Pharmaceutical and Medical Safety Bureau (PMSB), 250
Pharmaceutical patent cases, 304–309
  atorvastatin, 305
  escitalopram, 305–307
  levofloxacin, 308–309
Pharmacodynamics (PD), 219–223
  β-adrenoreceptor antagonists, 220–221
  calcium channel antagonists, 221–222
  enantiomers of propoxyphene, 220
  racemic mixture, 219–220
Pharmacokinetic-pharmacodynamic modeling, 226–228
Pharmacokinetics (PK)
  absorption, 206–208
  age-related differences, 217
  diet/lifestyle, 217–218
  distribution, 208–211
  drug-drug interactions, 218–219
  excretion, 216–217
  gender-related differences, 217
  metabolism, 211–215
Pharmacologically active eutomers, 278
Phenoxymethylquinoxalinone II, 178
1-phenylethyl isocyanate (PEIC), 85–86
Phosphate, 150
Phosphorus, 19
Piperazine-N,N-bis(2-ethanesulfonic acid) (PIPES), 150
PIPES. see Piperazine-N,N-bis(2-ethanesulfonic acid) (PIPES)

Pirkle-type CSP, 95, 96, 101–102
Plasma protein binding, 182–199, 208–210
  age and, 198
  analytical developments in, 190–196
  and disease states, 198–200
  drug-drug interactions, 218–219
  drugs interaction with, 183
  overview, 182–183
  species difference and, 197–198
Plasma proteins
  AGP, 182, 186–189
  binding (see Plasma protein binding)
  HSA, 182, 184–186
PMDA. see Pharmaceutical and Medical Devices Agency (PMDA)
PMDEC. see Pharmaceutical and Medical Devices Evaluation Center (PMDEC)
PMSB. see Pharmaceutical and Medical Safety Bureau (PMSB)
Polarimetry, 167
Polymeric CSP, 109
Polymeric surfactants, 157
Polysaccharides, 154
Prilocaine, 223
Prima facie case, obviousness, 299–300
Prior art disclosure
  anticipation of enantiomers by, 296
  law on obviousness of species in, 297–298
Prior art genus
  anticipation of species, 296
Propoxyphene, enantiomers of, 220
Propranolol, 77
Protein CSP, 106
Proteins, 154
Pseudoephedrine, 81, 85

QSAR. see Quantitative structure-activity relationships (QSAR)
QSER. see Quantitative structure-enantioselectivity relationships (QSER)
Quantitative structure-activity relationships (QSAR), 42–43, 178, 193
  3D, 43–45
Quantitative structure-enantioselectivity relationships (QSER), 42
Quinidine, 216

Racemate, 28, 246–247, 294–300
  development of, 222
  single enantiomer development and, 256–257
Racemic acid, 6
Racemic mixture, 27, 42, 49, 59–66, 69
  chromatographic separation of, 75–76
  greater activity with, 219–220
  properties/resolution of, 52–66
(R)-(+)-α-methoxy-α-(trifluoromethyl) phenylacetic acid (MTPA), 84, 86
(R)-amphetamine, FLEC and, 80
Reduced folate carrier (RFC1), 208
Reexamination Certificate, 295
Regulations, stereoisomeric drugs, 240–269
  guidelines on chiral drug development, 264–269
  international guidance, 248–254
  overview, 240–241
  requirements
    in EU, 241–248 (see also European Union (EU))
    in Japan, 259–263 (see also Japan)
    in United States, 254–259 (see also United States)
Renal clearance, 216
Reversed-phase high-performance liquid chromatography (RP-HPLC), 77, 87
RFC1. see Reduced folate carrier (RFC1)
Rifamycin B, 153
(–)-(R)-isoproterenol, 278–281
  binding properties of, 283
  effects on AC activity, 286
  interaction
    with β$_2$AR, 286–288
    with β$_2$AR/β$_2$AR$_{CAM}$, 282–286
Ristocetin A, 153
Rocking tetrahedron model, 37
Rotation
  optical, 19–20, 28, 52
RP-HPLC. see Reversed-phase high-performance liquid chromatography (RP-HPLC)
(R)-propranolol, 33
(R,R)-DANI. see (1R,2R)-1,3-diacetoxy-1-(4-nitrophenyl)-2-propyl isothiocyanate ((R,R)-DANI)
(1R,2R)-1,3-diacetoxy-1-(4-nitrophenyl)-2-propyl isothiocyanate ((R,R)-DANI), 77, 78–79
  applicability of, 88
R,S-naproxen, 69
R,S-1-phenylethylamine, 69
Rules Governing Medicinal Products in the European Union, 243

Safety Division, PFSB, 260
Sanofi-Synthelabo, 99
Sanofi-Synthelabo v. Apotex, Inc., 299
Scalemic mixture, 27, 28
Scatchard analysis, 191
Sedatives, 222
Selectivity, solvents, 124–125
Sequence rule, C.I.P. convention, 26
Serum albumin, 194
SFC. see Supercritical fluid chromatography (SFC)
Shave recycle, chromatography, 131–132
Simulated moving bed (SMB) chromatography, 133–140, 269
  advantages of, 136
  basics of, 134–136
  combination with SFC, 144
  development of, 136–140
  examples of, 140
  parameters, determination of, 139–140
  principles of, 134
  processes, 140
(+)-(S)-isoproterenol, 278–281
  binding properties of, 283
  effects on AC activity, 286
  interaction

INDEX

with β_2AR, 286–288
with β_2AR/β_2AR$_{CAM}$, 282–286
SLC transporters. *see* Solute carrier (SLC) transporters
SMED. *see* String model for enantiorecognition (SMED)
SmPC. *see* Summary of Product Characteristics (SmPC)
Sodium ammonium tartrate, crystals of, 51
Solubility, solvents, 125–126
Solute carrier (SLC) transporters, 172–175
Solutes, 114–117
 in-column band shape, 116
Solvents
 selection for enantioselective screening experiments, 122–126
 selectivity, 124–125
 solubility, 125–126
 viscosity, 123–124
SPR. *see* Surface plasmon resonance (SPR)
(S)-propranolol, 33
(2S,4R)-N-heptafluorobutyryl-4-heptafluoro-butyroxy-prolyl chloride (HFBOPCl), 84
S-shaped isotherm, 117
(S,S)-N-3,5-dinitrobenzoyltrans-diaminocyclohexane-isothiocyanate (DDITC), 77
SSR separation. *see* Steady-State Recycling (SSR) separation
4S,6S-stereoisomer, 258
Steady-State Recycling (SSR) separation, 133
Stereocenter Recognition model, 36
Stereochemical terms, 17, 28
Stereochemistry, of drugs, 264, 265
Stereoisomerism, 17
Stereoisomers, 19, 28, 206–228
 chiral, 20
 defined, 20
 development of, 269
 factors affecting disposition, 217–219
 fenoterol, pharmacological effects of, 280
 of flupentixol, 23–24
 interaction with β_xARs, 278–281
 isoproterenol, 278, 279–280
 mixtures of, 27
 naming of, 24–27
 pharmacodynamics (*see* Pharmacodynamics (PD))
 pharmacokinetics (*see* Pharmacokinetics (PK))
 regulatory perspective on, 240–269
 selectivity of, 268
 types of, 20–24
Stereoselectivity, 36, 38
 activity/toxicity, 222–223
 in drug metabolism, 211–215
 transport of drugs, 171–178 (*see also* Transporter, drug)
Stereospecificity, 27
Stern-Volmer analysis, 186
String model for enantiorecognition (SMED), 41
Sudlow site II, 186
Sulfasalazine, 178
Sulfur, 19

Summary of Product Characteristics (SmPC), 242
Supercritical fluid chromatography (SFC), 99, 140–144
 additives in, 142–143
 combination with SMB, 144
 disadvantage of, 143–144
 method development in, 142–143
 scale-up in, 143
 schematic of, 141
 screening in, 142
Surface plasmon resonance (SPR), 190, 195–196
SWORD trial, 269
Symmetry, 17–19
 operations, classes of, 49
Symmetry element, 49

TAGIT. *see* 2,3,4,6-tetra-O-acetyl-α-D-glucopyranosyl-isothiocyanate (TAGIT)
Talinolol, 216
Tartaric acid, 1
 optically active, 14
TASA. *see* Total apolar surface area (TASA)
Teaching, suggestion, or motivation (TSM) test, 299
Telmisartan, 178
Temperature
 effect on enantioseparation, 154–155
Testing, for chiral drug substances, 251–252
Tetrahydrofuran (THF), 124
2,3,4,6-tetra-O-acetyl-α-D-glucopyranosyl-isothiocyanate (TAGIT), 77
 DANI and, 79
2,3,4,6-tetra-O-acetyl-β-D-glucopyranosyl-isothiocyanate (GITC), 77
Thermodynamics, 38–41
 analysis, 40
 free-energy differences, 41
 functions, 39
 parameters, 39
THF. *see* Tetrahydrofuran (THF)
Thinlayer chromatography (TLC), 168–169
Thioridazine, 189
Thioureas
 formation of, 76–80
Three-point interaction rule, 101
Three-point models, chiral recognition, 31–33
Threose, 25
TLC. *see* Thinlayer chromatography (TLC)
Total apolar surface area (TASA), 42
Trans-4-hydroxy-2-nonenoic acid (HNEA), 87
Transmembrane proteins, expansion to, 171
Transporter, drug
 ABC, 175–178 (*see also* ABC transporters)
 intestinal absorptive/secretory, 207–208
 solute carrier, 172–175
 stereoselectivity in, 171–178
Trimipramine, 189
Tri-O-methyl-β-cyclodextrin, 42
Tris, 2-(N-cyclohexylamino) ethanesulfonic acid (CHES), 150
Tromethamine, formation of, 58

TSM test. *see* Teaching, suggestion, or motivation (TSM) test

Ultrafiltration, 190
Unichiral reagent. *see* Chiral derivatizing reagent (CDR)
United States
   FDA (*see* Food and Drug Administration (FDA))
   implementation of ICH in, 250
   patentability perspective, and development of chiral drugs, 294–300
   regulatory requirements for pharmaceuticals in, 254–259
Ureas
   formation of, 85–86
U.S. v. Ciba-Geigy Corp., 300

Vancomycin, 153, 178
Verapamil, 185–186, 213, 216
Vincristine, 176–177
Viscosity, of solvents, 123–124

Warfarin, 185, 198
Whelk O1 CSP, 101, 102

Xanthosine 5'-triphosphate (XTP), 282, 286
X-ray crystallography, 34, 38, 168
XTP. *see* Xanthosine 5'-triphosphate (XTP)

Zafirlukast, 178
Zonal elution, 191
Z stereoisomer, 23